WHAT DOES
THE HONEYBEE SEE?
AND
HOW DO WE KNOW?

A CRITIQUE OF SCIENTIFIC REASON

WHAT DOES THE HONEYBEE SEE?

AND

HOW DO WE KNOW?

A CRITIQUE OF SCIENTIFIC REASON

ADRIAN HORRIDGE

ANU
THE AUSTRALIAN NATIONAL UNIVERSITY
E PRESS

ANU
E PRESS

Published by ANU E Press
The Australian National University
Canberra ACT 0200, Australia
Email: anuepress@anu.edu.au
This title is also available online at: http://epress.anu.edu.au/honeybee_citation.html

National Library of Australia
Cataloguing-in-Publication entry

Author: Horridge, G. Adrian.

Title: What does the honeybee see and how do we know? : a critique
               of scientific reason / Adrian Horridge

ISBN: 9781921536984 (pbk) 9781921536991 (pdf)

Subjects: Honeybee.
               Bees.
               Insects
               Vision
               Robot vision.

Dewey Number: 595.799

Cover design and layout by ANU E Press

Cover image: Adrian Horridge

CONTENTS

ABOUT THE AUTHOR

Adrian Horridge was appointed in 1960 as Director of the Gatty Marine Laboratory, St Andrews, Scotland, while completing the two-volume work with Professor T. H. Bullock of the University of California on *The Structure and Function of the Nervous Systems of Invertebrates* (1965). At St Andrews, he started a research group that worked on the eyes of crustaceans and insects. In 1969, he brought this group with him to Australia, when he was appointed as one of four foundation professors of the Research School of Biological Sciences at The Australian National University. Since that time, he and his numerous students and postdoctoral collaborators have discovered much about the vision of insects. In 1969, Horridge was elected a Fellow of the Royal Society of London and in 1972 a Fellow of the Australian Academy of Sciences.

PREFACE

The unique and self-contained topic 'The vision of the bee' is of interest in its own right as the best-known example of what a typical medium-sized insect can detect with its eyes. It is also a topic of philosophical interest because it raises centuries-old questions about perception, consciousness, sentient beings, human uniqueness and insect-like robots. What does the bee really see? How does the small brain of the bee see so well? How does this influence our ideas about perception, automata and future practical applications?

There are many ways to answer these questions. Bees assist us because they can be trained to come to objects or patterns, and trained bees will answer questions put to them in simple tests. The bee's visual system is also open to analysis by optical, anatomical and electrophysiological methods, by tracking the bees while navigating and route finding and also by studying visual flight control as they pilot themselves between obstacles. This book includes a historical survey of how scientists have approached, experimented with and argued about insect vision for 100 years, finally rejecting anthropomorphism and solving some critical questions.

One of the features is the (still-imperfect) coverage of the German contribution to the subject. Until about 1966, insect vision was scarcely mentioned in books in English. The older work was ignored, while the more recent contributions were controversial and unrelated to each other. There was little study of the topic in England and textbooks made a hash of it.

This particular science is grounded in observation and logic. There is little mathematics, chemistry or physics involved or need for great learning. Bees are found worldwide and are reasonably representative of all large insects. Research on bee vision can be very cheap. The results have mostly been published but the story is not generally known or even available to the educated public. Here now is an account of what bees really detect with their eyes and how scientists found this out.

The accounts of earlier experiments on this road of discovery, and how inferences were made from the data, make a fascinating account of the arguments and counterclaims of contending professors. The approach here is out-and-out support for experiment, backed by the logic of John Stuart Mill and the philosophy of scientific progress of Thomas S. Kuhn. The processes of designing the experiments and inferring the conclusions from the data make a miniature picture of scientific effort in several areas of biology, physiology and comparative psychology. They illustrate how the work was really conducted—not always amicably.

The basis of this study is the observation of the performance of bees. From the performance, an intuitive inference was usually made, as a stab at explaining the behaviour. Often, the inference was incorrect because the vision of bees was counter-intuitive in several ways. Explanations made by analogy with mammalian visual systems or drawn from the terminology of the cognitive sciences were usually found to be inadequate. There were many examples of excellent observations and reliable data, from which a wrong inference was made, followed by argument without new data, stubborn resistance to reinterpretation and refusal to accept advances made by others. Later, the mechanisms of the performance were analysed by extensive testing of trained bees and, after much thought, the counter-intuitive processing mechanisms of bees' strange visual behaviour were slowly revealed. The whole subject became an exposition of the stages of visual processing. Bees do not see shapes or objects; they detect parameters and recognise places. This story is told as an example of how early intuitive inferences have given way to the results of carefully designed tests of trained bees.

The book is intended for an audience who do not want a text crowded with references and every fact that exists. The aim is for it to be read more like an essay for anyone with some scientific background. It describes the process of scientific discovery on a limited theme, with two excursions into branches of the subject in more detail: the action of the retina and processing by the nerve cells (Chapters 5 and 6). For all chapters, there are sufficient names of researchers given for anybody who wishes to dig deeper, using the extensive bibliography. This book could be used by university students interested in subjects such as vision for physiologists, perception for psychologists, insects for entomologists or zoologists, robot vision for engineers looking for new ideas and scientific method for philosophers.

ACKNOWLEDGMENTS

First, I acknowledge with enormous gratitude the contributions of those ancient institutions, St Johns College, Cambridge, where I was for 10 years absorbing a myriad influences from curious role models, and the University of St Andrews, where I learned to stand on my own feet, and where the work on insect vision began during 13 years at the Gatty Marine Laboratory. At Yale and the University of California, I was introduced to research financed by grants. Then at The Australian National University, I was able to expand with my research group on my own broad topic that seemed scientifically interesting: insect vision.

Second, I thank the great number of colleagues, who worked with and near me, aided by technical staff, librarians and secretaries. I depended on their support. I owe a particular debt of gratitude to the numerous enthusiastic colleagues who made good use of the plentiful equipment, made scientific advances and mostly published their own work on insect vision. Their names appear throughout this book.

Some, in particular, stand out: Tudor Barnard, Malcomb Burrows, John Scholes, Steve Shaw, Rick Butler, Ben Walcott, Ayis Ioannides, Ian Meinertzhagen and David Sandeman. In Canberra: Simon Laughlin, Allan Snyder, Andreas Dubs, Randolf Menzel, Gert Stange, Benno Meyer-Rochow, Ted Maddess, Fred Doujak, Peter Lillywhite, Jenny Kien, David Williams, Roger Hardy, Yasuo Tsukahara, Keiichi Mimura, Mark Leggett, Dan-Eric Nilsson, Mike Land, Stjepan Marčelja, Willi Ribi, Martin Wilson, Richard Payne, Roger Dubois, Doukele Stavenga, David O'Carroll, Tom Matic, Shi Jian, Jan Dalczynski, Peter McIntyre, Qijian Sun, Yang En-Cheng, Danny Osorio, Joe Howard, Mandyam Srinivasan, Tony Heyes, Andrew James, Zhang Shaowu, Miriam Lehrer, Eric Warrant, Mike Ibbotson and Neville Fletcher.

There is especial gratitude to the assistants who worked with me, sometimes for many years—notably, Margaret Lang, Charlie Roemmele, Margaret Canny, Bob Jackson, Caroline Giddings, Miriam MacLean, Roland Jahnke, Jadviga Duniec, Ljerka Marčelja, Edyta Kucharska, Sasha Neist, Steve Lucock and Virginia Pierce—and to secretaries (before they became extinct): Veronique, Tess, Tiger Lily, Elizabeth and Margaret.

Finally, I owe an infinite debt of gratitude to Audrey, the loving woman I married 55 years ago, who cared for me and four children, in a family life that was full of interest, culture and generosity.

INTRODUCTION

Small in size but high among the wonders of nature, insects delight and amaze us with their skill in flight and their obvious ability to see. A pursued dragonfly will turn and twist but will not escape the one chasing behind, yet it can stop in an instant or catch a mosquito in midair. Unerringly, a bee will take the direct path to the hive and a hoverfly will stand still in the air with invisible vibrating wings. These actions are all controlled by vision.

We know, however, that when an engineer constructs a visual system it requires a huge computer behind the digital camera, and when we see things ourselves and make sense of them, we are using billions of nerve cells in our visual system, which occupies a large portion of our brain. So, it is a serious task to understand what insects see and how they do it, because anything that insects tell us about vision could be useful in constructing simple brains that see.

Medicine, psychology, law and religion are all subjects with a long written history loaded with confusions, contradictions and unjustified conclusions. One hopes it might be different in science, where there is supposed to be impersonal validation. Experience shows, however, that the original scientific literature is scarcely read, the all-important authors' summaries can hide the weakness of the data, the conclusions often turn out to be invalid, the titles of the papers are often misleading and textbook writers cannot know it all. Few go back and study the original design of the experiments and the data. In fact, experts are so few in the world that often there is only one research group at the cutting edge—and they have baggage and axes to grind.

The quest for scientific literacy and deep understanding is hampered by vast arrays of facts, huge reference lists and gigantic projects. The good life in science is achieved by a new technique, a chance discovery and having time to think. A student with nothing to offer but enthusiasm has two options: to join an established group or to find a little local topic that can be expanded later.

What the bee sees is an accessible topic large enough to show how we observe natural events, use experiments to discover how things work and validate the conclusions, and bee vision is a small enough topic to grasp as a whole. Bees are available worldwide and they are easily trained to select objects, patterns or any easily manipulated stimulus to receive the reward of sugar. The research is very cheap and anyone can set up shop in the hope of making discoveries.

Discovery of visual mechanisms has been sufficiently slow and controversial to make a good story. The subject is not threatening or overdone, such as climate change, nuclear power, oil or food supply. There are no high voltages, nasty chemicals, radioactivity, risks of infection or heavy lifting. The experiments are not dangerous, except to those few people who are allergic to bee stings. Moreover, the experiments are a pleasure to watch as the bees make their choices, show their abilities and reveal the errors in existing theories before your very eyes.

During the nineteenth century, 'what bees could see' was an unfathomable topic. In the twentieth century, there was haphazard progress with lots of good data, many erroneous conclusions, internal contradictions and controversies. It was a topic in crisis looking to harmonise two paradigms that seemed true but incompatible. Then suddenly, about 1990, the way forward became clear.

Examination of the process of discovery leads me to the conclusion that there is no philosophy for scientific advance. Many practical strategies are essential. We do not need full understanding to be able to make discoveries, but it is a good idea to read the literature to know what has been done. Observe nature with an informed mind. Repeat old experiments that look interesting. Look for unresolved controversies and inconclusive experiments. Design original new apparatus and techniques. Plan for the next decade of discovery. Think about the problems all the time. Go for mechanistic analysis and make an observation every day. Never accept a conclusion that is merely compatible with the data; always devise a test to check it in a different way. Avoid noology—that is, science based on thought, not observation—and avoid computer models until the very end. In fact, all models are dangerous. Learn the lesson that Darwin demonstrated by collecting lots of data before venturing to conclusions. Keep making theories but put no trust in them. Look for real mechanisms but look out for anthropomorphism.

The factor that most governs the advance of understanding is undoubtedly the ability to produce the correct thought. At any time, there is a network of mutually consistent concepts that explains most of a topic. To break out of this paradigm requires a lucky observation, a stroke of imagination or a new technique that leaps ahead. When bees were first trained to come to a pattern, it was believed that they saw and remembered the pattern. When trained bees accepted unfamiliar patterns, it was believed that they generalised certain patterns that had a parameter in common. The technique of shuffling the patterns was thought to make the bees look at them, but it also taught the bees to ignore looking anywhere else, so they were left with only one landmark to identify the place of the reward. For about 100 years, bees were trained to look at patterns before it was realised that they were not interested in the pattern, only the cues, and that they did not have foveal vision to look at things, only 300-degree vision to triangulate on a few cues at wide angles to each other to locate a place.

The factor of next importance is the ability to spot the unjustified conclusion, find its source and change it. Unfortunately, most experts are committed to their own variety of the truth as they see it, and most scientists cannot understand most scientific papers. A researcher has an idea, tests it with an experiment, finds the results are compatible with the idea, writes it up as proven and turns to another topic. A textbook writer takes the result from the title of the paper, incorporates it in a wider theme and consolidates the error in a broader theme to a larger audience. Sadly, peculiar results are not rejected, just not mentioned, like family black sheep. It is impossible to change ideas, though ideas are changed. Researchers should publish a list of known errors to guide future textbook writers. At all times, progress has been achieved by hard slog, training and testing bees in hundreds of experiments—not by insights of genius.

In scientific research, almost everything that you need to know must be learned on the job: doing research, observing nature, building experimental equipment, and so on. Existing theories can help in planning only when you notice that they are out of line with your observations.

I have tried to push back anthropomorphic concepts—such as shape, topology, fitness, generalisation and cognition—and replace them with mechanistic principles derived directly from experimental tests of theories in turn derived from earlier experimental results on bees. In a topic such as vision, an analytical mathematical theory is not appropriate, but synthetic computer models are quite valid and might lead to useful applications in computer vision. They are single channel and iterative, however, and so are not models of the visual system, which is heavily parallel with successive arrays of adaptive filters.

However deep our understanding, we will never know every detail of the bee's visual system or any other simple brain, because even if we describe every detail of the nerve cells and synapses, and record the activity of all simultaneously, we still omit the essential settings of the gain, noise, time constants, feedback loops, ionic changes and hormonal effects, as well as the processes of growth and decay that all contribute to neural activity. Anyway, vision is active; the motion of the bee activates the visual system.

Contemporary culture often appears to be in the grip of numerous contradictory beliefs, conclusions, theories or truths—call them what you will. One doomed belief was the idea that the more science became testable by being experimental, the fewer ridiculous beliefs would persist, but science is now so extensive it has become a race between the speed of education and the mortality of the experts. One sensible view was that science was what was in the textbooks, but the more that science became testable, the faster the textbooks went out of date. *Wikipedia* might keep some topics up to date, but it will struggle to provide a consensus view because experts always disagree at some level of detail.

Meanwhile, keen, hardworking undergraduates believed that science was a process of steady discovery. Their older brothers doing research knew it to be a trail marked with the corpses of previous theories and conclusions, but they still persisted in doing some experiments. Then, out of the air, they would guess a new conclusion that was consistent with their new observations. At this point, problems multiplied. First, the conclusion might have been incorrect. Second, conclusive experiments can be impossible to do. Third, secular fundamentalist professors—all German in the case of insect vision—continued to teach their erroneous beliefs, as long as they lived, as Kuhn (1970:151) quoted from the autobiography of Max Planck: 'a new scientific truth does not triumph by convincing its opponents and making them see the light, but rather because its opponents eventually die, and a new generation grows up that is familiar with it.' My job is to inform that new generation.

In the mid-twentieth century, when scientists and social scientists alike believed that (apart from minor impediments such as communism, the Pope or a belief in Father Christmas) science and society had to build success on success to strive to arrive at truth or utopia, we thought of Reason versus Faith. Some said that the rise of nuclear power destroyed our innocence; some blamed the post-modernists; some blamed the break down of consensus societies when travel revealed their shallowness; some blamed the pluralist society with its mutually incompatible beliefs living in peace together. It is hard to accept, however, that an agreement to differ in our beliefs is a way of conducting our sciences.

Therefore, something must be decided, otherwise we dissolve in a soup of ideas. Modernisation has been a shift from a state of belief to one of choice. The way to restore much needed certainty is signposted by observations of the real world and turns out to be a perpetual seesaw between new theory and experimental demonstration.

At a dinner in the College Hall, I found myself asking the distinguished-looking woman next to me what was her attitude to the interaction of science and the arts. 'As for myself,' she offered, giving me a stern warning look, 'I am post-permissive—that is, I used to be permissive but am no longer' (I wish all women were as honest). She was referring, however, to a debate of the time called the 'science wars'. One side said that scientific theories were reliable conclusions from experiments based on earlier validated theories, and nothing else was to be believed. The other side said that theories were unreliable personal whims based on experiments that someone thought might be helpful or possible. Each accused the other side of being chicken-brained and pig-headed. On one side we have the idea that scientific objectivity is impossible because opinions and conditions in different societies differ and are volatile. This position is as impossible to sustain as the view that absolute truth exists. In all subjects, claims about truth come in all sizes and shades of grey, and fill the time of the chattering classes.

No doubt there is a difference between those who passed their science exams and those who didn't. Nonetheless, a compromise can emerge from the study of the development of ideas as we pass from one practitioner to another in the history of a topic. The realists had in mind the hard sciences of metallurgy, geology, thermodynamics, physiology, molecular biology, all three branches of chemistry, and so on, which were closed books to the radical philosophers. Their critics had in mind the so-called soft sciences such as anthropology, psychology, psychiatry and lesser superstitions such as media studies, business ethics and aromatherapy. They presented as being mutually exclusive, but they were both right. Bee vision has both hard and soft material and my effort has been to harden some of the softest parts.

This account of what honeybees see could be cannon fodder for this battle. There is an abundance of explosive material and generations of potential victims. The ideas, even in the hard sciences, are always personal but might not show it. In the hard sciences, however, they are mostly reasonably testable ideas, and some concerning bee vision have now been tested. There are interesting themes along the roads of discovery, and the foundations are strengthened by knowing their historical development.

The history of science teaches us that new ideas are thinly scattered and winners are not easy to spot. Also, we are notoriously lazy in consolidating the new into the old and especially in working out the logical consequences of new findings and deleting sections of old teaching. So, progress is slow. To pass even the most elementary examination on inshore navigation in several advanced countries, it is still essential to learn Newton's seventeenth-century theory of the tides, although the data to change it lie in every book of tide tables.

Like all science, the topic of insect vision is dotted with forgotten unsolved problems and theories that have turned out to be inadequate. The original observations were usually valid, but on the evidence available at the time the authors drew erroneous conclusions that were later overgrown with new ideas. Ageing experts become unwilling to revise their basic beliefs and nowadays it is almost impossible for anyone to finance a promising heresy. Knowledge staggers along in search of simple statements until a theory fails in too many ways and is replaced. 'Accepted, if not disproved' is not, however, sufficient. There must be no sensible alternative to the accepted story. Testing a theory means testing it with the facts against all other theories that can be imagined—an impossible task. The non-stop nature of all science soon puts books out of date and the patchy nature of publication in journals never forms a complete account. Generations of students are puzzled by inconsistencies in old accounts or they search for help in recent journals. They continually need a new synthesis. I hope that I can provide one, for a very small part of science.

CHAPTER SUMMARY

1. Early work by the giants

In the nineteenth century, progress on what insects saw was slow and contentious, with fierce argument about whether bees found places by odour or by sight and how they navigated to places, identified them and returned home. The pioneering experiments were done in Germany from 1896 to 1940. None of this research has been previously summarised in English for the scientifically interested layperson. The chapter concludes with a discussion of the young bee's transfer from using odours in the dark hive to vision outside as she matures.

2. Theories of scientific progress: help or hindrance?

Science is validated by logic, experimental observations and commonsense, starring Aristotle, Francis Bacon, J. S. Mill, T. S. Kuhn and Karl Popper, with discussions of empirical laws, mechanistic analysis and computer modelling. The visual system presents us with a special problem because there are arrays of processing channels in parallel, and many layers of these arrays, but no identifiable end point in the brain. There are also multiple causes for every observed effect. One reason why bee vision was only slowly understood was the lack of methods of analysis of a multidimensional system without an existing map.

3. Research techniques and ideas, 1950 on

New postwar electronic techniques for the study of the nervous system gave a fresh lease on life to a topic ideal for the age of laboratory experimentation. Persistent research in a small number of labs produced detailed information about the functioning of the retina and optic lobe neurons. The modelling of motion perception of the fly in Tübingen, Germany, was upset by the discovery of rapid eye movements that controlled piloting—much as they do in humans.

4. Perception of pattern, from 1950 on

Returning to his job after the defeat of the Nazis, Karl von Frisch appointed a youthful group of researchers in Münich, and when he retired, they began afresh in Frankfurt, with new techniques for the behavioural study of pattern vision in trained bees. They worked with huge patterns and obtained entirely new results that dominated the topic but could not be integrated into the rest of the literature. This led to much argument about images in the brain of the bee and a hiatus until new work started again in 1987.

5. The retina, sensitivity and resolution

An account of the structure and function of the retina of the bee is filled out with some material from the more detailed work on the fly. The eye catches light and processes the image, the signals are transmitted to the neurons below and vision is limited by the noise in the signals. There are simplifications for the non-specialised reader, but it is not a primary school account.

6. Processing and colour vision

Tedious probing with sophisticated equipment has revealed how nerve cells respond and collaborate. Based on the recent electrophysiology of identified single neurons with microelectrodes, this chapter continues the description of how the visual image is processed, transformed and summarised as it passes from the retina to the memory and initiates behavioural responses. The insect visual system is one of the best-known parts of the central nervous system in the animal kingdom. There is no evidence of reassembly of patterns in the brain. This technical account is essential for understanding the machinery of vision.

7. Piloting: the visual control of flight

This chapter is a description of how insects fly by visual control—a topic that was expanded by work in Canberra in 1987. There are accounts of keeping a straight course, avoiding collision, how to turn without getting in a knot, how to counteract sideswipes from gusts of wind, how to measure altitude, range, speed over the ground and distance travelled, and how to make use of the parallax caused by one's own movements. This work has led to significant practical applications for self-guided flying vehicles.

8. The route to the goal, and back again

The work of Karl von Frisch, from 1914 until the 1960s, slowly brought together the previously unimagined navigation and dances of the honeybee. The use of the sun as a compass in the sky—already known from detailed work on ants—required an internal clock, also known previously. Aristotle knew about the bees' dance, but its function in directing foragers was discovered in two stages by von Frisch in the 1920s and 1940s. The pattern of polarisation of blue in the sky also acted as a compass. Bees can learn to negotiate a maze, which involves the use of a sequence that is stored in memory—like the recognition of landmarks along a track.

9. Feature detectors and cues

New people, new apparatus, a new research theme, new ideas and generous funding spawned a hive of activity in Canberra from the mid-1980s, when the world's best bee trainer, Miriam Lehrer, arrived as a seasonal visitor. We brought Zhang Shaowu from Academia Sinica and started on pattern perception. First, the orientation cue was isolated. Later, the feature detectors for edge orientation were shown to act independently and were only 3 degrees long. The cues from modulation, radial and tangential edges, bilateral symmetry, position of the centre of black and position of hubs of radial and circular symmetry were also demonstrated and placed in an order of preference by the bees.

10. Recognition of the goal

To a bee, a panorama or a very large target displays parameters that overlap several local eye regions, so several landmark labels are learned and recognition of the place is relatively certain. The simultaneous responses of numerous small feature detectors form a cue, of which there is one of each kind in each local region of the eye. A coincidence of cues in a local region forms a label that identifies a landmark. The label is the unit of memory like a signpost on a route. The local regions are distributed around the eye, so that a place is recognised by the expected coincidence of a few labels at large angles around the head. As the bee nears the destination, she heads in the direction that changes these angles towards their expected values. This task makes good use of the 300° coverage of the compound eye.

11. Do bees see shapes?

From the beginning there has been contention about this question. Some think that bees see separate objects distributed in the panorama, with corresponding spatial representations in memory in the brain. They propose that bees recognise abstract properties such as triangularity, squareness or shape in general. These ideas originated from earlier theories of human vision. Careful testing of trained bees reveals no evidence for spatial representation, object or shape recognition, but only the recognition of coincidences of the cues already described. This idea is supported by the neuron responses, by efforts to make artificial vision and by numerous recent training and tests of bees.

12. Generalisation and cognitive abilities in bee vision

There has also been disagreement for a century as to whether bees can learn one pattern and then accept other patterns because they have a concept of a general likeness or difference, called generalisation. However, they also accept many quite different patterns. The explanation is that when trained on targets that are moved about to make the bees look, the bees learn insufficient cues to enable them to distinguish all other patterns. When they generalise, they are simply confused. There have been many claims that bees detect generic categories such as symmetry, topology and categories such as faces, but when the trained bees are carefully tested, it turns out that they have learned the particular cues required for the single task at hand. Trained bees accept an unfamiliar pattern if it displays the cues that they learned to expect—and no extra ones. The idea that they generalise in a cognitive way is founded on poor data, an inadequate variety of tests, failure to consider the cues and intuitive use of terminology borrowed from the cognitive sciences.

GLOSSARY

The *parameters* **outside the eye** display features, such as colour or edges, which are detected by *feature detectors* of several kinds **inside the eye**. A *cue* is the sum or count of the responses of one kind of feature detector in a local region of the eye and is therefore an abstracted part of the local region of the *image*. A cue, like a neuron, has its own quality (referring to the feature detected), a quantity (from the size of the sum) and a position on the eye. The bee detects the cue, not the original feature detectors. There is an order of preference to the known cues. The coincidence of the several different cues in a local region of the eye is remembered as a *landmark label*.

For humans, the *centroid* is the unique position of the centre of gravity on which a pattern balances on a pivot. For bees, the centroid is the position of the centre of gravity of the sum of the feature detectors that compose one cue in a local region of the eye. Bees learn centroid positions.

Configurational means laid out spatially like a picture.

Disruption of a pattern is roughly equal to the total length of edges. The motion of the eye over the disruption generates the modulation of the receptors.

The *feature detectors* are the units of perception of modulation, edge orientation, black, white or colour. They are small, about 3 ommatidia long on the retina, and all respond independently in parallel. The bee detects the cue, not the individual feature detectors, which are lost in various summations to form cues.

The *field* of a filter or neuron is the region in space and time within which a signal is detected.

A *filter* is a stage of processing in a model. It usually represents a neuron or group of neurons broadly tuned to detect a feature or cue. It can be represented as a mathematical operator that is multiplied or convolved with an input pattern to yield a signal that is passed on to the next stage of processing.

Fixation is a rigid holding of one position in stationary flight, usually with a high frequency of wing beat while apparently looking at a small spot, a contrast or a hole to fly through.

A *fixed* pattern—as opposed to a *shuffled* one—has the pattern fixed, as seen from the choice point of the bee.

A *generalised parameter* is one that is recognised in a context other than in the training pattern. Originally, it was merely in a different position on the target, but later it was in a different pattern.

A *hub* is the centre of a pattern of radial or tangential edges.

The *image* from the bee's point of view is the pattern of excitation in the array of receptors.

The *label* is the group of cues in a local region of the eye by which the bee recognises a landmark.

The *layout* of the image, features, cues or labels means the arrangement in space.

The *modulation* of a receptor is the change in the light intensity in the receptor and the consequent electrical signal. Modulation is directly related to the length of edge in the field.

Orientation of an edge is usually the angle to the vertical in a vertical plane. The orientation cue is the sum of the responses of the edge orientation detectors in a local region of the eye and, like all the cues, is independent of other properties of the pattern such as shape, continuity of edges or division into separate areas. Within the local region of the eye, the average orientation has a retinotopic position that bees can be trained to remember.

Orthogonal edges are at right angles to each other.

A *parameter* is a scalar or vector measurement of some aspect of the pattern outside the eye—for example, the area or total length of edge.

The *patterns* are displayed on the *targets* during training and tests.

Place for bees is a geocentric term, like the place on a map; *position* and *direction* are usually retinotopic terms on account of the radial arrangement of visual axes. *Location* or *position* refers to the position of a cue on the target, a shift in position of a pattern or a shuffle of the locations of boxes, targets or bars during training and tests.

Point of choice is the place where the bee detects a cue and makes a choice by moving away from or towards the next target.

A *retinotopic* memory is one that is laid out behind the retina to correspond with the layout of the pattern.

A *sign stimulus* is an older and more general term that is not restricted to vision—for example, it can be used in relation to the call of a bird. It is the human idea of the essential stimulus, not the parameters or the cue detected by the feature detectors.

A *template* is a hypothetical mechanism that detects a fairly complicated pattern. It can be innate or learned. In vision, a spatial copy is usually implied.

01 EARLY WORK BY THE GIANTS

Palaeolithic humans—with excellent vision and endless opportunity—must have examined honeybees busy at their work and wondered what the insects saw and what they were doing, just as somebody else, watching the humans in turn, wondered what on earth they were doing studying bees. For a social animal, it was important to know who was watching what.

The powerful obstacle to understanding what bees see—and they obviously see something—is that the human mind reads itself into the minds of others, even into bees. We call it anthropomorphism. We imagine that the bees are seeing things. We see the bee and the bees see us, which becomes obvious if we steal their honey. So, for about two millennia since Aristotle, the general opinion seems to have been that the bees see things and organise their affairs very much as we do. They see flowers; they collect honey; they fly home; they defend their store; the drones pursue the virgin queen. It is like a play based on the human world, just as Shakespeare described:

> For so work the honeybees,
> Creatures that by a rule in nature teach
> The act of order to a peopled kingdom.
> They have a king and officers of sorts,
> Where some like magistrates correct at home,
> Others like merchants venture trade abroad,
> Others like soldiers armed in their stings
> Make boot upon the summer's velvet buds,
> Which pillage, they with merry march bring home.[1]

We now know much more, but a veil will always obscure our understanding of what bees see because we are not in a position to observe it ourselves. The collapse of confidence in the reality of human perception began with René Descartes (1596–1650) and proceeded through Bishop George Berkeley (1685–1753) and the French rationalists of the eighteenth century.[2] Now it is the turn of the bees. The interest now lies in how the bees manage to do so much with a tiny brain, how humans have evolved scientific techniques to investigate them and how and what they see.

Nineteenth-century beginnings

The giants—or rather, gentlemen of leisure—on whose shoulders we stand began systematic observations and the experimental approach only in the nineteenth century. Sir John Lubbock (1834–1913), a polymath, who invented, among other things, bank holidays and the term 'Palaeolithic', published his own careful observations of the behaviour of ants, bees and wasps about 1873.[3] He found that ants detected ultraviolet (UV) light and that blue was their preferred colour. When he carried bees away from the hive and gave them honey, they rarely returned for more (they were probably not foragers). When he fed them in an upstairs room, they failed to recruit other bees (they could not remember heights above ground). His writing illustrated his difficulties because he could not refer to a body of reliable observations and there were no relevant theories.

Auguste Forel (1848–1931), a remarkable medical professor in Zürich and best known as a psychiatrist and expert on Hymenoptera, aimed to eliminate anthropomorphic ideas from the study of insect behaviour.[4] One of his targets was Felix Plateau (1841–1911), the Professor of Zoology at the University of Ghent, Belgium, who had the misfortune to be the son of a very famous mathematician. Between 1885 and 1899, Felix studied how bees found flowers and published many papers.[5] Unhappily, he repeatedly produced the wrong answers, so they caught the eye of one or two critical scientists.

Plateau tried to attract the bees with paper flowers that were carefully painted with natural colours. He also hid dahlia flowers behind paper but the bees went under the paper, so he concluded that the shape and colour did not attract them. Because the bees ignored his flowers, he concluded that bees recognised flowers by their odour, not by vision of shapes or colours. He was unwilling to concede that the bees remembered the place of the reward by use of landmarks.

On the numerous works of Plateau, Forel (1908) lamented:

> It is with reluctance that I have decided to undertake the criticism of this author, not, indeed, that it will be difficult, but because of the space which it demands, and because it is painful to me to have to bring to light the false conclusions of a colleague whose patience, work, honour and good faith I esteem.

Forel then launched without mercy into 50 pages of objections, supported by his own experiments.

2

Plateau worked with large artificial flowers that were scarcely distinguishable from real flowers. The bees passed them by, so he supposed that the match of the colours was unsuited for the bees. Forel repeated the experiment with crude artificial flowers, laced with honey and found that bees would not visit them unless they found the honey by chance, or had it pushed at them. They would then return repeatedly to the same artificial flower. Plateau persisted with his contention that the bees used their sense of smell, but other published work had shown that bumblebees returned to their flowers when their antennae, palps, mouth and pharynx (that is, the seat of the sense of smell) had been removed.

After 50 pages of fierce criticisms, Forel accepted the correct data but not the false conclusions:

> I must make an excuse for my long criticism and my long series of controls, as much to M. Plateau as to the reader. But it was necessary. In using the experiments of M. Plateau himself to show the errors of judgement that he draws from them, I fully render homage to his scientific honesty. And it is precisely this honesty which has allowed us to follow the author step by step, and to pick up, by the help of his faithful narration of facts, the thread of their actual connections [how the facts relate to each other] and their agreement [that they are mutually consistent]. Thanks to this, our study will not have been a sterile polemic, for it has brought us to see more and more clearly into the very question which occupies us. (Forel 1908)

Modern science has lost this art of pulling the rug from under one's opponent.

The main obstacle to this research was that the bees had already arrived at their destination, with a memory of what they expected to find, so, in any experiment, they were likely to be frustrated by any change and would either start to relearn the place of the reward or simply go away.

Forel was one of the first to use individually marked bees effectively in a variety of experiments. Confirming earlier work by Lubbock, he stressed that the bees did not follow an experienced bee, but they were attracted to a number of bees feeding and they remembered the place where they found food. To Forel, these observations showed that feeding bees made little use of their sense of smell. Forel also concluded that bees distinguished the contours of objects poorly and that they returned to any shape that offered them honey at the expected place. 'Vision of form, colour, dimension and distance…guides the bees by means of visual recollections associated to those of taste and smell' (Forel 1908).

In the nineteenth century, public criticism was more robust than we enjoy today. Serious scientists flung identifiable mud at one another's conclusions and sometimes at the experiments themselves. For example, Forel again:

> The publications which have appeared on the subject before us are very
> numerous, but they consist for the most part of theoretical dissertations
> only, of hypothesis, and, as Lubbock (Linn. Soc. Journal vol 12,
> Observations on bees and ants) has very well remarked, of oft-told tales
> of ancient experiments, borrowed, through more than a century, from
> one 'authority' to another, without attempt at control or checking.
> (Forel 1908:5)

The philosopher of scientific method, Karl Popper, would have appreciated this
approach: not advancing on the shoulders of others, but shooting them down.
In the end, Rabaud (1928) covered the literature in French but did not refer to
Plateau's numerous works.

Professor Albrecht Bethe (1872–1954), father of Hans, the physicist, was a
versatile physiologist and anatomist of the nervous system, sometimes called
the 'conscience of German biological science' of the time because he brought
attention to the errors of the other professors—not a bad idea, actually. For
Bethe, all comparative psychology of animals was an absurdity. His paper of
1898 illustrates the conflict between the general belief in bees' cognitive powers
versus the experimental evidence of their extreme stupidity. He replaced his
beehive with an empty hive with an open back, so that returning bees flew out
through the back and continued repeating this manoeuvre. When he moved
a normal hive back by a metre, the bees stopped at the former position of the
entrance as a cloud in the air, failing to recognise the hive. These observations
were old hat to beekeepers, illustrating the isolation of professors from artisans.

Bethe's belief in mechanistic analysis guided him to do experimental tests. His
experiments proved to him something beyond the science of the day: the bees
did not locate their hive by scent and could do so after their antennae were cut
off. When a hive was closed at night, and opened the next day in a new place far
away, at first the released bees made short exploratory flights and returned to the
hive. They remembered the new position of the entrance with great precision
and appeared to be guided by something external to the hive. Similarly, when
carried in a box for up to two kilometres, either they flew upwards and headed
in the direction of home or they flew in a circle and returned to the box. After
many such experiments, Bethe logically concluded that the bees obeyed a force,
absolutely unknown to us, that carried them back to the place in space from
whence they came. At that time, radioactivity, radio waves and x-rays were in
the news and Bethe must have been disappointed with the hilarious reception
of his hypothetical force.

Accounts of insect behaviour of the nineteenth century had pages of detail of
how the flowers were arranged or how the bees appeared to do this or that,
repeated with variations in other papers and whole books. We find voluminous
accounts later in the works of Karl von Frisch (1886–1982) and others, but from
1920 onwards the professional journals gave only bare accounts of experiments

and tables of results. Finally, the detailed measurements disappear also and we moderns are left with boring summaries of methods, results, condensed diagrams and long reference lists. Unfortunately, the loss of innocence—and incidentally, disproving the other fellow—was not replaced with a better design of experiments, greater significance of the results, more critical polemics or lucid logic in the conclusions.

Scientists of the nineteenth century, such as Lubbock and Romanes, who understood the experimental method, struggled to separate the mechanistic analysis from the descriptions of performance. They tried to interpret their observations on the vision of honeybees, ants and wasps, but sought in vain for explanations. They had no idea of peripheral processing by connections of neurons in parallel. Anyway, the neurons were only just being discovered by new techniques. We find the same in every science, every sphere of activity. At first, understanding is slow to start because there is no map to guide us through the jungle of unrelated observations. In the case of honeybee vision, the early analysis was documented by Lubbock, Romanes, Plateau, Bethe and Forel, provided with an anatomical substrate in the histological works of Grenacher and Exner, and the arrays of neurons in parallel were described by Cajal, Sanchez and Zawarzin.

Forel ejected the nineteenth-century 'astrologers with their ancient rubbish' (von Uexkull 1908) with many trenchant comments of his own:

> As we have seen, the causes of the erroneous judgements with which Plateau has obscured the question at issue are errors of interpretation, inadmissible and continual generalisations, and the almost total omission of the psychical faculties of the insect, especially with regard to memory and association. (Forel 1908:193)

He found that the bees learned to return to the place, not to the flower, and shape was of no significance, but he could take the analysis no further. Later, in 1910, he noted that bees would return at a time of day when they were regularly given a reward, and so started the study of their time sense. The reward for the bees was marmalade at breakfast time—a novelty in Switzerland, adopted from the first English mountaineers.

Figure 1.1 Apparatus with vertical presentation for visual discrimination experiments with honeybees. a) Patterns on boxes that are shuffled in position. b) Defining a range by partitions. c) The apparatus introduced in 1967; before they enter, the bees hover with the pattern subtending an angle of about 130° directly in front of them. d) The Y-choice maze, with baffles; the bees enter at the front into a choice chamber from which they see both targets; they select one side, reach the reward hole, then when satisfied, exit by the way they came; the targets and the reward change sides every five minutes.

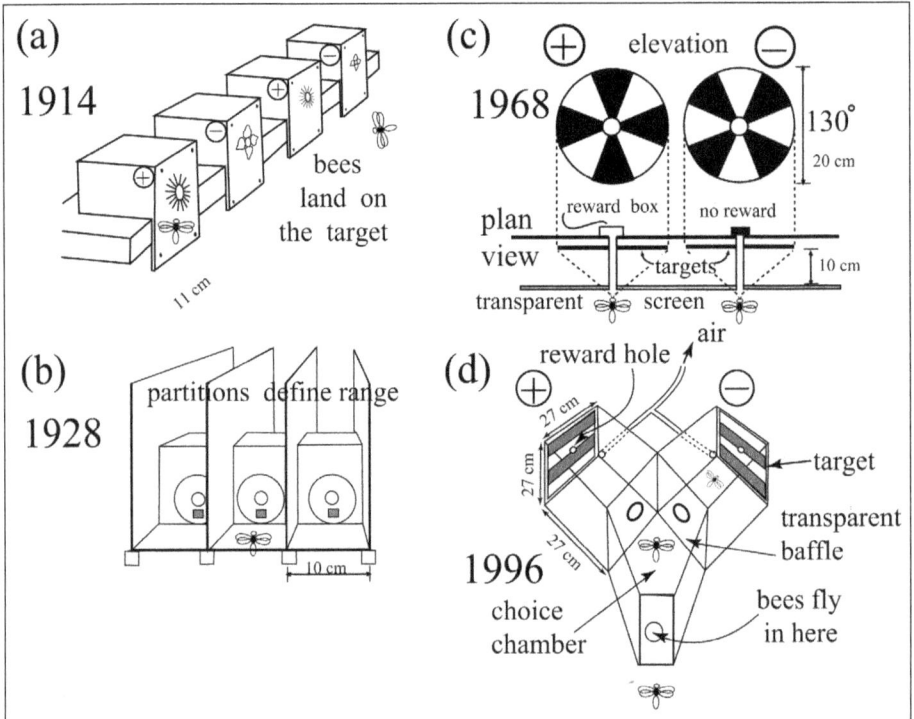

Sources: (a) after von Frisch (1914); (b) after Baumgärtner (1928); (c) after Wehner (1981).

The early twentieth century

In 1914, when von Frisch published his results with trained bees, he had been working at his family's country house in the Austrian Alps on the colour vision of freshwater fish, but he turned to bees to give a demonstration of learning when fish were not available.[6] He used Sigmund Freud's principle of association by simultaneous presentation. On the balcony of his alpine summer house, von Frisch copied from a remarkable African-American naturalist Charles Henry Turner (1867–1923) the method of putting a reward in one of several small cubical boxes, with a different pattern on the vertical front of each box (Figure 1.1a). Access to the inside was through a hole in the middle of the pattern (Turner 1911). von Frisch put a reward of odourless sugar in the box that displayed the pattern to be learned and nothing in the other boxes. The positions of the boxes were shuffled in a row all facing the same way, so that the bees could not learn where to go, but were obliged to look at the patterns to find the reward. Without realising it, von Frisch (and all others after him who interchanged the targets) trained the bees to ignore the landmarks that indicated the exact place of the box. This was an important ingredient in the experimental design because it made possible the acceptance of unfamiliar patterns that displayed the same combination of cues.

The criterion of success was the bee landing on the correct reward hole, and therefore the angular sub-tense of the target was not known at the moment of the bee's decision, but could be very large. As later demonstrated by Baumgärtner (1928) and Friedlaender (1931), the bees took special notice of the region immediately below and around the reward hole.

Figure 1.2 Results of early experiments. The pairs of flower-like patterns in (a) and (b) were discriminated from each other in the vertical plane, but those in (c) were not when presented together. The patterns in (d) laid out flat were not discriminated, similarly those in (e), but those in (d) were discriminated from those in (e).

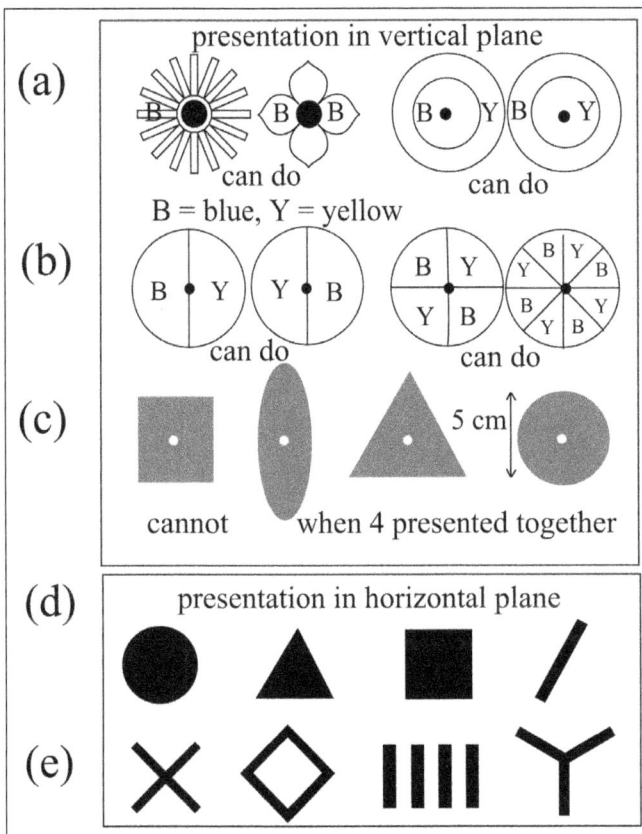

Sources: (a–c) after von Frisch (1914); (d–e) after von Frisch (1965).

In the first half of the twentieth century, bees were trained to discriminate between two or more patterns. von Frisch found that bees easily learned the difference between a flower shape with many small petals and one of the same size and colour with a few large petals (Figure 1.2a). Also, bees easily discriminate between pairs of flower-like shapes in which an area of blue and one of yellow have different positions in the two training patterns, even a left/right reversal (Figure 1.2b). He demonstrated that bees distinguished yellow and blue from all shades of grey. He was interested mainly in the bees' abilities and he had no particular theory in mind to guide his experiments.

von Frisch proposed that bees were able to distinguish flower-like patterns because their vision was adapted to their normal repertoire. The idea of pattern processing being adapted to the normal function was not examined further until the 1990s. von Frisch's bees, however, could not learn to discriminate between a blue square, triangle, disc and a diamond shape (Figure 1.2c) although trained for five days. This failure would have caused confusion in the literature but for some reason it was almost ignored for 90 years. Much later, in the decade between 1995 and 2005, it was discovered that bees learned quite quickly to discriminate between pairs of these shapes. Probably von Frisch failed with four shapes because he displayed too many in one task, but that excuse needs further testing.

von Frisch's students and younger colleagues found that bees discriminated between pairs of many varied patterns, but there were other pairs they confused for no apparent reason. His student Baumgärtner (1928) found that bees could distinguish between flowers (taken in pairs) with three, four, five or six petals, if one but not the other had a petal immediately below the reward hole. He was interested primarily in how well bees detected the place of a coloured patch relative to the reward hole, and in the angular resolution in behavioural tests compared with the angles between the receptor axes. Bees could distinguish between a blue and a yellow 20mm by 20mm square in bright light if they subtended at least a minimum of 3° at the eye—a result that was ignored for 70 years. Two 5mm by 5mm squares of different colours could be discriminated from the same squares that were exchanged in position only if they were located immediately below the reward hole. This was the region that would be visible to both eyes and outside this area this particular task was impossible (Figure 1.3). In 1999, I showed that the discrimination of the exchange of positions of two coloured spots in the horizontal direction was possible only when there was contrast with the green receptors, which was essential to stabilise the position of the eye in the horizontal direction.

Figure 1.3 Discrimination between small yellow (Y) and blue (B) squares, with vertical presentation and landing on the reward hole as the criterion of success. a) Two examples of the targets. b) The region around the lower lip of the reward hole where discrimination of the colours is successful (black squares) and unsuccessful (white squares). c) The overlap between the two eyes in angular coordinates.

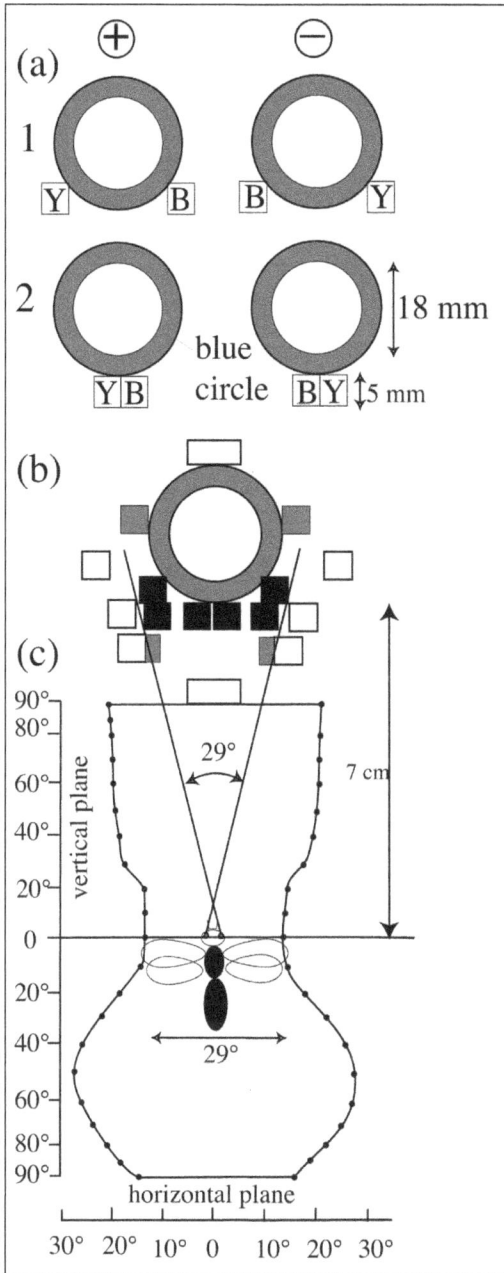

Sources: (b) after Baumgärtner (1928); (c) after Seidl and Kaiser (1981).

Mathilde Hertz (1891–1975), daughter of the physicist Heinrich Hertz and acquainted with many German scientists, studied bee vision from 1925 to 1936 at the Institut für Bienenkunde in Berlin (where von Frisch worked). She caused more puzzled brows among reviewers and students than any other bee researcher in the twentieth century. Her method was to lie a number of patterns on a white table and place a reward of sugar solution in a glass dish next to the rewarded pattern. She stressed that the bees could use different parts of their eyes to discriminate correctly. The patterns were shuffled in position at intervals to make the bees look for the rewarded one. The bees were therefore trained to ignore the exact place. On the flat table, with shuffled target positions, the bees did not discriminate edge orientations or relative locations. Later, we found that only patterns that were salient for the bees would be learned in these conditions. Examples are areas of spots, coloured patches, radial patterns, concentric circles and patterns rich in black/white edges (Chapter 9).

Knowledge about any visual system was in a sorry state at the time. Hertz used a great variety of training and test patterns, following any idea that the results suggested and, from 1926 to 1933, discovered many interesting details, most of which have been neglected for three reasons. First, she wrote in an obscure style that was difficult to translate. Second, the bees did not correlate edge orientations with the directions of their sun compass as they flew in all directions over them, although Wiechert (1938) later showed that bees used edge orientation on patterns laid flat when restricted in their direction of approach. Third, having no general paradigm outlining how insect vision operated, Hertz interpreted everything in terms of the Gestalt theory[7] for human vision, as in Wertheimer (1923), which was briefly expressed as:

> There are wholes, the behaviour of which is not determined by that of their individual elements, but where the part-processes are themselves determined by the intrinsic nature of the whole. It was the hope of Gestalt theory to determine the nature of such wholes. (Wertheimer 1924)

As will be seen, the idea of 'global' vision had great influence on later research on bees.

On the one hand, Hertz analysed patterns into low-level parameters, such as area, length of edge and circular versus radial contours, which looked as though they were fundamental—much of which was later verified. On the other hand, patterns were classified by arbitrary global characters such as symmetry, disruption, isotropy, smoothness, texture, variability of patch size and separation into parts. There was, however, no demonstration that these categories really had any meaning for bees.

Figure 1.4 The basic separation of figural intensity (disruption) in columns and figural quality (shape) along rows, when patterns are laid flat. Bees easily discriminate the patterns on the top row from each other, and with greater difficulty within each row going downwards. The further the patterns are apart within each column, the better the bees discriminate them. The numbers indicate the relative lengths of edge.

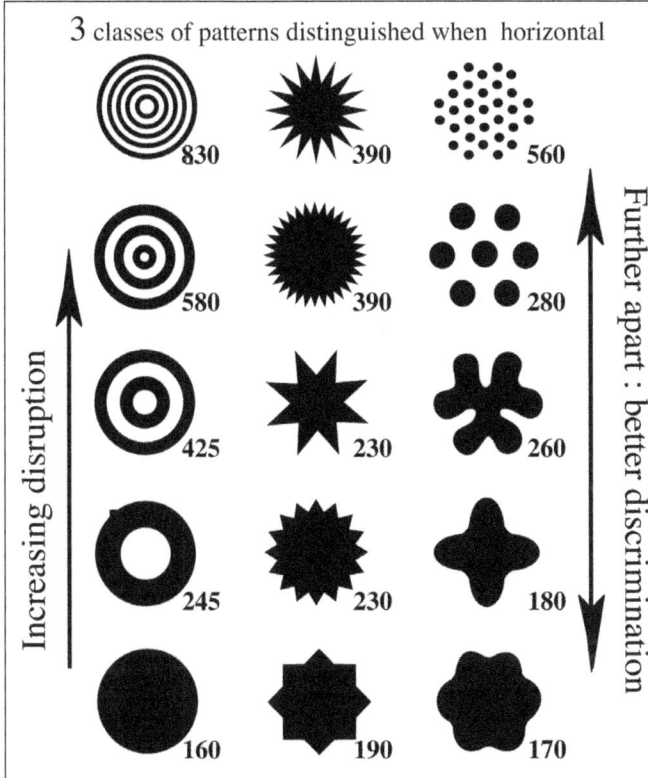

Source: After Hertz (1933).

Although the language and terminology were frequently obscure, the actual data were not so bad. Hertz discovered that the bees discriminated area or size and total length of edge (Figure 1.4). The last is sometimes called the disruption of the pattern and has been mistakenly identified with spatial frequency. The bees detected radial symmetry about a centre and some major types of pattern such as blobs, groups of spots, gratings of parallel bars, concentric circles and radial patterns of bars. In the first 60 years of experiments, only one other parameter was discovered: the orientation of edges (Turner 1911; Wiechert 1938).

Figure 1.5 An example from Hertz (1933). Bees trained on (a) accepted (b) equally well. Bees trained on (e) accepted (c) and (d) equally well. Bees trained on (d) or (f) avoided (a) and (e). These results were a puzzle until it was realised that the bees learned the modulation and avoided rings and crosses unless trained on them.

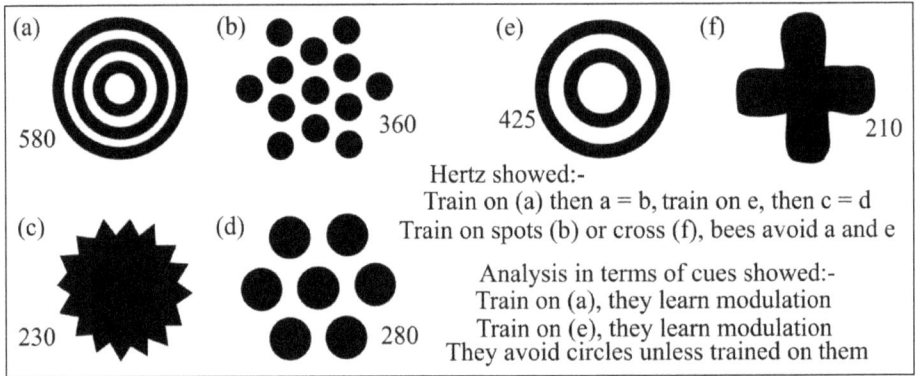

(a) 580
(b) 360
(e) 425
(f) 210
(c) 230
(d) 280

Hertz showed:-
Train on (a) then a = b, train on e, then c = d
Train on spots (b) or cross (f), bees avoid a and e

Analysis in terms of cues showed:-
Train on (a), they learn modulation
Train on (e), they learn modulation
They avoid circles unless trained on them

Gertrud Zerrahn (1933), working in Heidelberg but obviously in touch with the contemporary work in Berlin and München, also presented flat patterns on a white table. Although the patterns were different, Zerrahn's conclusions duplicated some of Hertz's work, and Zerrahn also showed that the preferences of untrained bees for patterns of various types mirrored their ability to learn to discriminate them.

A student of Professor Otto Koehler in Königsberg, Marianne Friedlaender (1931) used the same type of reward boxes as von Frisch. Bees were trained to discriminate between a rewarded yellow square cross (with bars 40mm by 8mm) and a square (24mm by 24mm) of the same colour and area, both on a white background (Figure 1.6a). In a significant advance, the trained bees were given several tests. Both the cross and the square were accepted as the original when turned through 45°. Movement of either shape within the target had little effect on the discrimination. The radial pattern of the square cross had salience for the bees because they detected it even when it was moved a short distance or rotated. It was 60 years before the next steps were taken (Chapter 9).

Friedlaender found that bees could not discriminate between a target with a rectangle of grey on the left of the reward hole and one with the rectangle on the right (Figure 1.6c). When the bees had been trained with radial spokes adjacent to the rectangles, however, they could do this task and they retained the discrimination of the locations without the spokes (Figure 1.6b). Moreover, patterns that included radial spokes could be moved up or down on the targets without spoiling the discriminations. Her explanation was that the position of the centre of symmetry of the spokes provided a salient reference point. The bees in flight scanned continually in the horizontal plane and they failed to remember the position of an image that was not stabilised on the retina.

Figure 1.6 Early analysis of the effect of a change of location. Each new group of bees was trained with the pairs of targets on the left and tested with the pairs on the right. The targets were fixed in position and the criterion of success was landing on the reward hole, so the patterns were huge at the moment of choice. a) The cross and square are discriminated although they are changed in orientation and moved relative to the reward hole. b) Radial rays stabilise the eye and the discrimination persists although the patterns are moved relative to the reward hole. c) These patterns were not discriminated. d) A blue panel on the left and a yellow one on the right, with green contrast where they meet, are discriminated from the mirror image. In tests with the panels moved up, discrimination persists. e) Preference is reversed when the panels are moved to the right. The cue is the colour adjacent to the reward hole. f) Single coloured panels on opposite sides of the reward hole are discriminated, but not in tests with the reward panel moved up.

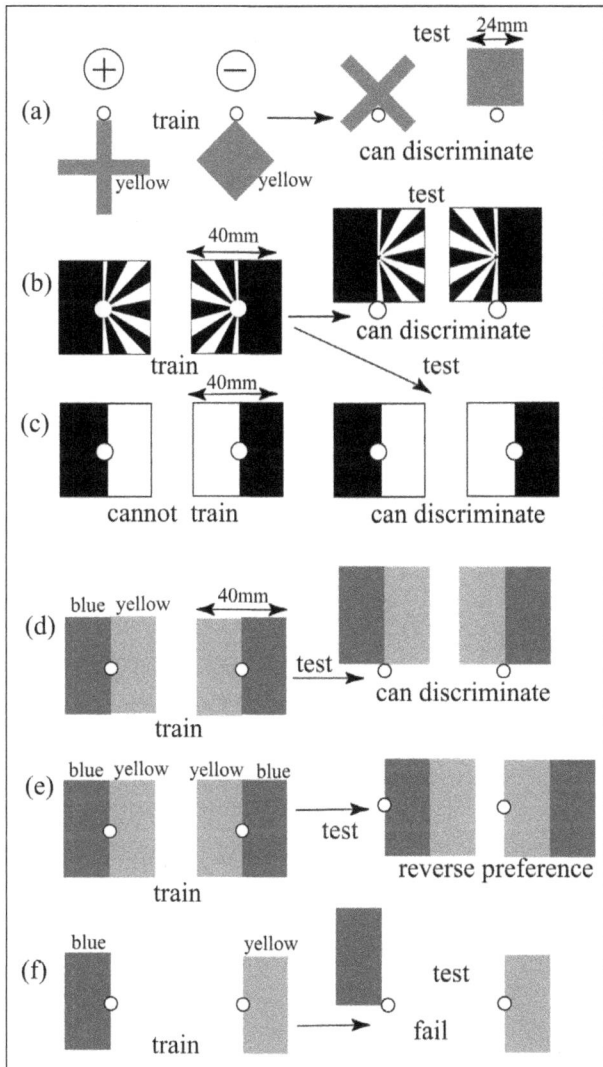

Source: Redrawn from Friedlaender (1931).

Next, bees were trained to discriminate a pattern with a blue patch on the left of the reward hole and a yellow on the right versus another pattern with the colours reversed (Figure 1.6). The bees failed in tests when the position of the reward hole was shifted or one of the patches was omitted. When the bees were trained with a yellow patch on the left of the reward hole versus a similar patch on the right of it, moving a patch in the horizontal direction relative to the reward hole in tests had little effect, but moving a patch in the vertical direction spoiled the discrimination (Figure 1.6c). These important clues to the bee's spatial world were forgotten for 70 years (Chapter 9).

Elsbeth Wiechert (1938), another student of Koehler, showed that relative positions of two rectangles of different colour were discriminated when displayed in the horizontal plane as well as in the vertical plane, if natural obstacles restricted the direction of approach of the bees.

Conclusions by mid-twentieth century

By 1950, the training patterns had been selected at the whim of each experimenter and there was still no general idea of bees' visual mechanisms or how to design useful experiments. With the aid of hindsight, we can draw out a few useful generalisations that have been clarified since that time.

One parameter that the bees detected was the disruption of the pattern or the total length of edge in it. In the oft-copied Figure (1.2d) from the work of Hertz (1929–31), bees could distinguish between any of the figures in the top row only with difficulty, but all figures in the top row were easily distinguished from those in the lower row. The explanation was that the bees learned the modulation of illumination of the receptors as the eye moved across the figure. In fact, this result does not apply when the eye is stabilised on very large targets (as illustrated in Chapters 4 and 9).

It was never asked what was the area in which the length of edge was measured. The training patterns were always isolated simple shapes, so it seems to have been believed that the bees first identified the rewarded shape and then measured its outline.

A second generalisation, also from Hertz, was the easy discrimination of radial patterns of edges and avoidance of circular patterns. Bees trained to go to a group of spots would not visit a pattern of concentric circles, but those trained to go to the circles would visit the spots (Figure 1.5). In addition, Friedlaender found that the centre of a square cross would act as a reference point (Figure 1.6a). Radial bars enabled the bees to discriminate a shift in the position of a black square (Figures 1.6b and 1.6c). Radial patterns of edges or a dark spot on a light background were salient to the bee's eye. Together with the spontaneous preference and ease of learning of radial patterns, the salience suggested that the bees' visual processing was adapted to flower-like forms, as von Frisch (1914) had suggested. The mechanism, however, remained a mystery.

Third, size was discriminated, so, when working with other parameters, patterns must be of similar size. Baumgärtner, with vertical presentation, had shown that if a coloured patch was to be discriminated, it must subtend a certain solid angle at the eye so that many facets were stimulated.

Fourth, bees discriminated easily whether a black spot was above or below the reward hole. Several authors had shown that bees could learn whether a yellow patch was to the right or to the left of an adjacent blue patch when both were close to the reward hole. The reward hole acted as a landmark. No-one, however, commented that when the reward box with the training pattern on it was regularly moved to make the bees search for the rewarded pattern, the bees were still able to discriminate the positions or relative positions within the training patterns. We now know that shuffling the positions of the reward boxes during the training causes the trained bees to ignore the local landmarks. They still used more distant landmarks and learned the range of relative positions of the parameters in the vertical direction at the places where they were displayed during the training.

By mid-twentieth century, it was well known that bees knew the approximate place to come for a reward by using landmarks at different distances and directions, so it was clear that they also learned the positions of areas of black or colour relative to a point of reference.

As part of the wider field of experimental psychology, it was thought at the time to be useful to plot the stimulus/response as a function of a variable in the tests, or a curve that showed the progress of learning. The percentage of correct responses, however, depends on the training conditions and duration. For the bee's visual system, many such empirical relations were published, but they were almost useless for the analysis of the visual system because there was only one experimental treatment and no tests of what the trained bees really detected.

Hiatus in Germany

All this early twentieth-century research on bee vision was done in Germany. Almost all the scientists were Jewish or had Jewish relatives. After 1938, research under von Frisch continued on navigation, route finding and colour and odour discrimination, but the work on pattern vision ceased until revived in the 1960s. Wolf and Zerrahn emigrated after 1933 and joined Professor Selig Hecht at Columbia, New York, from where they published many empirical relations about the vision of the bee and other animals. Again, they were hampered by lack of a sufficient body of bee neuro-ethology to interpret the measurements.

Others, such as Harald Esch and Rudolf Jander, later moved to the United States—not because of persecution by the Nazi regime, but because they were persecuted by German professors.

As a Jewish scientist in Berlin, Hertz was ruined after 1933, and from 1936, she and her mother lived in Cambridge, England. She was befriended by Bill Thorpe, a Cambridge don (like Lorenz and Tinbergen, an intuitionist). In Thorpe's (1956) otherwise excellent and comprehensive book on insect behaviour, he scarcely mentioned the discrimination of patterns by trained bees, as though he could make no sense of what was available to him. His lectures in 1949 to a Cambridge zoology class revealed to me no insights into the mechanisms or methods of analysis of insect vision.

There was an enormous number of descriptions of insect taxes, kineses and tropisms in response to light, but almost all the experiments reported in the influential textbooks by Willi von Buddenbroch (1937, 1952) and the handy little book by Carthy (1958)—which was almost all that was available in English—were unrelated to each other or to a comprehensive theme. The details of the neurons of insect optic lobes, so beautifully described by Cajal and Sanchez (1915) or Zawarzin (1913), were ignored as possible substrates of visual behaviour. The growing knowledge of responses of neurons and the physiology of synapses was also ignored.

To each generation, honeybee vision simply did not make sense. The reason for the failures and the confusion that followed, we now see with the benefit of hindsight, was due only partly to the lack of abundant data and theories to guide new experiments. In particular, it was never asked whether bee learning was like wax—which could be moulded to any external shape or pattern— or like a set of innate boxes that could be ticked when a limited variety of parameters appeared. Consequently, the right experiments were delayed. There was also the dead weight of the anthropomorphic belief that bees really saw the things they looked at. There was no theory of how or what insects really saw. The main problem, however, was the criterion of success (landing on the correct pattern or reward hole) in the training of the bees, so the experimenter could not control or infer what parameters the bees were using.

The primacy of odour

Worker bees are helped out of their sealed cells by nurse bees and soon take up a task within the hive, feeding the grubs, filling storage cells with pollen or nectar, fanning to ventilate, cleaning out old cells or tending the queen. All this is done in the darkness of the hive, so in the busy life of the young bee every action is governed by the sense of smell. The bee's antennae have thousands of odour receptors and the main association areas of the brain serve olfaction. Bees also detect the direction of gravity. The whole world of the young bee is one of odour, up/down, touch and taste.

After two or three weeks of working in the hive, the young bees are like those unenlightened human beings living in a dark cave, as described by Plato in Book 7 of *The Republic*.

The mouth of the cave is open, but the unenlightened slaves see only shadows of themselves, like shadow puppets on the opposite wall, never their true nature. Similarly, the young bee has sensed the real world of flowers, pollen, foliage and the passage of the days only by odours. Reality for young bees is nothing but sequences, mixtures and memories of odours. If they communicate with each other, they must do so in a language of odour or touch.

One day, when about a month old, a bee finds itself with a load of rubbish that must be thrown outside the hive and there is no experienced bee to take it, so it approaches the entrance. In his book, Plato suggests that in the first stage of liberation, the bright light is distressing and nothing is understood from the unfamiliar visual sensations. Gradually, the young bee will use the familiar odours to educate the visual system and soon will find that it can detect another bee or a flower at a distance without relying on the odour. For a time, as Plato describes, the shadows (odours) appear more real than the visual world. At first, the bee detects best the motions of contrasts, then the motions and ranges of the objects themselves, until finally the bee gazes on the blue sky and the sun in the proper place for that time of day and learns that the sun and the sky move in a regular way from horizon to horizon.

In Plato's words, the enlightened bee does not return willingly to the drudgery within the hive but would rather explore the outside world and return to the hive loaded with food for the unenlightened ones left behind. So, an education for new roles is required. As Plato says, to function in the upper world, the upwardly mobile 'must use the motion of objects and astronomy':

> The spangled heavens must be used as a pattern, with a view to higher knowledge…The sun will be seen to be the universal author of all things beautiful and right, parent of light in the visible world, and this is the power upon which the eyes of the rational are fixed.

Two thousand years later, we have to confess that we know little more than Plato about the process of the transition of the young bee from the dark world of odours to the bright world of vision. It seems obvious, however, that the new visual memories must be correlated at some point with the existing odour memories. The mushroom bodies in the brain, which are thought to be the centres of memory, certainly take the majority of their inputs from the odour pathways, with smaller tracts from the optic lobes (Figure 6.4). It is likely that the higher-order optic neurons establish connections with the existing higher-level odour centres before the new visual associations can be remembered. Similarly, the existing gravity detectors of the vertical, which are so essential for the ordered construction and business of the hive in the dark, are somehow associated with the new uses of vision to detect the orientation of an edge and the pattern of the sky and horizon in relation to the sun.

Endnotes

1. Quote from *King Henry V*, Act 1, Scene 2, speech by Canterbury on the strategy to defeat France.

2. Among many others, early ideas about human vision are summarised by Richard Gregory (1981).

3. Lubbock wrote *Prehistoric Times* (1865) and *The Origin of Civilisation and the Primitive Condition of Man* (1871), as well as many books and papers on natural history.

4. The points made by Forel foreshadow many of those in the rest of this book.

5. Plateau did some good work on the physiology and anatomy of insects.

6. This story is recounted in his autobiography.

7. Gestalt theory tried to solve the problem of the apparent wholeness of sensations by proposing that there was a corresponding representation in the brain. See Gregory (1981:Ch. 12) and a full account in Koffka (1935).

02 THEORIES OF SCIENTIFIC PROGRESS: HELP OR HINDRANCE?

We expect to have a problem in understanding what honeybees see because they have a tiny brain combined with a very wide view of the world: *multum in parvo*. We must draw conclusions from the way they behave. We reasonably expect that they detect only relatively simple parts of the scene, but at first we are unable to imagine how they see anything in a moving panorama. To make progress, we have to devise ways of asking questions of the bees so that logical conclusions can be made from the way they react. This chapter is about making firm conclusions.

Unlike most experimental science, there is no need for equipment, purified chemicals or electricity. Anyone with some patterns and sugar solution could have inferred most of what we know about bee vision at any time in the past centuries. That did not happen. Progress was excruciatingly slow, although the bees were eager for lessons once they learnt that sugar was available at school. Why the delay?

At each step, progress was limited by error and the slow development of ideas, so it took a long time to formulate each next appropriate question. At first, the questions put to the bees never produced sensible answers. The bees were observed, their responses to experimental change were a mystery and the proposed explanation was just a guess. There were many unsuspected factors and guesses that were not tested became facts. The errors blocked the imagination of those who followed. As a result of this patchy acceptance of a mixed bag of insight and error, there was no acceptable answer to the interesting question 'What does this insect actually detect?' The question was not asked.

Let us examine the development of scientific theories to see whether the ways of thinking about explanations—what we call the philosophy of science—have been of any help.

Early theories of scientific advance

There is a long and fascinating road that winds through history and explores how the natural world was elucidated. The problem faced by the great innovators of the philosophy of science in the past 3000 years was to find a general method that would apply to any problem, although as things turned out, this was a bad place to start. The process is not direct because the best questions to ask become obvious only when the answers are anticipated. The process starts with collecting facts of interest long before any moment of truth arrives. We have to observe and think at the same time, followed by a dissection of the subject into components, an effort at analysis to see what causes what, and then we must assemble the tentative mechanisms into a coherent story.

Aristotle,[1] an ancient Greek philosopher, taught that we should accumulate facts and look for generalisations about them. In coming to his own conclusions, however, he was usually short of facts and relied on the primitive assumptions of the day. As a result, he was knowledgeable but often mistaken. He was unwilling to abandon his general principles, although, on the topic of the reproduction of bees, he admitted that there were insufficient facts to warrant any conclusion at all.

Although real experiments had been done for millennia—for example, helpful and fatal trials with medicinal herbs—the idea of the experimental approach and the concept of an *experimentum crucis* was first systematised by Francis Bacon[2] at the end of the sixteenth century. The idea was to invent a crucial experiment that allowed the observer to decide between two alternatives. We now know that it is a rare piece of luck to find such an experiment that is conclusive. It can be wrongly conceived, so that the result cannot be interpreted, or there might be more than one explanation, or new facts emerge later. Bacon stressed that the gathering of facts must be steady and progressive, with conclusions at each stage, and this advice was followed by great scientists such as Charles Darwin, but Bacon had no idea how a scientific concept or theory was formed in the first place. He advised us to be suspicious of first principles (meaning Aristotle's principles), but we still find them getting in the way today, such as the belief that there is something special about symmetry or the idea that insects see things, even if blurred.

Bacon was aware of the danger of proposing a theory and then inventing experiments to prove it, but he could not stop the practice. You have an idea out of the blue, then enthusiastically rush around proving it. Sadly, it was equally useless to rely on Bacon's pet method: induction. This is the formation of a general principle that is consistent with a number of separate facts. Induction depends on regular occurrences and the uniformity of natural events. It is, however, boring to collect facts without knowing why. The opposite of induction is deduction, in which the observations are logically deduced from data and general principles.

Alone, or even together, deduction and induction are not strong enough, or even sufficient, to generate useful experiments. The two missing ingredients are imagination and a caution about multiple causes. Like Aristotle, Bacon argued as though a phenomenon could have only one cause. The visual system—with numerous receptors in parallel, multiple pathways to the brain and numerous superimposed arrays of nerve cells, always changing with time—would never be understood if single effects always had single causes.

The classical and medieval minds tended to work in terms of rather rigid categories with sharp boundaries. Something was either this or that. They respected categories as though they had been created with the universe and had an independent existence. Classifications also had a value of their own. The categories ruled the discussion without being questioned themselves, directed the next venture and diverted attention away from unexpected but significant novelties. Observations were suspect, as Galileo was firmly told by the Church. We still see the pleasure enjoyed in an armchair argument about concepts and the definitions of terms.

More recently, we have been urged to think not only of alternative causes, but of all the intermediate stages between them. Categories also become blurred. I prefer to assume that the visual system operates with parallel pathways, each with a definite function. As a first approximation, I assume that each type of component and pathway can be analysed separately with yes/no answers if the appropriate tests can be devised. So far, it has worked.

In the late seventeenth century, John Locke (1632–1704) traced the origins of knowledge, while David Hume (1711–76) analysed ideas about causation in the mid-eighteenth century. Bishop Berkeley doubted the evidence of the senses but still relied on learning and commonsense. These English philosophers were more empirical than their continental colleagues and, in the early nineteenth century, the differences were sharply intensified in the battle between John Stuart Mill (1806–73) and Sir William Hamilton (1788–1856), who in general accepted as valid anything that was intuitively obvious, especially the rules of reasoning and the evidence of the senses. Hamilton imported these ideas from Germany. For centuries, induction had also been relied on, with little criticism. As already mentioned, induction is the method of inferring the general rule from the particular instances. The more general statement that applies to many situations is derived from a number of less general statements that apply to only some cases. Induction is based on two principles: that nothing can happen without a cause, and that the same combination of causes is always followed by the same effect.

The methods of scientific induction—noted mainly because they were effective in the industries developing in the foundries, potteries and factories, and were indispensable for progress in physics and chemistry—were summarised by William Whewell (1794–1866) in an influential book *The Philosophy of the*

Inductive Sciences (1840).[3] With one foot still in the past, Whewell accepted the ancient view that the rules of thought, including the intuitive recognition of categories in visual perception, were built innately into the human mind. They were not to be questioned. This was in line with German philosophers, of whom Immanuel Kant (1724–1804) was the most influential in the early nineteenth century. Kant assumed that reason was not subject to space or time. Basically, a reasonable cause that was proposed on intuitive grounds was accepted until further observations made it untenable.

In the early nineteenth century, accompanying the further development of mathematics and the exact sciences, empirical philosophy was strongly promoted by an intellectual prodigy, John Stuart Mill. Mill put the arguments of the English empiricists of the previous century—Berkeley, Hume and Locke—into a systematic framework. He replaced intuition with learning from experience, particularly by relying on numerous observations and deducing their logical consequences. A lack of independent checks infuriated Mill, who, in 1865, wrote a long condemnation of Kant's support for intuition. To Mill, the combination of induction and intuition was the way to errors of thought, and the German philosophers were a threat to right thinking.

Apart from governing India from a distance, writing a stream of articles in favour of freeing slaves, the liberation of women and guiding the social conscience, Mill's contributions were crucial for the development of experimental science, especially biology.[4] In his book the *System of Logic* (1843), Mill laid down the rules for the inference of causes from effects. I recommend them as a guide to any budding investigator. Mill distinguished between necessary causes, sufficient causes and possible causes. He accepted multiple causes operating in parallel and repeated Newton's advice that 'no more causes of natural things are to be admitted than such as are both true and sufficient to explain the phenomena'—a principle that is usually called 'Occam's razor'. A necessary cause is one that is logically required. A sufficient cause is one that is adequate but there might be more to be said.

Second, Mill did not accept anything just because it appeared to be intuitively so or was a reasonable guess. Even Mill's most abstract works were aimed against the German a priori school, called 'Intuitionism' and best known in the works of Kant. Mill denied any ability or performance that was reckoned to be 'innate' and instead derived all human knowledge from human experience: 'The notion that truths external to the mind may be known by intuition or consciousness, independently of observation or experience, is, I am persuaded in these times, the great intellectual support of false doctrines and bad institutions' (Mill 1843). To him, all causes, inferences, conclusions or categories were obtained by making bare observations, noticing regularities and then deducing the causes. A century before Jean Piaget, therefore, the development of the human mind was an exercise in self-education by trial and error.

Mill was well known and influential among the scientific community in London, where he became Secretary of the India Office and later MP for Westminster. Mill, once said to be the cleverest man in the world, also demonstrated that there was more to science than observations, empirical laws and rules for scientific investigations. As a result of his efforts, the teaching of philosophy in England was deflected from the path led in Europe by Kant and saved England from the Gestalt and holistic psychologists of Vienna and Zürich.

At the time, these ideas had no effect on research on the vision of the bee. Instincts were proposed and accepted as innate as explanations of behaviour. An experiment was an observation, followed by a guess about the cause. Most of the philosophers of the nineteenth century and more recent times were of no further help, being engrossed with the meanings of words and the theoretical basis of physics, astronomy, mathematics and the relation between mind and matter. Towards the end of the century, however, Mill's methods were taken up in the United States by C. L. Morgan (1890), Edward Thorndike, Margaret Washburn and J. B. Watson, who opened the subject of experimental psychology, but little of this spread back to Continental Europe. We can detect Mill's influence in the work of Herbert Spencer Jennings (1868–1947), who concluded that the detailed behaviour of lower organisms was controlled largely by learning by trial and error (Jennings 1906).

Mill's rules allow us to make deductions from observations and experiments, provided we do not ignore some hidden cause. Unfortunately, we can never list all possible causes. Another difficulty has always been to arrange sufficient examples so that a common cause is established. A third difficulty is that a number of facts might be totally unrelated but we might still derive a theory from them. A fourth difficulty is that we might be totally ignorant of the type of system being studied—for example, whether memory is a solid-state molecular transformation, a wet chemical reaction or a rearrangement of connections between nerve cells—so no lasting conclusions can be made. Perhaps the most common error is to waste time on facts that prove useless. The most dangerous error is to postulate a hidden cause, give it a name, raise it to the status of reality and then validate it by devising an experimental proof, while still missing the real cause. This is called 'misplaced concreteness'. These potholes produce errors of deduction and account in part for the hesitant and meandering progress in every branch of science, but especially in vision and analysis of the nervous system, where we start with multiple causes in parallel but no map.

For experimentalists, Mill's best contribution was originally Newton's idea that a postulated cause must be capable, realistically and mechanically, of producing the effect that was observed. He went further and advised that the nature of the postulated cause should be demonstrable by an independent means. We can translate this as 'list the components and find out how the mechanism works, then confirm it experimentally'. How few students of animal behaviour even bother to list the components! How easy to label the performance 'cognitive'!

Figure 2.1 a) Two patterns of similar area and position on the targets that are easily distinguished by bees in the apparatus shown in Figure 1.1d. The patterns are fixed, not shuffled in position, but are interchanged every five minutes to ensure that the bees look at them. The square is the rewarded pattern. b) The intuitive idea that the bees 'compare the stored image with the current image of another shape' by the areas of overlap and non-overlap when they are superimposed.

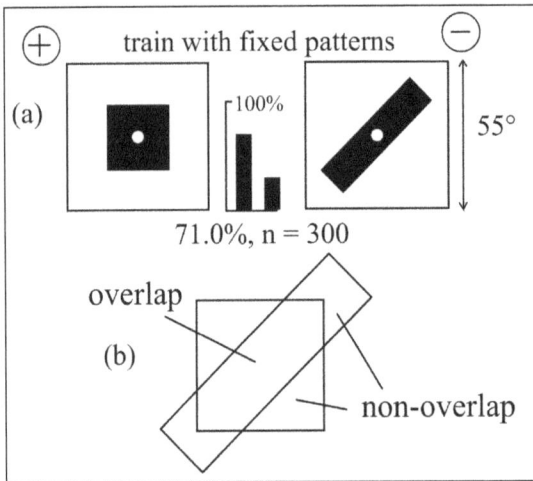

Source for (b): After Wehner (1981:Fig. 86).

How Mill's logic was applied to bee vision

Now we come to the part that requires a little concentration. Figure 2.1a shows the success rate when bees are trained to select the rewarded black square, but those choosing the oblique rectangle receive nothing. The two patterns were in the apparatus in Figure 1.1d, of the same size at the same centre, to give the bees a fair choice. The two targets were interchanged every five minutes to teach the bees to look at them and not simply learn to go left or right. The bees clearly learned this task, but the real score means little because it depends on the length of the training, the hunger of the bees, and so on.

Until recently, the success of the bees would have been explained by the difference in shape of the square and the rectangle. It was proposed that the bees measured the region of overlap and the regions of non-overlap (Figure 2.1b). This is a general explanation that could apply to all patterns, but we have no indication that it is the correct explanation. In fact, it was a misleading guess.

Figure 2.2 a) Training patterns; the bees avoid the bar. b) The trained bees avoid the bar when displayed versus a bar that is moved down. c) They avoid the bar when displayed versus a bar with modified edge orientation. d) They distinguish the edges of the original bar from the square. e) They confuse the edges alone with the original bar, which they fail to recognise. These results show that the trained bees recognised only the orientation of the bar edges in the expected position. They say nothing about the square.

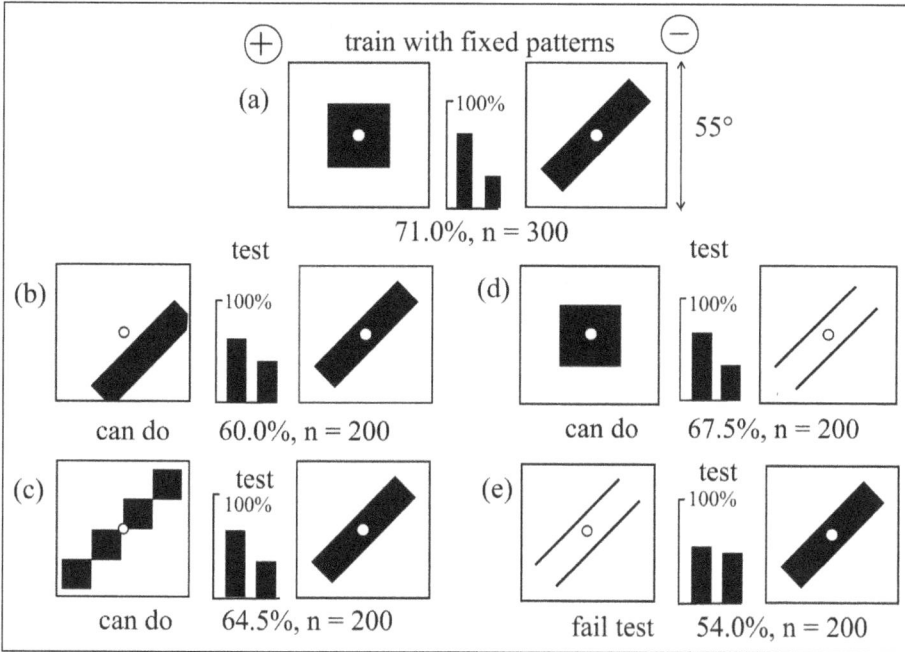

Now turn to Figure 2.2, where the trained bees are given four tests to reveal what they have actually learned. They distinguish the original bar in the training from the same bar moved down (Figure 2.2b), so they are sensitive to bar position. They distinguish the bar from a similar shape with stepped edges (Figure 2.2c), so they detect the orientation of the edges. They distinguish the square from a hollow bar (Figure 2.2d), which by itself tells us little; however, the trained bees have equal preference for the hollow bar and the original bar (Figure 2.2e). In this case, whatever makes the bees avoid the oblique black bar is displayed on both targets in Figure 2.2e, so in the training they have learned to avoid the orientation on the edges of the oblique bar.

Figure 2.3 a) Training patterns, as before. b) The trained bees fail to distinguish the square from a horizontal bar; or c) from a bar with stepped edges. d) They also fail when the square and the bar are both moved; and e) when a black spot is added. They do not recognise the square in any of the tests.

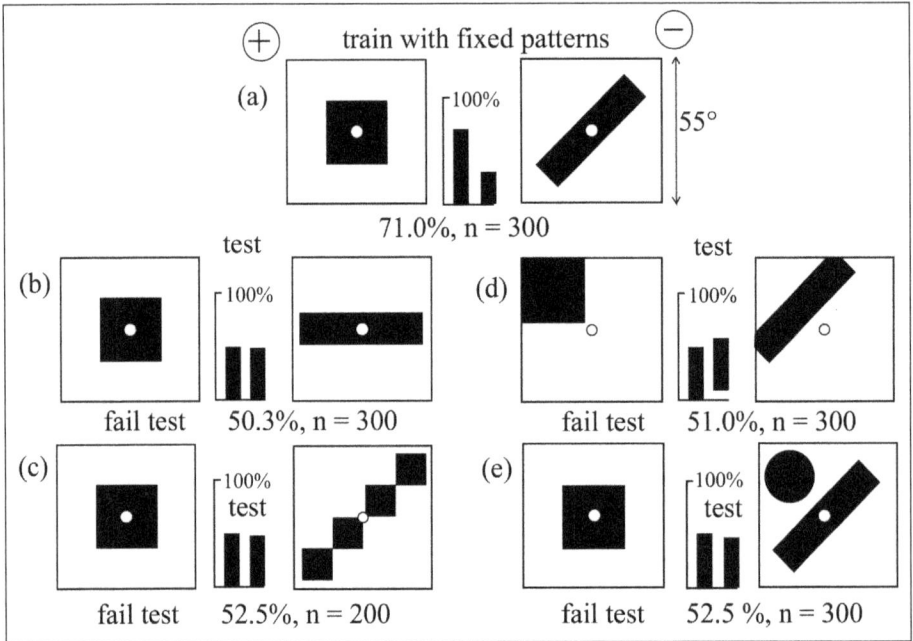

Figure 2.3 shows what they did not learn. The trained bees were tested with the original black square in the training versus other patterns. The square was chosen equally with the rotated bar (Figure 2.3b), or with the bar with stepped edges (Figure 2.3c), because the orientation cue was not displayed on either target. The square was chosen equally to the bar when both were moved (Figure 2.3d) because the expected parameter was not in the expected place. Finally, the original patterns were chosen equally when a large black spot was added (Figure 2.3e), because this additional parameter was not expected, so the bees failed to recognise the place. Clearly, the bees had not learned to recognise the square or the bar, only a simple cue in the expected place. This experiment does not exclude the possibility that bees can recognise some patterns that were untested.

In my search for the cues that bees use, I tried a large number of pairs of patterns. Some they discriminated, others they did not. Between every pair of patterns that bees are able to discriminate there is obviously a difference, or more than one, which the trained bees detect. When we find a number of successfully discriminated pairs of patterns (Figure 2.2), and the pairs have only one common factor or common difference, we have probably found the common cause—that is, the cue that they detect, if we have persisted in looking at enough examples. This is Mill's 'Method of Agreement' in the search for factors. In the above example, the parameter was the edge orientation at the expected place.

Although thoroughly criticised by later writers, Mill's rules are still useful in the crucial design stage of experiments and in making inferences from them. When there are groups of pairs of patterns that bees cannot be trained to discriminate, there is likely to be a common factor that is missing from all of them (Figure 2.3). This is Mill's method of 'Agreement in Absence', which is very decisive if the number of instances is large. In all cases where an effect is missing and one possible cause is consistently absent, there is a strong presumption that these two circumstances are cause and effect in spite of possible multiple causes. There might, however, be several reasons why the bees fail. We saw this in the failure caused by the addition of a spot, which was in fact a strong salient parameter that was easily recognised by bees.

When bees fail to learn to distinguish a pair of patterns, or trained bees fail to discriminate in tests (Figure 2.3), we call it negative evidence. It is good evidence of the inability to perform, which is not the same as absence of evidence. Although it can be observed and confirmed, the bees' failure to discriminate still gives us a problem. We have to be careful that the same bees can discriminate other similar pairs of patterns, so that we are sure that they are not failing for some trivial reason. The bees might be unable to learn to discriminate because they fail to notice the patterns or the cues displayed, the two patterns might display the same cues or the bees might be unable to stabilise their eyes on the targets. Similarly, when trained bees are tested, they might fail because they detect an unfamiliar cue on the test targets or because the preferences for the available cues are balanced on the two targets (Figure 2.3e), so it cannot be assumed that one cue has been omitted. The ambiguity makes the research harder and longer but the situation can be resolved with a sufficient number and variety of tests. The solution to the difficulty is to take two patterns that bees prefer equally despite extensive training—that is, that they cannot learn to distinguish (Figure 9.12)—then add a parameter that bees recognise. The resulting learning is then a positive demonstration that the parameter is effective when the patterns are not.

Similarly, when the bees succeed, we must devise control experiments to show that there is not some other irrelevant cue, such as an odour, a difference in size, range, position or illumination, which enables the bees to 'cheat'. When we find two patterns that bees easily discriminate, but with no known cause, we can suspect that there is a previously unknown parameter.

In the study of bee vision, I assume that the bees detect certain features in the patterns—called parameters—to which their feature detectors are innately adapted. They remember something in the brain derived from the feature detector responses—called a cue—that is a small part of the whole pattern. There is no evidence that the bee brain is able to reassemble the visual inputs to make a more complete picture. This is 'absence of evidence' and not conclusive. There are many pairs of patterns that look different to humans, however, which bees are unable to discriminate in training experiments (Figure 2.3 and Chapters 9 and 11), so I infer that they generate no cue or no difference in cues. Clearly, the bees do not distinguish or recognise them. This is 'evidence of absence of recognition' in these examples. Now, to emphasise the effectiveness of the empirical method to investigate the matter further, look again at Figure 2.1b. The intuitive idea of overlaps and non-overlaps of shapes never surfaced in the exposure of the parameters. It was just a guess.

Problems of applying theory

With logic defined, why was progress so slow?

In the late nineteenth century, apart from the efforts related to Darwin's theory of evolution by natural selection, biology produced almost entirely what we now call 'natural history'. It was a period of belief in the progressive improvement of understanding, but the biological sciences were mostly descriptions of species, anatomy, development, fossils, geographical distributions, physiological systems and the chemistry of some processes in animals and plants. At the very end of the century, detailed anatomy, histology and the physiology of the nervous system became established. The analysis of insect vision took off in Germany with the works of two giants, Grenacher (1879) and Exner (1891), who studied stained sections of the insect retina with newly invented compound microscopes and provided the basis of our modern knowledge of the compound eye (Chapter 5).

The advances in the nineteenth century were made, for the most part, by men who were thoroughly conversant with the knowledge of their time and who made more than one outstanding advance. Romanes was famous for his work on the nervous systems of primitive animals. Forel was a distinguished entomologist. von Frisch dedicated his whole scientific career to the study of the honeybee. At the time, the interest lay in describing and making sense of the performance. Bees were trained and the brain of the bee was recognised as adequate for its own dedicated tasks, but there were no techniques to reveal mechanisms. Possible mechanisms were not mentioned until the late twentieth century. The scientists knew how to plan an experiment, and the experimental equipment was available, but there was no fund of knowledge of what tests the bee had previously passed or failed. The early experiments show that the early scientists simply did not know what to do.

A number of properties of the visual system contributed to these difficulties.

Systemic hurdles

1. Diversity

A drag on the advance of ideas was the enormous diversity of insects with miraculous behavioural patterns. The fascinating descriptions concealed the lack of analysis. The first stage was the assumption that sign stimuli were used in the recognition of mates, food or predators. The sign stimuli, however, are the consistent signals such as colours, movements or markings, which humans intuitively assume the animals detect. This work is still going on. It is far more difficult to identify the real features that enter the nervous system. They are much smaller than the sign stimuli and are probably common to all insects. When the feature detectors have been identified from behavioural experiments, they can be sought by electrophysiology.

2. Parallel processing

As Grenacher, Exner and Ramon y Cajal showed in great detail about the turn of the century, the insect retina has thousands of photoreceptors in parallel and feeds into several arrays of pathways in parallel that pass through successive layers of neurons. The first requirement is to list the anatomical components—a formidable task not yet complete even in the most studied insect, the fruit fly. Finding the rules of action requires hundreds of tedious physiological recordings of the individual nerve cells during the continuing behaviour, as well as decades of obsessive study of their connections. Having done all that, there is still a long way to go, because the nervous system functions by the coincidences between activities of neurons in arrays and they learn, so they are not constant.

To analyse a system of elements in parallel it is useful to know what the mechanism cannot do or what is not there. For example, if a genome is known not to contain a particular gene, it is known that that gene is not essential for any remaining process in the living animal, and this fact helps the analysis of the gene's function elsewhere. Whether or not he had a motive, if the suspect proved his absence, he could not have been the agent. This is the correct use of the evidence of absence. In law, it is the theory of the alibi. Further, if there was no motive to kill grandma, no sign of violence and there was no-one around at the time, grandma probably died of natural causes, so a positive deduction might be made. Positive evidence of absence is perfectly valid and is essential when there are many possible causes that can be eliminated one by one. Critics of negative evidence, please note.

There are other ways to separate multiple causes. In the 1920s, R. A. Fisher (1935) developed novel experiments randomising different treatments of agricultural crops with different amounts of fertilisers, soil types and water, to separate the effects of possible multiple causes. Right from the beginning of our work with trained bees, we eliminated some possible causes by randomising those aspects of the stimulus that we wanted the bees to ignore, as in the papers by Lehrer et al. (1988) on the measurement of range irrespective of size or position and van Hateren et al. (1990) on the discrimination between horizontal and vertical bars irrespective of bar width or position. The method appeared pathetically late in the analysis of insect vision.

From 1909 on, bees were trained to discriminate a pattern or between a pair of patterns displayed on movable targets that were shuffled in position to prevent the bees learning their locations (Figure 1.1d). This had the unexpected consequence that the bees ignored local landmarks outside the training targets (Chapter 12).

From 1926 onwards, the trained bees were also tested with patterns that were related to the training pattern or various parts of it. The method was progressively refined after 1990 as more cues within the bees were recognised and their limits found. It was essential that some of the test patterns contained the necessary cues and others did not. Results obtained using all the methods, in the hands of many researchers using a variety of training and testing techniques, were eventually all explained by the same few simple cues in parallel, as listed in Chapter 9.

3. Feedback loops

One of the great hurdles that had to be surmounted in the analysis of the nervous system was the control of the action by a part of the action itself. The feedback loop makes the system look purposeful. In the nervous system, a feedback loop can be within the animal—for example, the sensors at our joints and in the muscles keep our limbs in constant positions under a varying load. We would not be able to stand up without these feedback loops. The loop can be outside the body. We guide our hands and feet with our eyes. We hear our own voice and adjust it under control. Feedback loops were not really understood until cybernetics became popular in the 1960s, and it took decades before the experimentalists found even poor methods of analysis by breaking or clamping the loops, or more refined methods by replacing the natural loops with artificial ones.

4. Preconceived ideas

Despite Mill's efforts, a trust in intuition and acceptance of causes that looked reasonable were serious sources of error in every decade. Intuition was represented in every aspect of bee vision in the twentieth century—not only by the Gestalt psychology that influenced Mathilde Hertz and, through her, other studies of bee perception. We will find numerous examples in the coming pages, even though Morgan (1890) advised readers to 'endeavor to distinguish observed fact from observer's inference' and 'to apply Occam's razor, especially to proposals of cognition in animals'.

Almost universally, it was assumed that the behaviour of insects was genetically innate and that a sign stimulus initiated a predetermined act. Quite the contrary: if the insect is allowed to initiate its own movements, it quickly modifies them by rapid learning that eliminates errors and selects the effective responses. This was described by Jennings (1906) as 'reaction by varied or overproduced movements, with selection from the varied conditions resulting from these movements'. We now know that this applies to simple motor movements and to relearning the effects on sensory inputs by all parts of the central nervous system, even in the posture of the legs and the supposedly fixed opto-motor response.

The most insidious preconception, damnably difficult to escape, is to read human sensations into animal systems—and in this case to assume that bees 'see'. For most of the twentieth century, this anthropomorphism led to an assumption, derived from the study of primate brains, that in the brain of the seeing bee there was a structural layout of the image—a bad guess that was not excluded until 2005 (Chapters 11 and 12).

Backsliding into intuitionism

From about 1900 onwards, two theories of general application and wide influence appeared in studies of vision. First, Wertheimer (1912) found that the mechanistic concepts of his time failed to explain why human perception seemed to proceed from an assessment of the whole to the recognition of the parts. Nowadays, there is abundant evidence of human top-down processing, as in visual search and size constancy. Wertheimer was followed by Koffka (1924) and Koehler (1925) and the Gestalt theory spread—notably to bee vision. The word 'Gestalt' is translated as 'configuration of a whole'. The modern equivalent is the recognition of configurational layout.

In human vision, the Gestalt theory included numerous laws and factors. The most important is that the human visual system transforms the stimulus into the most perfect that the situation allows—for example, a ring of dots becomes a circle. Characteristics such as regularity, inclusiveness, symmetry, simplicity and unity are detected intuitively, even if present in an imperfect form. Similar patterns or objects that move together are seen as groups and are classified into categories. An enclosed surface is seen as a shape. This was the accepted theory when Hertz researched bee vision. Mill would have suggested that Gestalts were all learned.

The second corrosive influence was the development of ethology by Konrad Lorenz and Niko Tinbergen, who used intuitive concepts such as 'fear' and 'drive'. In the 1930s, Lorenz published numerous observations of the behaviour of free-living animals and, with Tinbergen, produced a comprehensive theory of instinctive behaviour. The responses were 'innately controlled movements' initiated by a 'specific releasing mechanism' (internal) or a 'sign-stimulus' (external). They were driven by 'reaction-specific energy' that could accumulate and also be 'depleted' by the activity itself. Actions were directed with an intention of obtaining food or a mate or driving away a rival. Behaviour was basically stereotypical and inherited, but could be modified by learning. The theory was developed into a hierarchical model with many layers and parallel pathways.

In my terminology, this was mostly detailed description of performance, however wonderful and colourful, followed by an intuitive guess about causes, followed by a redescription of the performance in terms of the supposed causes. There seemed to be an intellectual block to empirical analysis of mechanisms. In a biting critique, Lehrman (1953) found serious flaws:

> It involves preconceived and rigid ideas of innateness and the nature of maturation. It habitually depends on the transfer of concepts from one level to another, solely on the basis of analogical reasoning. It is limited by preconceptions of isomorphic resemblances between neural and behavioral phenomena...Any instinct theory that regards 'instinct' as immanent, preformed, inherited or based on specific neural structures is bound to divert the investigation—from fundamental analysis.

Mill would have loved that.

After World War II, ethology became a powerful force in Germany and spread with Tinbergen to Oxford, then moved to Madingley Hall, Cambridge, with Thorpe, Hinde, Bateson and many others. Even in my student days, however, Gray and Lissmann (1946) were taking pot shots at innate behavioural patterns. Throughout the 1960s and 1970s, the ethologists said they made an enormous effort to understand mechanisms by studying behaviour, but they in fact redescribed it in other words. The available theory could never be satisfactory because the amount of information in a behavioural pattern was dwarfed by the vastly greater information required for its explanation.

In his *Biographical Memoir of Niko Tinbergen*, Robert Hinde (1990) summarised the conflict between ethology and comparative psychology, with an effort to paper over the cracks: 'both sides were partially right.' This was whitewash and rubbish. Gradually, the battle died down, however, as it became obvious that ethology was descriptive natural history of whole animals, mainly vertebrates. Those with the techniques to study neural mechanisms of behaviour formed their own subject, neurobiology, with their own Congress of Neuroethology, but that alone did not help to identify the mechanisms of bees' visual discrimination, because most of the researchers on bees had been educated in the Continental system based on the philosophy of Kant and his successors. The whole subject was dominated by German biologists who somehow thought that empiricism was immoral.

Empirical laws

For the past 200 years, physical scientists have studied a system by changing the input progressively and recording quantitative data from the responses. Investigators looked for regular relationships between inputs and outputs, which were called empirical laws. The search for an empirical law was useful because a theory might be found to be consistent with it. When they were not able to see why such a law should exist, they hesitated to extend the law to cases varying much from those that were in fact observed. Mistakenly, the search for empirical laws was taught as a way to proceed scientifically. I still remember those never-to-be-forgotten practical physics classes at school in which we measured the volume of air at different pressures and the length of a copper bar while it was heated. We then described the empirical relationships as laws that were explained in terms of the motion of molecules, which became more and more real as we listened to our enthusiastic physics master. In fact, we had no evidence of molecules in our experiments—and they did not come into science by this route.

In the case of vision, the exact relation between input and output is even less useful. In the bee, these searches for exact relations between the features of the pattern and the percentage of correct responses hampered progress. Dozens of such papers by Crozier and his colleagues at Harvard and by Selig Hecht at Columbia were published in the *Journal of General Physiology* in the 1930s. This approach was futile when applied to a system such as vision, with many separate channels of transmission in parallel.

Starting on a new tack, Lindauer and Wehner in Frankfurt looked for empirical laws from 1965 until about 1973. Wehner (1967) trained bees to come to a square cross, or alternatively to a regular striped grating, and then plotted the choices between the training target and the same target rotated by various angles. A theory could be found for one set of results or the other, but not for both (Chapter 4).

In one favourable example, Wehner (1969) trained with a broad single bar and then plotted the scores of the discriminations between the original bar and the same bar at different angles. The scores fitted the increasing mismatch between the positions of the areas of black on the training and the test targets (Figures 2.1b and 4.5). To explain the results, Wehner proposed that the bees remembered a copy of the training pattern and compared it with each test pattern, but this guess led to the erroneous conclusion that the image was remembered in the brain of the bee.

Explanations consistent with the data

Most experimental data are explained by making an intuitive hypothesis that is compatible with them, such as postulating that a cat ate the missing cream. It is not good enough, however, to accept a theory just because it is consistent with the data. This was, however, the commonly accepted way to proceed. The cat starts as a guess and, even if never seen, it can become an explanation of other events. Later authors refer to the cat as though it was seen eating the cream. If there really was a cat, however, it might have been innocent. Much of the discussion about theory in the scientific advances of the twentieth century was an effort to cover up the sad consequences of accepting erroneous explanations that were compatible with the data. In fact, like sex, everyone was doing it, and it was regarded as an ultimate necessity. Everyone hoped they had the true theory and that they would not be responsible for erroneous concepts.

Popper and the method of disproof

There is a way to make progress. If there is no cat, we can look elsewhere for the cream. This is the 'theory of disproof' attributed to Karl Popper (1935), whose ideas were popular with students about 1950. In Popper's view, progress requires a definitive experiment that excludes the cat. In the study of honeybee pattern vision from 1914 to 1989, the cat was the idea of parameters as parts of the image and memories within the bees, or alternatively the idea of a complete retinotopic memory of the image, but for some strange reason, the critical experiments to exclude either of these popular proposals were never done.

The contribution of Popper—the idea that theories could never be verified, only disproved—partially resolved the otherwise difficult acceptance of induction. Theories can never be proved because unforeseen observations can appear in the future. After Popper, theories that were consistent with the facts could be accepted as valid for the time being, but only if the effort to disprove them had been made. The disproof might have been useful if the discipline had in fact been applied when a theory presented itself, but none of those working on bee vision ever validated their theories. Also, until the 1990s, different theories were presented separately without reference to one another. So much for Mill, Popper, Kuhn and the rest of the philosophers of science.

The new paradigm

In the mid-twentieth century, explanations based on intuition or empirical laws gave way to a new attitude, based on fiddling with the mechanism; it was the era of Meccano, mending a bicycle or a radio, which led to understanding signals and machines. With the development of electrophysiology and cybernetics after 1945, the idea developed rapidly that sense organs acted as filters in parallel pathways. Recordings from neurons in the visual systems by Kuffler (1953 on the cat), Barlow (see below) and Burtt and Catton (1952 onwards, on the locust) revealed examples that responded to complex stimuli such as a moving spot of the right size, but not to a simple flash of light. In 1952, at a supposedly private meeting in Cambridge while I was a research student, Horace Barlow gave a seminar on the responses of frog retina ganglion cells, including one type sensitive to a small black spot, with an inhibitory surround. An American spy at the meeting distributed a report to all holders of US Office of Naval Research (ONR) grants in the United States (Mollon 1997) and Ted Bullock certainly knew of it. Much later, however, Lettvin et al. (1959) created widespread interest when they published similar work. That year, I visited Jerry Lettvin on the way to California to work on a book with Bullock and was less than impressed by their equipment, methods and flash in the pan.

In fact, the promise of this work was not fulfilled, partly because in invertebrates the neurons recorded were not representative and not identified; the real cues were not well defined and neuron specificity was not really demonstrated. The main reason, however, was that recording from only a single neuron at a time simplified the data, but blinkered our understanding about arrays in parallel and coincidences between neurons. In the years 1955–62, innate edge detectors were found in huge arrays in vertebrate retinas and mammalian cortex. Jander (1964) was stimulated by these studies of vertebrates. He combed the literature and proposed the general hypothesis that small feature detectors were the basis of vision in ants and bees; but, without the necessary equipment or techniques, it was only a guess. Later, McCann and Dill (1969) recorded from edge detectors in the optic lobe of a fly. Those working on insect visual behaviour flatly ignored these indications. There is some excuse for heading the wrong way when signposts are undeveloped, but no excuse when signposts from related disciplines emerge.

Forty years later, we are no further advanced in identifying the part played by any neuron for anything subtle in the visual repertoire of the honeybee (or other invertebrate). There has been a huge effort on the illusory unit motion detectors of the fly, but negligible electrophysiology of pattern vision of any invertebrate because no-one knows what features have been detected.

Kuhn and progress

Thomas Kuhn (1922–96) argued that scientific topics passed from an initial chaotic pre-paradigm stage into a productive period of growing understanding and satisfying results that fit the theories of the time. There is nothing there to object to. Ultimately, perhaps because new methods or technology are found, there is a crisis because anomalies accumulate and the accepted theory is no longer tenable. At this point, a new experiment or theory points the way to the next period of advances, and the cycle repeats.

At first, a new theory should be simple—that is, it should propose few new postulates or variables with acceptable relations between them. To survive, the theory must pass every test and there should not be a better theory. The new theory must not account only for previous results and anomalies, it must predict further developments. Kuhn observed that from time to time a new technique—a new means such as the availability of electric power or a real discovery such as of x-rays—altered the way facts were collected and explained and there was then a shift away from the old towards the new. The new paradigm would not be acceptable to all, especially to those whose career was based on the old one. All 'theories' became transient and replaceable.

The route was familiar to Mill. Collection of data led to an inference or guess of causes. As a paradigm developed, a new theory would become more and more confirmed. At best, this was induction; at worst, mere intuition. Next, it was essential to design experiments that would probably establish or perhaps exclude the new theory. No philosopher, however, explained how to design the right experiments, which required intuition, lots of thought and extensive knowledge of the literature.

The new territory, filters and neurons

In vision, the way forward was indicated by the arrays of single neurons in the primary visual cortex in mammals that each responded to an extremely simple feature—notably, the orientation of a moving edge (Hubel and Wiesel 1959). In this case, every neuron could be represented as a spatio-temporal filter with its own field, but the meaningful signal was carried by the coincidences of many neurons in parallel. The artificial-vision fraternity enthusiastically embraced the concept—as reviewed later by Hinton et al. (1986)—but robot vision needed much more.

Filter theory, like electrophysiology or neuron anatomy, is not sufficient to explain vision. First, it is essential to discover how many different kinds of parallel paths are active at any one time, exactly what is the meaningful part of the signals they carry, the excitation in each, how they are interconnected, the field sizes, the destinations, the time delays and finally the central reckoning. This leads slowly to an understanding of the way the parts work together. A systems analysis with interacting boxes (Figure 7.1) can be made only after the difficult behavioural analysis has been done—not before. Second, the visual system relies on the visual feedback from the movements that it controls. Vision is active, so the pinned-down preparation is only a beginning.

The new postwar paradigms in physics—notably, information theory, cybernetics, feedback, signal-to-noise ratios and modulation transfer functions—and the expectation that useful artificial vision would quickly follow, sustained the enthusiasm for recording from amphibian and arthropod visual systems. The studies of mammalian vision in medical schools led the way. In the period from 1966 to 1974, it became established that at the front end of vertebrate vision there were arrays of simple filters in parallel that detected local intensity, directional movement of local contrasts and orientations of edges and contrasts.

A few pioneers enthusiastically proposed that behaviour would be understood in terms of nerve-cell interactions. As more and more results emerged from a greater variety of preparations, however, it slowly became obvious that this ideal would never be achieved, because the information content of behaviour was so low compared with that of neurons. Even in 1981, Rudiger Wehner, in his long review, could not reconcile the new work on the neurons of the optic pathways with the visual performance known at the time. It was also essential to direct research to the analysis of behavioural responses that might explain the activity of the neurons.

In the 1970s, new techniques became available for identifying single dye-marked neurons that were also recorded with a microelectrode, so theoretically, the physiological interactions could be reconciled with the anatomical pathways. There was some excitement at the prospect of working inwards from the retina towards the brain to discover what features of the image were detected. There was no final understanding, however, even of the directional motion detector neurons of the fly's lobula plate, which were exceptionally large. Even in this case, the results were at first misleading because an animal with a fixed head could not initiate the movements that were part of its vision.

In the insect optic lobe there are dense arrays of neurons in parallel and in series and it has proved impossible to track a sufficient fraction of the synaptic connections or decide which are inhibitory or excitatory. Many of the neurons—and perhaps all the small ones—function by graded potentials and have no action potential. Although it was possible, through the second half of the century, to record from many of the large ones in the insect optic lobe, behaviour was not explained by neuron responses. The reasons why are that the units of visual behaviour are not known; bees are freely moving and their vision is active, with continual learning from visual feedback from their own movements; they respond to only a small part of the image seen by the human eye; only a few parameters have been identified; and sensory integration depends on coincidences. Moreover, the behaviour of interest—visual recognition—was stored in an unknown language and appeared only briefly.

When we recorded neurons in the deep optic lobes of insects 45 years ago (Horridge et al. 1965), we found that very simple stimuli such as flashes and moving edges were adequate and the field sizes were very large. Some responded to sound, touch or body movements. How could a group of neurons with these properties carry the signals for vision? Certainly, the reassembly of an image would be impossible. Only much later, I gradually learned that the parameters that bees detected and the cues within their processing system were also very simple and were summed in local eye regions, and that they functioned as a parallel array. In brief, large arrays of a few types of small feature detectors are summed together to form a few cues in each local region of the eye. The relation between behaviour and neuron activity lies in the coincidences of the responses of neurons with different inputs that lie in parallel in local regions. Again, the idea of multiple causes can be traced back to Mill.

Endnotes

1. Aristotle (384–322 BC) wrote the *Organon*. See Westaway (1937).

2. Francis Bacon (1561–1626), an English essayist, philosopher and politician, wrote the *Novum Organon*. See Westaway (1937).

3. Whewell also wrote *Novum Organon Renovatum*, as well as *History of the Inductive Sciences* (1837). See Westaway (1937).

4. Recounted in his autobiography (Mill 1873).

03 RESEARCH TECHNIQUES AND IDEAS, 1950 ON[1]

Several powerful new techniques became available as a result of wartime research. First, the electronic equipment developed for sonar and radar—particularly the availability of low-noise high-impedance amplifiers, the oscilloscope, tape recorder and automated cameras—encouraged a burst of effort to record everything of interest, particularly in the visual system. This was undertaken by Roeder at Tufts; Burtt and Catton at Newcastle; Parry and Pringle in Cambridge, in the United Kingdom; McCann at Cal Tech; Bishop and Keehn at the University of Southern California; Kuiper at Gröningen; and Kuwabara, Naka and Eguchi at Fukuoka. It started with only a few devotees with electronics experience and access to a workshop; much of the equipment was modified surplus from the war. Because neuron activity was visible on a luminous screen, they were exciting experiments for students and the techniques quickly spread.

Second, in the 1950s, the wartime interest in operational research spilled out into biology (and economics and elsewhere). Systems analysis brought new ideas about control systems with feedback loops. The first sign was a rash of block diagrams, in which postulated causes in boxes—such as the motion of a target—were linked by arrows to effects, such as eye motion. Next arrived a thin sprinkling of quantitative data about these boxes and arrows, and some mathematical relations that fitted the data and were supposed to summarise the interactions.

Third, many universities purchased a transmission electron microscope, soon joined by a microtome for cutting thin sections of tissues embedded in araldite (another wartime production). About the same time, various techniques for staining individual nerve cells with silver salts, originally developed by Golgi and Cajal, were revived by William Holmes in Oxford and passed on to a number of English students—notably, J. Z. Young, Brian Boycott, David Blest and Nicky Strausfeld. Correlations between the anatomy and the physiological responses of neurons and synapses began to appear, culminating in the individual identification of the recorded neurons by injecting dyes through the microelectrodes. Theoretically, it became possible to identify every neuron, record its activity and plot its various fields of sensitivity. Recording from the neurons, although hard ground to plough, was accepted as the only way to understand the control of behaviour.

Exciting new research

After World War II, Cambridge, England, like Cambridge in the United States, was a focal point for experimental analysis of the nervous system, neurobiology and animal behaviour.

There was indeed a revolution in all these fields, and of course many others too, combined with intensity of action, because those returning from war duty were catching up for lost time. They had ideas to unload, new techniques to carry them out and, for a time, they were building a better world.

In physiology in Cambridge, first-year lectures on the nervous system were given by Lord Adrian, who had been awarded the Nobel Prize in 1932 for his demonstration of nerve impulses and the discovery of the frequency code in single axons. In the practical class, activity in nerve fibres was made visible with an amplifier and cathode ray tube put together by Sir Bryan Matthews, who had perfected the string galvanometer that did the same job in the 1930s. Second-year lectures on the nervous system were given by Willie Rushton, who discovered the principle of univariance in vision, Andrew Huxley and Alan Hodgkin, who, with Jack Eccles, were awarded the Nobel Prize for medicine in 1963. At the time, they were working actively on the squid's giant axon membrane. In line with the time, the result was a set of empirical equations that fitted the data. Only later did Richard Keynes measure the ionic fluxes with radioisotopes.

The Zoology Department was stuffed with Fellows of the Royal Society and was committed to the experimental approach. Vincent Wigglesworth had a strong group working on insects. Professor James Gray was interested in the reflex control of movement. Eric Smith was an expert on the nervous system of annelid worms and echinoderms. Carl Pantin was the world expert on the nervous system of the sea anemone, such as it was.

On my arrival with a PhD studentship, I was given a key to a room on the second floor. It was quite empty and I had no particular topic. It was still vacation so few people were around. My nominal supervisor, Carl Pantin, had gone to Brazil for 15 months, so I went to the Marine Laboratory in Plymouth on my motorbike. I had been there on a course in marine biology in my second year at university. This time, I walked in and asked the director, Freddie Russell, if I could work there for a while on the Cambridge research table. There, quite by chance, I was instructed in the art of staining nerve fibres with methylene blue by a remarkable old professor. J. S. Alexandrowicz was a Polish refugee, once

Minister of Health and Director of the Medical Institute in Lwów. He had been a medical officer in the Polish and then the Russian armies, escaped through Persia (now Iran) and found sanctuary in England and the Plymouth laboratory after an adventurous war. Coming from the School of Zawarzin, Orlov and other Slavic neuron stainers, Alexandrowicz had, in 1932, published the classical account of the innervation of the crab heart. In the library, it was he who showed me where to find all the literature on the nervous systems of invertebrates and he thought nothing of translating verbatim from six or so European languages.

Back in Cambridge, as a research student, the situation was equally favourable. I happened to bring back to Cambridge some jellyfish, *Aurelia*, from the Norfolk coast and examined them with a newly invented phase-contrast microscope that Victor Rothschild had purchased to look at frozen bulls' sperm. Rothschild had been in charge of security of research during the war; it was he who had stolen a highly secret magnetron from Sir Mark Oliphant's lab and then returned it next day with a stern warning about security. Immediately, when I used his fancy new microscope, a network of large nerve fibres was made visible in the living state. This was something that no-one had previously observed, so I had a discovery in the bag that had to be exploited. That is why I became an electrophysiologist. I built all the electronics for recording from nerve fibres and inherited a massive old mahogany oscilloscope camera from Jerry Pumphrey, who had left for Liverpool in 1949. Pantin was still in Brazil, but he was so shy of students that he would run out of his bolthole if a student knocked on his door.

Before the war, Pumphrey and John Pringle had been visiting fellows with Bronk and Hartline at the Johnson Institute in Philadelphia, and had returned to Cambridge with the most advanced techniques of recording from nerve fibres. In 1938, Pumphrey had published the first recordings from chemoreceptors of the frog, while Pringle had a series of papers on recordings from mechanoreceptors in insects. In 1941, Pumphrey had been transferred to Admiralty Signals at Witley and was made responsible for calibrating radar for early warning of approaching aircraft. Pringle had been sent to the Teddington Radar Research Establishment (TRE) and for a time was in charge of all airborne radar work in Britain. He developed radar responder devices that led bombers to their targets. Alan Hodgkin had also been in radar research and had improved the device for detecting the small reflected signal. Many other brilliant biologists had been boffins in the war and it was no accident that when they all returned at the same time, Cambridge became a beacon of progress.

At the end of the war, when huge amounts of surplus equipment became available for next to nothing, they equipped their laboratories in a manner that would otherwise have cost a fortune. They also acquired large amounts of exactly the right kind of junk to make any other equipment that was needed. I remember finding in the basement of the Zoology Department dozens of American power packs, transformers, oscilloscope tubes, boxes of assorted capacitors and resistors, automated gun cameras, pentodes and double triodes for push–pull amplifiers. Plenty more could be purchased cheaply in London in Tottenham Court Road and that was how I equipped the Gatty Marine Laboratory for electrophysiology in the 1950s.

Looking back, it is clear that the analysis of the nervous system was driven by the technological possibilities, as the electronic and optical instruments became available in the hands of those who understood them.

The Cambridge laboratories all had excellent workshops, with managers who provided tools, sheet metal and nuts and bolts, and willingly showed research students how to build whatever they needed. Superficially, Pringle was rather formidable, and did not supervise research students, but he answered my questions readily enough and allowed me to examine his equipment.

To be sure of plenty of jellyfish, I spent the next summer at the marine laboratory at Millport, on the Clyde estuary in Scotland. By chance, Gray and Rothschild were both there for the salmon fishing further north and I was able to show them the first recordings from a coelenterate nerve net. I spent the rest of my 10 years at Cambridge shuttling between marine laboratories, working on a range of fascinating animals, such as sea slugs, sea mice, sea pens, comb jellies, hydromedusae and corals on the reefs of the Red Sea. Later, while we were both at the Stazione Zoologica di Napoli in 1955, Pantin wrote to his friend Mick Callan, Professor of Zoology at the University of St Andrews, Scotland, who appointed me to the Gatty Marine Laboratory as a lecturer. I left Cambridge because there were no marine animals there and I needed a job.

In the three decades after the war, there was also enormous growth in support for universities and in particular for research. Cash for equipment and technical help was available for the asking. In 1958–59, I spent 15 months in California, working on a book with Ted Bullock, and visited most of the invertebrate electrophysiology labs in the United States, absorbing American methods of funding research, so that I was able to assemble a group quickly on my return to St Andrews.

The retina

Arriving back at the Gatty from the United States, my research fell by accident into the topic of the compound eye. In 1962, Burtt and Catton at Newcastle published in the *Proceedings of the Royal Society* a ludicrous account of the optics of the compound eye of the locust as a diffraction grating with summation of rays at different levels in the receptor cell layer. This was impossible because the cells were full of black screening pigment. Three newly arrived students, John Scholes, John Tunstall and Steve Shaw, decided to tackle the insect retina with intracellular microelectrodes. Adopting the best techniques that I had seen in Cambridge and the United States, we set up new heavy, steady benches, designed and built our own flat-bed electrode pullers, operated by a spring, and copied the Bak pre-amplifier with compensated input capacity and neutralised grid current. At the height of the Cold War, we imported large, heavy Russian copies of Leitz micromanipulators with a grease plate and extra-fine screw. We designed and built our own Cardan arms, using parts from tank gun sights, with a rapid-release screw, as on a sextant, with adjustment to one-tenth of a degree. We purchased narrow-band interference filters to arrange in a filter wheel for rapid changes of wavelength. At first, we used very small white pin lights close to the eye and only later moved to a xenon arc. Following the techniques published in 1961 by Ken Naka, we described the receptive fields of the locust photoreceptor cells by intracellular recording.

Because he worked through the night, when the locust eye became 1000 times more sensitive, Scholes discovered the 'bump' potentials caused by capture of single photons for the first time in an insect eye. Tunstall showed that the fields of the locust retinula cells were uncomplicated and Shaw explored the lamina monopolar cells. With Callan's electron microscope in the Zoology Department, Tudor Barnard described the palisade that appeared in the dark-adapted eye and altered the light-guiding properties of the rhabdom. After all, there were no peculiar optics. A summary of the results appeared in the *Stockholm Symposium on the Compound Eye* (Bernhard 1966).

As published in the same symposium, Pete Shepheard discovered that a stationary crab eye remembered a retinotopic projection of the positions of surrounding contrasts, even with a brief exposure. Rudiger Wehner, who was just beginning his period of training bees to come to black bars on a vertical surface, noted this performance. The crab and the bee detect black/white edges quite separately from broad areas of black, with corresponding types of input channels, phasic for edges and tonic for areas. The crab eye responds at angular velocities less than earth speed (15° per hour) and the accuracy is much better than the interommatidial angle. The results were totally at odds with the Reichardt model (see below) of motion detection by the compound eye, as was the behaviour of the freely moving crab eyestalk in tremor and when recovering from a voluntary eye movement.

After the war, there was rapid growth at Baltimore and then Harvard, with Steve Kuffler, Furshpan and Potter. Bullock and Hagiwara set up an electrophysiological laboratory at the University of California at Los Angeles and found the crab heart ganglion to be the perfect preparation for synaptic interactions and spontaneous rhythms in single identified neurons. At the same time, Arvanitaki, in France, showed the way to record intra-cellularly from the giant nerve cells of *Aplysia* and set off another bandwagon that later took Eric Kandel on board.

The Germans took about five years to reach the level of expertise that I had seen in Cambridge. Hansjochem Autrum, who replaced von Frisch as professor at Münich, built an electrophysiological set-up and measured the spectral sensitivities of the three types of photoreceptors in the honeybee eye, as well as the angular sensitivity of receptors in the fly. He and his collaborators, Burkhardt, Wiedemann, Vera von Swehl and pupils, spread the techniques in Germany.

The 1960s was a rich new era for invertebrate physiological research. As recorded later in this chapter, a strong group developed rapidly in the Max Planck Institute in Tübingen. There, Scholes and later Kuno Kirschfeld studied the optics and recorded the noise in fly photoreceptors. In the Netherlands, Kuiper expanded a group at Gröningen, where Doekele Stavenga established his career on the insect retina. These, and a few others, went deeper into the optics and receptor physiology in the 1970s, which was the high noon of studies on the insect retina. Numerous new students graduated with advanced skills in recording, neuron identification, electron microscopy, the optics of light guides and online data analysis.

There was a new group around me by now, committed to a program on optics and electrophysiology of compound eyes. In 1967, I had a project with Gay Grimshaw, a physics student at Dundee University, who built a wax model of a locust ommatidium and shone radar waves down the axis. We had trouble with standing waves caused by reflection at the far end of the rhabdom, but managed to get some measurements of angular sensitivity. We also worked on superposition eyes of beetles for some years and Rick Butler, a student from Canada, found huge day/night changes in the receptor fields of the cockroach eye, but only two types of receptors, for ultraviolet and green.

The period between 1967 and 1969 saw the appearance of a number of official reports that foreshadowed hard times for disinterested research. The *Robbins Report* recommended putting the funds into teaching; the *Rothschild Report* recommended more contracts for more applied work, and direction from industry; the *Dainton Report* spelled out the inability of the State to support an ever-expanding university sector and suggested that too many PhDs were being produced in the pure sciences. The scientific fraternity noticed the clouds on the horizon. The new Principal of the University of St Andrews could not, or would not, be as generous as his predecessor, as shown by his letter to me dated 31 January 1967:

> When you were with me there was one point I did not raise about your £6,000 for the Gatty. We both were talking on the assumption that it could be nothing but good to accept £30,000 from the Science Research Council. What will inevitably, and rightly, be asked is how far the £30,000 from the Research Council will commit the University to a take-

over [of the] operation, and hence mortgage our future funds and pre-judge academic developments. It would help me if you could give me more information about this.
Steven Watson.

That was the moment I decided to leave.

I had spent much of the 1965–66 academic year at Yale, teaching and writing, and so renewed my contacts in the United States. The money I earned was banked in the United States. In 1967, Steve Shaw and I had a Grass and a Rand Fellowship to work at Woods Hole, Massachusetts, where we took our recording gear and families. We studied the retina of dragonflies caught at the Prosser family's pond, with a net made from one of Hazel Prosser's curtains. There I discovered fireflies winking at night in the bushes and collected them for electron microscopy of the light guides in eyes by day. About that time, I met Ben Walcott in Eugene, Oregon, who said that he would come and join us at St Andrews. 'How can we find the funds?' I asked. 'No problem,' he replied, 'I will sell my aeroplane!'

While working at the Marine Biological Laboratory in Woods Hole in 1967, I was invited to visit Australia to consider joining a new institute as a foundation professor. So, sadly, after 13 years building a research team at St Andrews, I resigned, on 5 February 1969. The attractions of selecting my own staff with adequate funding in a new environment in Australia were conclusive. St Andrews University also lost Professors John Burnett from botany and John Cadogan from chemistry—partly because at that time the university was not outward or forward looking and did not provide sufficient facilities for research. They could not even buy the essential journals for the university library. It is an important object lesson in the way that research groups rise and fall because individuals make use of opportunities when the time is ripe and go elsewhere when the funds dry up.

In 1969, I became a foundation professor of the Research School of Biological Sciences at The Australian National University in Canberra, bringing with me David Sandeman (on crab eyes), Rick Butler (on cockroach eyes), Peter Shelton (who later turned to the development of the insect eye), Agis Ioannides (on hemipteran eyes) and Ian Meinertzhagen (on retina-lamina connections), with Ben Walcott (on water beetle eyes) as a postdoctoral fellow—as well as their families and other staff from elsewhere. The numbers that became temporary emigrants tell plainly of the enthusiasm and commitment of that group.

As a vanguard, I sent Meinertzhagen ahead to Canberra to order equipment and get the labs ready. He needed more time to produce complete maps of the projections of axons from the retina to the lamina in various insects. Allan Snyder turned up about a year later, not knowing one end of a rhabdom from the other. His work on the optics of ommatidia provided the inspiration for his analysis of polarising monomodal wave guides and their application for long-distance transmission in light guides for communications.

The new scholars from 1970 until about 1985 were given generous four-year PhD studentships, which seemed to be always available, and which included an allowance for a spouse and return fares. The students made recordings from small photoreceptor cells in various eyes, including mayflies and spiders, and eyes with mobile receptors. Laughlin, Doujak, Wilson, Lillywhite, Hardy, Dubs, Howard, Payne, Matic, Shi and others counted photon arrivals in a variety of insect eyes. They were exhorted to do an experiment every day to retain the skill.

There was a memorable period in the mid-1970s when Stavenga, Snyder and Laughlin, aided by Pinter, Srinivasan and Howard consolidated the data (mostly from our own lab) on photon captures, interommatidial angles, field sizes, lens apertures and rhabdom cross-sections, to produce a comprehensive theory of design of compound eyes for the known range of ambient light levels. During this period, Dubs, Guy, Laughlin and later Hardy, James and Howard analysed the function of the large lamina monopolar cells, which responded with a temporal derivative of the photon flux, minimised the noise and compressed the signal (Chapter 5). This became the best-understood example of optimisation of synaptic transmission in any nervous system.

We studied the movements of screening pigments in the day versus the night eye, as well as the light-adapted versus the dark-adapted eye. Some of the retinula cells themselves make large movements between day and night states in night-flying beetles and moths. By day, highly refractive guides carry light from the cone tip to the retinula cells in many of the nocturnal insects, but at night these eyes have a clear central zone where light from adjacent facets is summed. Some diurnal moths reach the theoretical limit of resolution in a superposition eye; some nocturnal beetles have very poor resolution and integrate light over huge fields as a strategy to collect as much light as possible for flight in starlight. In fact, in 1985, Doujak showed that a single crab ommatidium could detect a single bright star. We found that, in some beetles, the cone changed shape to adjust the optics between a light-guide eye by day and a superposition eye by night. Gert Stange showed that the dragonfly ocellus controlled pitch and roll in flight by summing the illumination from horizon to horizon, and Martin Wilson showed that the locust ocellus detected the position of the horizon mainly by UV contrast. This tradition is still alive in Sweden with Eric Warrant and Almut Kelber at Lund.

Interest turned to the recurrent problem of how to analyse the several parallel processing pathways in the insect visual system. For years, we had tried recording in the optic lobe, but the puzzling properties of the neurons in clamped insects could not be explained by the poorly known visual behaviour. Single neurons responded to moving edges, spots and changing intensity, but little else, with no indication of vision of shape or pattern. Enthusiastic electrophysiologists soon discover that an animal's behaviour is more likely to explain its neuron properties than vice versa. Willi Ribi described retina/lamina connections by Golgi-EM, which was just the edge of the neural jungle. A notable advance was Jenny Kien's discovery of neurons in the brain of the locust that could measure the angular velocity of the flow field. Similar neurons were found in the crab eyestalk by Sandeman and Erber, although more peripheral optic-lobe neurons were tuned to a low temporal frequency of passing edges.

In another significant advance, in 1984, Maddess discovered that optomotor neurons were most sensitive to a temporal frequency that increased as they adapted to high frequencies—that is, motion detection became more sensitive to faster motion. Danny Osorio identified neurons of the locust medulla and, with Andrew James, started the difficult task of characterising them by spatial and temporal resolution kernels. Later, with Ljerka Marčelja, I showed that several groups of insects had slow and fast motion-detector neurons (just as they had slow and fast neurons at all levels). Therefore they have the information to measure angular velocity from the ratio of the excitations of these two types of otherwise frequency-dependent neurons (see Figure 7.4). By that time, Srinivasan was interested in how the flying bee measured the perceived velocity and range of surrounding objects.

The new work in the 1980s on the perception of range from the angular velocity of the surroundings by flying insects led directly to practical applications. We formed collaboration with the research officer of the Guide Dogs for the Blind, Tony Heyes. We conceived the idea that a person with damaged vision might be assisted by an artificial insect eye stuck on the end of a finger, with an output in the form of a vibrator on the wrist. The eye-on-finger successfully measured the range and direction of the contrasting edges in view. Unfortunately, we could not find an industrial collaborator because there was little profit in making gadgets for the blind. Eventually, when the design of seeing robots became an urgent requirement after the nuclear accident at Chernobyl, our efforts attracted the attention of the Fujitsu Computer Company, which gave The Australian National University $10 million for our know-how. Later, in the 1990s, the research was also supported by the US Air Force, the US Defense Advanced Research Projects Agency (DARPA) and NASA, which installed our artificial insect vision system into freely flying pilot-less helicopters and drone aeroplanes.

At the end of 1992, I found a topic for my retirement that required little equipment or expense: visual processing of patterns by trained bees. Since von Frisch had shown in 1914 that bees discriminate some pairs of flower-like patterns very well but fail to discriminate geometrical shapes of similar size, the subject made no sense, although plenty of good observations using vertical presentation were made before 1939. In the second half of the twentieth century, there had been a lot of published data on bee pattern discrimination but not much agreement about their interpretation (Chapters 9–12).

Figure 3.1 The optomotor response. a) and b) The beetle *Chlorophanus* was held by the thorax as it walked on a light maze of paper strip. A fixed black drum around it was pierced with holes that allowed stimulation of selected vertical rows of facets. Around this, the stimulating drum was rotated or oscillated. The situation was an 'open loop' because the beetle turned the paper, not itself. c) The open-loop situation. d) Land's experiment; the fly was fixed on a freely rotating pin and the head and body positions were recorded by an overhead video camera. The situation was a 'closed loop' because the fly could turn. e) Interactions in the closed-loop situation.

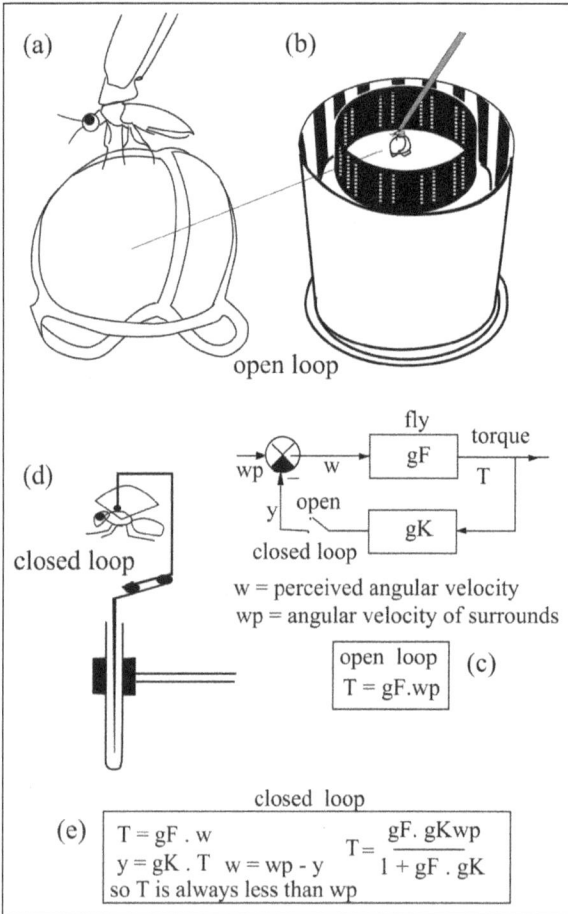

Sources: (a) and (b) after Hassenstein (1951); (d) after Land (1975).

Motion perception in the second half of the twentieth century

Up to the mid-twentieth century, there had been many descriptions of how particular insects responded to movement in the visual field, but little had been written about mechanisms. Long ago, Exner had pointed out that the detection of the movement of an edge necessarily involved the change in light intensity at one ommatidium followed by the same change at an adjacent ommatidium. The unit detector for motion would therefore have a preferred direction and a preferred angular velocity, with a peak response at the optimum coincidence rate and a fall-off of the response at either side of this peak. Such a detector, however, measures only the timing and therefore cannot distinguish between broad stripes passing rapidly and narrow stripes passing slowly. The motion detector gives a larger response to a more frequent passing of edges, up to a peak, declining at higher frequencies (Figure 3.3c). Also, any response implies either of two contrast frequencies on each side of the peak. With this mechanism alone, an insect would detect the directions of local motion in separate eye regions, but that is all—with nothing about pattern.

The optomotor response

We are all familiar with the way that our eyes follow the movement of the passing countryside seen through a train window. Many animals have a similar response when the whole visual field is moved around them unexpectedly. The eyes (if mobile), the head or the whole body follows the passive motion of the whole visual field on any of the three axes, roll, pitch or yaw. At its most dramatic, a hovering fly rotates in flight to track a patterned drum that is rotated around it. Typically, the eye lags behind the motion and there is a response over a range of low-oscillation frequencies.

The optomotor response introduced several important ideas into the exact study of behaviour. The directional nature of the motion detection was established a century ago, and a function was inferred to be station keeping when hovering or on the surface of flowing water. The stabilisation of a straight path in locomotion was uncritically accepted. The machine-like performance strengthened the ideas of input–output relations and reflex control of posture by sensory processing that were characteristic of the first half of the twentieth century. In the bee, the response—like all motion detection—was green sensitive and colourblind, which gave rise to some controversy in the early twentieth century after von Frisch demonstrated colour vision in the bee.

Figure 3.2 The interactions during visual control of locomotion with voluntary turning and 100 per cent visual feedback; 'g' is a measure of the gain in the internal loop. a) As usually portrayed for the fly. b) With the halteres included.

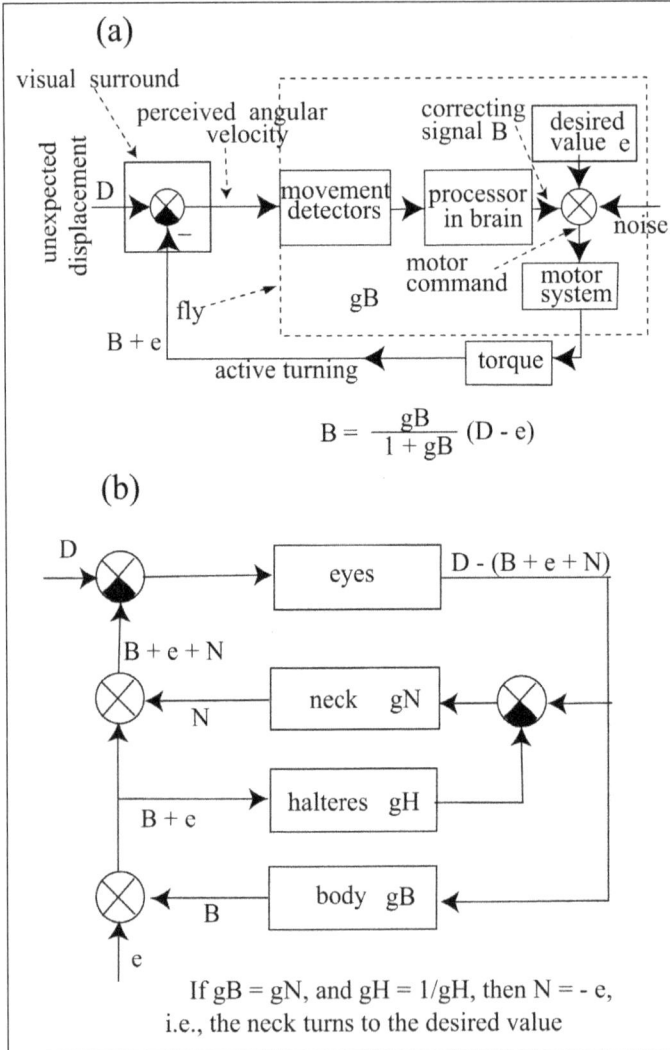

$$B = \frac{gB}{1 + gB} (D - e)$$

If $gB = gN$, and $gH = 1/gH$, then $N = -e$, i.e., the neck turns to the desired value

Source: From Land (1975).

Figure 3.3 The optomotor response of the bee, 1960s style. a) The bee was held by the thorax on a stiff rod attached to two small coils, the whole being hung on a thread within a large drum (not shown) and free to rotate. When the flying bee exerted a torque in response to the motion of the drum, the electrical signal flowed through both coils and controlled the electric magnet that held the bee in a constant position. The response was a measure of the torque. b) The usual posture of the bee, indicating slow flight. c) The normalised response to the rate of passing of bars on the drum, at four different grating periods, from 3° to 60°. d) The response at the optimum temporal frequency as a function of the period of the stripes. The points where zero is crossed indicate a spacing of 5° in the motion detector. e) The interaction between the bars of a grid and the array of ommatidia can result in perceived motion in the opposite direction to the stimulus, as in (d), caused by the Moiré effect between the spacing of the ommatidia and the bars.

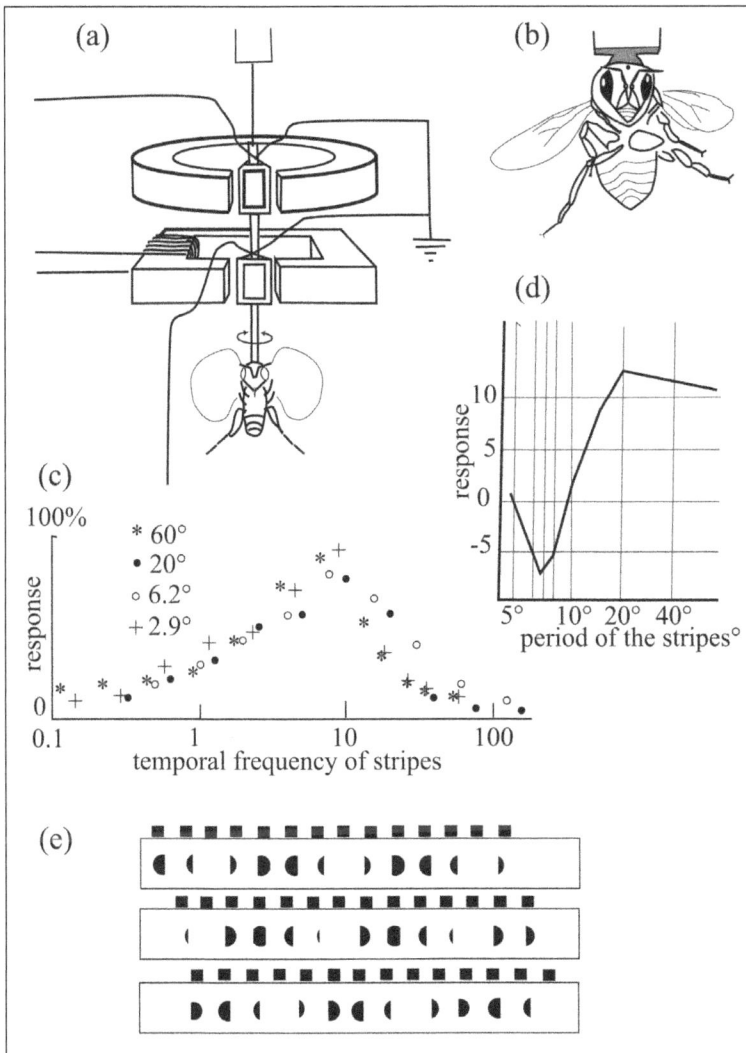

Source: (d) after Kunze (1961).

Various measures of the optomotor response have been made with a variety of insects—notably, the amplitude, the turning force generated by muscles of the neck or by the whole body in flight or the rotation of a ball or a maze of paper strip on which the insect walks (Figure 3.1a). When the body is fixed, the movements of the head as it follows the rotation of a surrounding drum are easily observed in many insects—for example, butterflies and locusts. At the same time, some neurons, particularly in the medulla, lobula and the neck muscles, are easily recorded and respond in a similar way to a moving striped pattern, so that several steps between the input and output can be recorded (Figure 3.5).

In a pioneering study in Königsberg, Lotte von Gavel (1939) plotted the optomotor responses of *Drosophila* to the movements of gratings at different periods at different light intensities. The central region of the eye had the highest resolution. The interesting point was that, as the period of the grating was reduced, the response reversed at a period near 9°. The reversal was correctly explained as a Moiré effect between the interommatidial angle and the grating (Figure 3.3e). The response beyond the reversal point showed that the resolution of the bars at the limit was better than that predicted by the interommatidial angle. The reversal also showed that the fly perceived the best correlation between the adjacent facets with the shortest delay, not the real direction of movement of the drum. At low light levels, the reversal occurred at a larger period of the bars, showing that the spatial tuning of the motion detectors had increased. In an earlier study, to explain why the period at the reversal point increased so much at low light levels, Hecht and Wald (1934) championed the improbable idea of a wide variety of receptor field sizes and sensitivities, and indeed interactions between sub-adjacent visual axes were later found (Figure 3.4). Even at this early date, there were sufficient data to show that insects detected the output of the motion detector as a vector without pattern, not the real image of the bars.

Most mobile animals make a predictable optomotor response when the visual scene is moved, yet this apparently strong reflex disappears when the animal moves itself. There had been many discussions about how the optomotor reflex was switched off during a voluntary movement. Using the popular new systems theory, Horst Mittelstaedt and Erich von Holst (1908–62) outlined the theoretical interactions that might combine visual stabilisation via the optomotor response with self-steering. About 1960, they defined the re-afferent signal, which was the sensory stimulus to the eyes as a result of head movement (Figure 3.2a). Their interactions between boxes joined by arrows influenced many subjects, from robotics to social science. Further, it was proposed that every central neural command to make a movement was accompanied by another command, called the 'efferent copy', which would exactly cancel the effect of the expected feedback, so that no optomotor response would follow.

Figure 3.4 The observed spacing of the interactions between ommatidia in the detection of motion in the optomotor pathway, with arrow thickness indicating the relative strength. There could have been other interactions in other channels that were not tested.

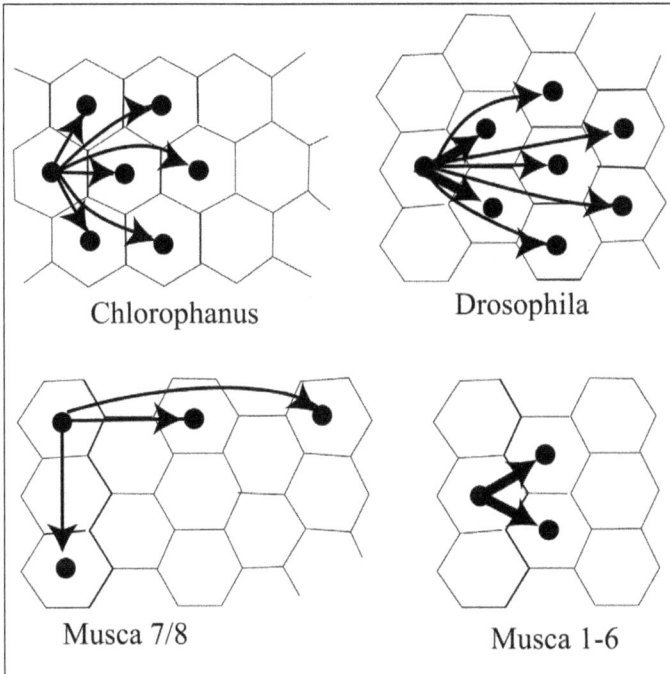

Source: After Buchner (1976).

With 100 per cent feedback at the speed of light as the animal moves, the optomotor response is an ideal illustration of the visual feedback loop (Figure 3.1). When the eye turns freely in the response it is said to be a 'closed loop'. The loop is broken when the eye is fixed and the output is measured as the turning force (mechanical torque) in flight or as the rotation of a ball held by the insect's feet. The preparation is then an 'open loop' (Figure 3.1a).

Immediately after World War II, one of von Holst's students, Bernhard Hassenstein, was a great enthusiast but so poor that he had to sleep in a laboratory cupboard. When I was in California, I remember that he had entered the state illegally through quarantine carrying a mud nest of a South American bird disguised as a pudding and a chocolate box containing luminous beetles. The connection between von Gavel and von Holst's student was probably the influential experimentalist Otto Koehler, who literally walked out of Königsberg when the Russian Army arrived in 1945 and escaped to the West.

In his classical experiments, Hassenstein fixed a beetle (*Chlorophanus*) by the head but allowed it to hold a paper ball that rotated beneath it as it walked at the centre of a horizontal drum (Figure 3.1a). Instead of the head moving, the rotations of the ball were observed, so avoiding the visual feedback in the open-loop situation. The fields of view of the receptors could be restricted by fixed vertical slits. He found that:

1. The turning tendency is zero at very low and also at very high temporal frequencies of passing of stripes (contrast frequencies), but rises as a bell-shaped curve at medium frequencies, peaking in the range 1–20 hertz (Hz) for most insects.

2. The shape of this contrast frequency response curve is independent of the stripe period (spatial frequency) and is not altered by rearrangement of the component spatial frequencies in the pattern on the drum. The insect does not give the same response to the same angular velocity of different patterns.

3. When the contrast is reversed between two successive positions of a regular grating, the insect responds in the opposite direction, apparently making the best correlation it can between the successive distributions of intensity.

4. Adjacent receptors in pairs provide sufficient input to infer motion.

In 1950, Hassenstein showed his results to Werner Reichardt, a former air force officer who had been involved in a plot to kill Adolf Hitler, but who was then working as an electronics expert. Together they introduced the idea that the motion and its direction were computed by cross-correlation of the outputs of each pair of adjacent receptors. This was the minimum mathematical relation that was consistent with the data. Based on the newly emerging signal transmission theory, the pathways of the visual input were considered as filters and the outputs as variable pattern generators.

The success eventually led to the formation of the Kybernetic Forschungsgroup at the Max Planck Institute in Tübingen, led by Hassenstein and Reichardt, who switched to work on the visual system of the flying fly. After 1965, Karl Götz and Kuno Kirschfeld joined the group. With a period of intense study of the optomotor response by Götz, Buchner, Reichardt, Varju, Wagner and others, work also covered many aspects of the fly's visual system. Later, Kirschfeld and Franceschini showed that just two adjacent visual axes were sufficient to give a response, and Buchner found that sub-adjacent axes also contributed (Figure 3.4). Götz initiated the isolation of the behavioural mutants of *Drosophila* that were later essential to the analysis of straight flight.

A student of Hassenstein, Peter Kunze (1961), gave an account of the optomotor responses of the bee, using the torque produced as the bee tried to turn in flight. The bee was clamped in the standard Tübingen apparatus of the time, which generated an opposite torque that compensated for the bee's efforts (Figure 3.3a). The responses rose to a peak with an increasing rate of passing of the bars on the drum and declined at higher temporal frequencies. The position of the peak near 10Hz was independent of the period of the grating (Figure 3.3c). At the optimum temporal frequency, the response fell to zero at a grating period of about 10° and was in the opposite direction between 10° and 5°. Adjacent and sub-adjacent ommatidia are involved. The mechanism therefore detects the direction of the modulation sequence, not the direction of the motion, and is certainly not concerned with seeing the bars or the grating.

In this work on the fly and the bee, the insect was clamped and could not make a real turn or a saccade of the head, so there was no visual feedback. The stimulus was controlled by the experimenter and was unexpected by the insect. This had the effect of isolating the optomotor response. The insect soon learns that its efforts are frustrated, so these efforts decline in strength and have to be measured as smoothed averages over many repetitions. Even the posture of the bee changes from that in fast flight to that in slow flight or hovering (Figure 3.1b), as though it is aware that it is going nowhere. Also, the measure of resolution is an average, because it is the value at which the opposite turning tendencies from different parts of the eye exactly balance.

The next step, taken with *Drosophila* and *Musca*, was to introduce into the drum a vertical bar that could be moved independently of the background pattern in the fly's visual field. Later, this became a random-pixel bar on a random-pixel background, called the figure/ground stimulus. The insect would attempt to steer towards the moving bar despite the simultaneous movement of the drum. The researchers thought they were studying the fly's control of steering towards an object moving over a moving background. In a purely mathematical formulation of the interactions in motion perception, the responses were modelled with the same inputs and single output as in the optomotor response, but with an extra layer of non-linear interaction between the same elementary motion detectors to represent the moving bar. Poggio and Reichardt (1976) wrote an account with 125 pages of mathematics and Poggio produced 25 mathematical papers about the Reichardt model. There were many international presentations and a great deal of confidence in this methodology, but also a lot of criticism from elsewhere.

In his remarkably apposite poem of 1958, Frederick Winsor expressed it so:

> This is the Cybernetics and Stuff
> That covered Chaotic Confusion and Bluff
> That hung on the Turn of a Plausible Phrase
> And thickened the Erudite Verbal Haze
> Cloaking constant K
> That saved the Summary
> Based on the Mummery
> Hiding the Flaw
> That lay in the Theory that Jack built.

More usefully, and ignoring the maths, Hausen (1982) and Hengstenberg (1982) described in detail the fields and arborisations of the directional motion detector neurons in the deep optic lobe of the fly (the lobula plate), which gave responses to motion like those of the whole animal. These large neurons sum the motion stimulus in three axes around the head within quite wide fields (Figure 3.5). They illustrate regional specialisation, overlapping fields and coding of yaw, pitch and roll by different overlapping groups of neurons, but none responds to small objects. This result elaborated the earlier conclusions of Götz, based on systems analysis, that the outputs of the motion detectors ran towards the flight-control mechanism. Later, Egelhaaf (1987) described in the fly some neurons that responded to motion of a small object but were inhibited by large-field motion, as known in moths, locusts and dragonflies. He combined them with the large-field motion detectors to explain the responses to the figure/ground stimulus with one homogeneous type of elementary motion detector.

Technical problems appeared in the late 1970s. The frequency range of the optomotor response was too low to apply to free flight. The response strength of the clamped insect was an arbitrary measure, a feeble shadow of that of the freely flying fly, and measurable only when summed over numerous repetitions of the stimulus. Smoothing the output prevented observation of the small jumps made by the eye (called saccades) that later proved crucial for flight control. Mittelstaedt's control theory (Figure 3.2a) omitted the saccades and was wrong for other reasons (see below). Even worse, the clamped insect quickly learned that its efforts to turn were in vain, so the response changed and waned. Work on locusts by Rowell (1971) and later with Reichert (Rowell and Reichert 1991) at Basel indicated several separate neuronal systems. The method of systems analysis, inferring interactions from observations, was suitable for a single channel, but could not separate or characterise several unknown pathways that functioned in parallel. The main weakness, however, was in the data, which recorded a weakened optomotor response but obscured the main agent, the saccades.

There was a persistent difficulty in finding the way forward, because the response plotted as a function of the angular velocity depended strongly on the pattern, being weak for a sparse pattern and strong for a crowded texture, and therefore was no use as a measure of the angular velocity. Such a mechanism was useless for an insect flying in a wind. The response in *Drosophila* was too slow, rising to a peak at about 1Hz and falling to zero at about 10Hz, irrespective of the pattern or angular velocity. In short, the motion-detecting system was unsuitable for a freely flying insect that had to avoid obstacles.

Figure 3.5 a–f) The peak directional sensitivities and approximate fields of the principal motion detector neurons on the optomotor pathway in the fly; V = vertical directions; H = horizontal directions; Greek letters show the coordinates. g) Responses as a function of the direction across the eye. h) Representative fields and their spatial relations.

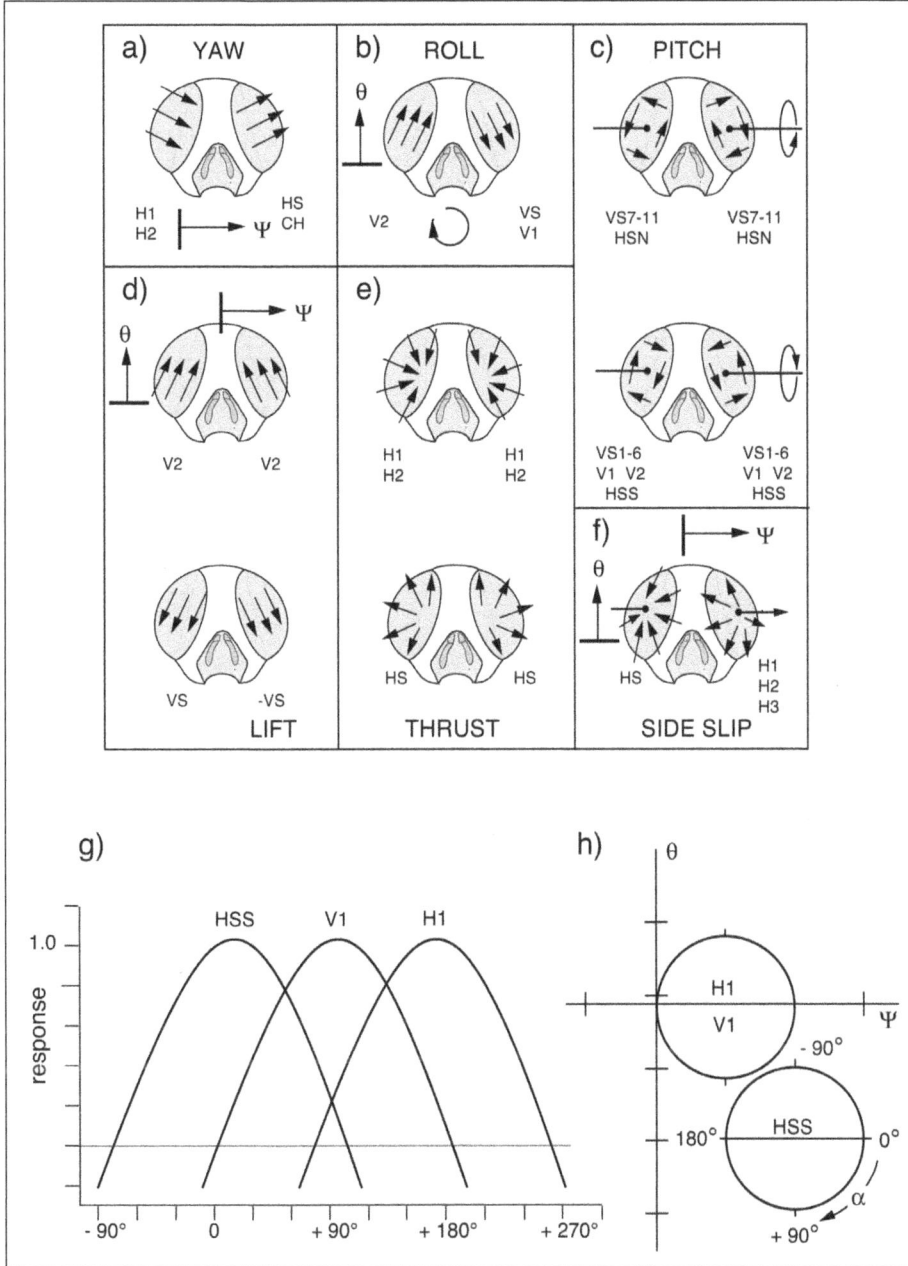

Source: After Hausen (1982).

Since the work of Kennedy (1940) on the free flight of locusts and mosquitoes, it had been clear that those insects detected their angular velocity and direction relative to the ground and controlled the direction and speed of their migrations according to wind and weather. The difficulty was highlighted by the discovery of neurons in the brain and ventral cord of locusts, and later in dragonflies and bees, which responded in such a way that they could measure angular velocity irrespective of pattern. The way that insects find the source of an odour is to turn upwind when they detect it and then follow the plume upwind, which requires detection of speed over the ground.

The rot really set in at the International Congress of Entomology in August 1972, when Mike Land presented new observations on the behaviour of a fly that was fixed by the thorax but was free to move its head and rotate on a pin (Figure 3.1c):

> While flying, but free to rotate, flies [*Calliphora*] show two kinds of head movements. (1) Rapid saccadic movements with amplitudes of up to 20° relative to the direction of the body axis and durations of c. 20 ms. These head movements…are accompanied by body turns which slowly bring the body axis back into line with the head. (2) Stabilization movements which tend to keep the axis of the head still with respect to the surroundings. (Land 1975)

In other words, the head makes a saccade that is so fast that it leaves the body behind and the optomotor system does not notice. The head is then held in the new position by the optomotor response while the body catches up. Human vision is rather similar, but more predictive.

Continuing Fred Winsor's (1958) version of the scientific method:

> This is the Space child with Brow Serene
> Who pushed the Button to Start the Machine
> That made (Hay of) the Cybernetics and Stuff
> Without Confusion, exposing the Bluff
> That hung on the Turn of a Plausible Phrase,
> And shredding the Erudite Verbal Haze
> Cloaking Constant K

For a decade, there had been doubt about whether the equations describing optomotor behaviour were related to the mechanism, but errors of thought and misuse of the mathematical muse were no longer the point when serious flaws were disclosed. The first flaw in the house that Jack built was that the fly's head had been clamped, so it could not see the effects of its own turns. The second flaw was the insensitivity of the equipment and the misleading averaged data that had omitted the saccades.

In studies of photographs of male flies chasing females in free flight, Land and Collett (1974) characterised the mechanisms by which the flying fly fixated on a small moving object, measured its angular velocity and turned to pursue it. Others separated the components. Srinivasan and Bernard (1977) found that the response of the fly to small moving objects, such as a bar, and to broad-field stimuli, such as the background, had different time constants, and Olberg (1981, 1986) distinguished the object-detector neurons of the dragonfly from the optomotor neurons.

Meanwhile, Strausfeld and his collaborators were anatomically separating the neurons specific for the chasing behaviour from those for the optomotor response. The connections were not to the flight muscles, as predicted from the Reichardt theory, but to the neck to turn the head. Later, we found that the directional motion detector neurons in several groups of insects were of two types—the slow optomotor ones and faster ones going up to 200Hz. As more electrophysiology and anatomy appeared, motion detection became more complex, there was not a single unit motion detector and the details required by the Reichardt theory were not found (see Figure 6.6). It was realised that mathematics could not be connected directly to behaviour and the whole idea of computational neurobiology was a house of cards. Of course, not everybody agreed, but books full of rubbish were published on the subject for years after.

Angular velocity

A flying vehicle must be sensitive to the perceived angular velocity of nearby objects and the surrounding panorama. Indeed, the early block diagrams used to summarise the optomotor response had in fact used angular velocity as the variable (Figures 3.1 and 3.2), but no-one had noticed. The importance of the whole gradient of angular velocity of the optic flow was taken by J. J. Gibson from classified wartime work on landing aeroplanes at Farnborough, England, and later published in an influential book on the subject.

One of my students, Jenny Kien, discovered two neuron types in the deep optic lobe of the locust, one (M1) sensitive to forward motion across the eye, the other (M2) to backward motion. The relation between response and velocity was independent of the period of the stimulating striped pattern. Neuron M1 received inputs from small fields 1–5° in diameter, with the maximum response to interactions between every fifth ommatidium along a horizontal row. The M2 neuron had maximum effect with every sixth ommatidium. The optimum excitatory interval corresponded with an angular velocity of 20° per second (Kien 1975). Although this mechanism solved the problem of velocity measurement by demonstrating different spacings of the inputs, the result was ignored for years.

In the 1980s, John Kennedy returned to the topic and, with David (1979a, 1979b, 1982), showed that freely flying *Drosophila* responded to the angular velocity of the flow field and to parallax, but not to the contrast frequency. In Canberra, we worked out the use of the angular velocity by the flying bee to measure range and avoid obstacles (Lehrer et al. 1988), and later Srinivasan et al. (1996) showed that bees measured angular velocity irrespective of direction and integrated it to measure the distance flown over the ground.

On mature consideration, it was obvious that after damage to their wings that spoiled the equal traction on the two sides, bees, dragonflies, butterflies and flies quickly adjusted and flew in a straight line. Also, insects can fly in a straight line inside a moving motorcar without turning towards or against the perceived motion outside the car. In the literature, there are many examples of how damaged insects rapidly learn to adjust their normal posture and movements— notably, when a leg is removed, and in the demonstration of learning of postural positions of the legs by headless locusts (Horridge 1962). A complex robot cannot survive in an unpredictable environment without rapid learning mechanisms to assist in the control of its posture and movement. Similarly, we rapidly learn to ride a bicycle or steer a boat with a tiller.

The control of straight flight in the fly

With little previous warning of its explosive contents, in 1984, Martin Heisenberg and Reinhard Wolf published a book about the way that the fly *Drosophila* controlled its flight path in the horizontal plane. As described further in Chapter 7, they showed that the fly responded quite differently to visual feedback from its own active motion and to passive rotation of the visual scene around it. They also showed that the saccades indeed controlled the straight flight of *Drosophila*, and later discovered that there was a dead reckoning of angular turns, so they kept a tally of all turns and could return to the starting direction. These discoveries meant that the averaged responses of a fly with fixed head were no longer relevant. For decades, the stabilisation of the flight of the fly on a straight track had looked like an inflexible optomotor reflex, but this revolutionary analysis done on *Drosophila* showed that if one wing was damaged or if the feedback loop was modified, the fly could quickly learn by operant conditioning how to reset the controls and steer once again on a straight course. The learning is called 'operant' because the fly makes voluntary movements and correlates its intentions with the resulting feedback.

Just as the fly can learn to fly on a straight course visually without an isolated beacon although its wings are not symmetrical, by making tentative test saccades in either direction, it can learn to direct its course towards a target that rewards it with scented or warm air. Flight speed (thrust and lift) and turning in the yaw plane can also be modified by learning. The fly finds its goal by checking the results of making saccades. More and more, it is apparent that the responses to visual motion are rapidly changed by other actions of the insect. The concept of an optomotor reflex has been replaced by a goal-directed action with continual operant learning of all motor outputs, providing active stabilisation against unexpected movements. This surprising result probably applies to all insects. For years, Reichardt and his colleagues at Tübingen ignored the early work of Kennedy (1940), the saccades later recorded by Land and the efforts by Heisenberg at Würzburg.

A research group with Martin Egelhaaf at Bielefeld has now taken up the enormous challenge of recording from the visual system and at the same time displaying to the eye the visual input in three dimensions (in open loop) that would eventuate if the eye were not clamped. They started by recording the exact motion of a fly, and its head movements, as it flew about in a box with a pattern on the inside surface. The pattern seen by the two eyes was then computed for that flight trajectory. The real stimulus at the eye was presented as the input to a computer model of the proposed motion detectors, represented by one of the large field detectors of horizontal motion of the lobula plate. Because the neuron is working outside its linear range at the saccades, the responses are insensitive to the choice of pattern in the visual field. It is still true that the responses of the fly's motion detectors are tuned to temporal frequency, not spatial frequency, and do not measure angular velocity. The way they interact, however, with the distribution of different spatial frequencies in the natural background, and an active control of gain, is dominated largely by the saccades, as in the two-dimensional situation in the horizontal plane. In other words, the optomotor component no longer dominates.

If we stand back in a critical mode, we see that the optomotor response is now just a mechanism for recovery from unexpected displacements during locomotion. The figure/ground stimulus was discarded when the separate fast and slow parallel channels with small and large fields were discovered. In the fly, the control of direction is dominated by the saccades, and insects have a memory of the retinotopic positions of outstanding contrasts at each place and of the accumulated angle turned. The motion detection is still consistent with the Reichardt model, as indeed any directional motion detection should be, and some synaptic circuits have been proposed, but not yet securely identified. Also, in the fly, the analysis will have to be extended to the halteres and the motor neurons (Figure 3.2b).

The bee often flies very slowly and appears to be different in its flight control, but it could turn out to be similar to the fly. The bee scans from side to side in flight, with small saccades of the head. In the natural environment, and in tunnels with patterned walls, the bee integrates the measured angular velocity of the flow field and remembers the distance travelled over the ground, but also detects the flight speed mechanically with head hairs and the antennae. The bee equalises the angular velocity of the flow field on the two sides and adjusts the flight speed to maintain a preferred angular velocity of the flow field irrespective of the pattern. Whether this is fundamentally different from the fly awaits further experiment, and it would be nice to have corresponding data from other insects to see how far the flight control is adapted to lifestyle.

It was a mixed blessing that most of the work was on the housefly and *Drosophila*. Being wandering scavengers, these flies have relatively simple visual behaviour compared with the bee.

The perception of motion by flying insects illustrates the fashions of the decades and how the Tübingen optomotor response was attacked in the work of Land and Collet, Goodman, Hausen, Hengstenberg, Möhl, Rowell, Strausfeld, Wehrhahn and many others, as the system became better understood, and then replaced by Heisenberg and Wolf. The mechanism of free flight of the fly is now being intensely studied by Martin Egelhaaf and colleagues. There is also a major effort in the United States to identify every synaptic connection between the neurons of the *Drosophila* optic lobe in an effort to separate all the arrays and systems with partially overlapping inputs and outputs. The story has not concluded.

Endnotes

1. A large number of personal names appear in this chapter. Further details can be found in the bibliography.

04 PERCEPTION OF PATTERN, FROM 1950 ON[1]

There were many earlier descriptions of how particular insects responded to lights or ran towards dark holes or contrasting edges, and several useful summaries of this kind of behaviour, categorised into taxes, kineses and tropisms (Fraenkel and Gunn 1940). The movements that insects make when stimulated by light, however, say almost nothing about the mechanisms of processing. Similarly, the outstanding work by Baerends (1941) and van Beusekom (1948) in Holland showed that wasps recognised their nest site by the memorised configuration of landmarks in different directions relative to each other, but this was about performance, not mechanism. In the second half of the twentieth century, starting with Bernard (1966), various symposia on insect vision were published. Wigglesworth's classical work on insect physiology, revised in 1965, gave a summary of many factual details but no mechanisms were in sight until the extensive compilation by Autrum (1979–81)—and even that was incomplete. Understanding was confused by the statements of Exner, Hertz, Autrum and Reichardt that insects distinguished shapes through motion, although there were plenty of observations that many insects hovered to take a better look.

When research started up again, the prewar (mostly German) effort was scarcely mentioned, even by the Germans themselves. From the autobiographical sketch of Karl von Frisch (1957), we can infer that National Socialism played a part in the suppression of the work of some of his students, and certainly after 1945 there was a strong tendency to deliberately turn away from all the events and literature of the 1920s and 1930s.

More fundamental reasons for ignoring the earlier results—except for a summary of Hertz—were numerous. One was a push towards quantitative data, with empirical mathematical relations to support a theory, as advocated by teachers of scientific method at the time (Chapter 2). Second, there was the cybernetic movement that sprang from the operational research groups of World War II, as illustrated by Reichardt's group at Tübingen and the journal *Kybernetic*, which he founded. There was also the destruction of libraries, a wish to forget the past, a new generation of students in a hurry, a general thrust away from natural history towards mechanisms and the particular interests of the key players.

In 1965, a view through Russian eyes was translated and edited by Timothy Goldsmith at Yale. Georgii Mazokhin-Porshnyakov, a professor of entomology at the Institute of Information of the Academy of Sciences of the USSR, had published numerous papers on the compound eye and trained bees to discriminate patterns. Mazokhin-Porshnyakov (1969) thought that bees responded to the 'the totality of the object's characteristics and not—to individual parameters such as shape, size, and so on'. He observed that '[t]he fact that insects can distinguish shapes is taken for granted by the majority of contemporary authors'. He ignored the fact that every researcher up to that time had found that trained bees detected parameters such as disruption, area, colour and radial symmetry, not shape.

Let me quote further:

> Insects, in particular honeybees, are able to distinguish shapes and recognize simple figures, such as circles and triangles, and complex figures like stars. Highly decomposed figures are perceived by insects through the flickering of light produced by motion of the retinal image over the receptors...Insects are even able to distinguish solid objects from plane ones, and can estimate depth. They distinguish colors and make large use of color vision in their normal patterns of behaviour. (Mazokhin-Porshnyakov 1969)

There is a very telling translator's footnote:

> The reaction to various figures exciting the eye to a different extent is not proof that insects really distinguish shapes. Insects can move back and forth along the edges of black bands and distinguish these bands simply as a darkening in their visual field and not as a field decomposition (shape). (Mazokhin-Porshnyakov 1969)

There is also an interesting footnote on page 123: 'Hertz's (1929–31, 1933) analysis of figure perception and identification in bees is not substantiated, and we will not consider it here.' Hertz was of the opinion that her bees did not see shapes that were laid flat.

Mazokhin-Porshnyakov himself accepted that earlier authors had sometimes failed to train bees to distinguish different shapes, but he believed that they had used patterns that were too large. His successful patterns contained the same shape on two spatial scales. From a distance, his 1965 patterns looked like an empty triangle, square or circle, but the edges as seen from a distance were composed of many small triangles, squares or circles that could be resolved close up. He convinced no-one because there were none of the necessary controls against the many possible cues. The emphasis was on what insects could see—

that is, successes in training—but successes alone leave us with the impression that 'insects really receive and make use of a variety of visual information and accordingly behave more or less like vertebrates do' (Mazokhin-Porshnyakov 1969). This is terrible stuff—mostly guesswork.

A related observation was made 30 years later by Campan and Lehrer (2002), who successfully trained bees to distinguish a filled triangle, square and circle, and who mentioned that 'A. mellifera tended to scan mainly those contours whose direction differed between the two shapes'. They concluded that the shape discrimination 'is based on the use of local parameters situated at the outline of the shape, such as the position of angles or acute points and, in particular, the position and orientation of edges'. These proposed parameters, however, were also guesswork, and there were no tests to identify them. Even into the new century, the majority of authors apparently continued to believe that insects saw shapes, despite the lack of evidence for the idea, and plenty of evidence against it.

A virtue of the condensed little book by Carthy (1958) was an account of some of the German, French and English literature on reflexes and other behaviour. Even so, while some of the findings of von Frisch, Hertz, Opfinger and Wolf on trained bees were mentioned in English, the crucial ones by Baumgärtner, Friedlaender (see Figure 1.6) and Wiechert were missing. It provided a partial impression of what was available in the mid-twentieth century, but no explanations. There was no useful theory at that time and little to assist students with their essays.

This was the sorry state of the art when Scholes, Tunstall, Bennett, Shaw and I started recording from insect retina and optic lobes of the locust at the Gatty Marine Laboratory, Scotland, in 1962. Besides showing that the field sizes of the receptor cells were nothing special, we found groups of large neurons with large fields that responded in a variety of ways to the motion of edges or black spots. The neurons were excited by very simple stimuli, but the large field sizes precluded the separation of features and the responses could not be related to different patterns presented to the eye. Insect vision remained a puzzle.

At the time, it was commonly stressed that recording the properties of neurons would eventually explain or even predict simple reflexes or even complex behaviour—and many still believe this. Our introduction to the problem of explaining insect vision by this route was so disappointing, however, that for 25 years we abandoned high-level neurons and tackled the retina first. In a system of pathways in parallel, it is impossible to know which neuron relates to which behavioural response or what is happening in pathways other than the one that is recorded. At the time, we did not understand that the behaviour could explain some properties of the neurons, not the other way round, so the behaviour must be analysed first.

The mobile eyestalks of the shore crab were also ideal objects to study, because the responses to motion were reliable and informative. In 1966, I published a series of papers with Peter Shepheard and then with Malcomb Burrows showing that the shore crab *Carcinus* had a peculiar sort of memory when placed in an illuminated arena with a few vertical black bars. After the crab has seen the stationary bars for a few minutes, the light is put out and the crab's surroundings are rotated by a few degrees while it is in the dark. When the light is switched on again, the crab's eye moves until it points towards the same position as before, relative to the bars, showing that there is a memory of their former positions on the eye. The precision was much better than the angle between receptor axes (2°). Some individual crabs respond as though they remember only the former positions of black areas, but others remember the former positions of edges irrespective of which side of the edge is black. Later, the separate memories of positions of edges and black areas turned out to be similar in ants and bees.

The meaning of these observations surfaced only when it was found that crabs were aware of the direction of their burrow at all times, and insects stabilised their walking or flight positions by exact memories of the positions of surrounding contrasts.

A promising new start

In von Frisch's busy institute in Münich, Rudolf Jander, Una Jacobs-Jessens (1959) and several students assiduously studied the navigation and visual system of the red wood ant (*Formica rufa*), which recognised some landmarks visually. They measured spontaneous preferences to various black shapes on the sides of a white arena and trained the insects to discriminate between the same shapes. Whereas a protozoan or a barnacle might have a single detector and gives the same response to almost any visual stimulus, the vision of ants and bees was treated by Jander and colleagues as a collection of simple detectors guiding a variety of responses. By 1963, Jander and his student Christiane Voss at Freiburg inferred that ants had detectors for: a) a dark area, b) horizontal versus vertical stripes, and c) disruption of the pattern. Untrained ants prefer solid black, closed figures to disrupted patterns, which can be reversed by training, and they prefer vertical stripes to horizontal ones, which is unaltered by training. A black triangle was distinguished from a black disc or square (Figure 4.1e), but possibly the centres of gravity of the figures were at different heights.

Figure 4.1 a) Patterns frequently used to illustrate edge length or modulation as a cue. It was supposed that each pattern (laid flat on a white table) could not be discriminated from those in the same row, but any pattern in one row could be discriminated from those in the other row. This does not apply if they are presented vertically. b) and c) The feature detector proposed for ants and bees. d) and e) Pairs of patterns that were discriminated, but note that the centres are at different heights.

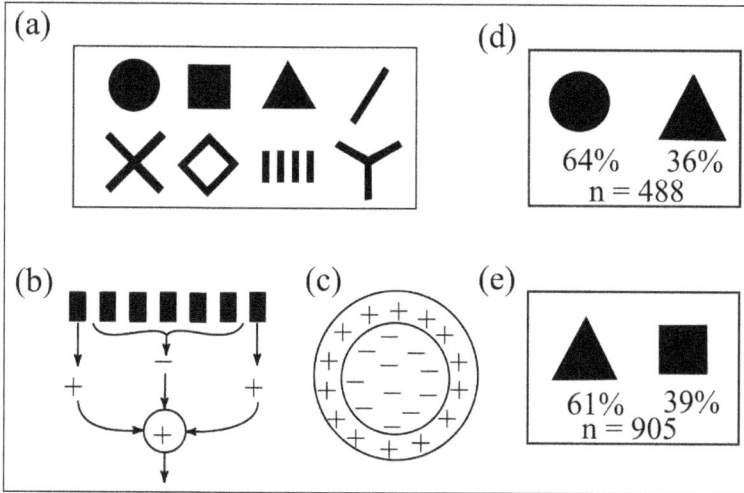

Source: (b) and (c) Jander (1964); (d) and (e) Jander et al. (1970).

In 1964, Jander proposed that insects had modulation detectors that acted as filters and that ants and bees learned features that they later detected in unfamiliar patterns. The feature detector proposed for insect vision was symmetrical, with a centre/surround structure (Figures 4.1b and 4.1c) that would detect a spot or an edge irrespective of which side was black or white. The responses of trained wasps to the orientation of edges were explained by strings of these detectors (Jander et al. 1970). Such detectors, however, would be useless for areas of black or colour. The concepts and terminology could be traced back to seminal papers of 1950–62 by Barlow and Levick, Kuffler, Lettvin, Maturana, Mittelstaedt, Hubel and Wiesel and others who found that arrays of single neurons in the peripheral visual system of vertebrates responded to simple visual features such as edges or spots.

In Frankfurt, Voss (1967) also showed that wood ants detected the orientation of black and white gratings with a minimum period of 1° although the interommatidial angle was 9–10°. This remarkable sub-pixel resolution, also found in several carnivorous insects that hunt for prey, must rely on modulation differences, not pattern. The ants distinguished between areas of black and edges, irrespective of which side was black. They also detected the difference between a sharp black/white edge and one that had the edge orientation spoiled by cutting it into square steps. Voss demonstrated that the ants measured the height above the horizon of a black spot of constant size (14° diameter), with a maximum spontaneous attraction at 30° up. There was no hint of shape discrimination. Although these findings were scarcely ever given due credit, they recurred in many later papers by others on ant and bee vision.

The quantitative approach

In a search for empirical laws, Schnetter (1968) at Würzburg used black patterns of four-pointed and six-pointed stars with circular symmetry, presented flat on a white table, as Hertz had done 40 years before. It was a mistake to start with such a complex image. He trained with one star that was duplicated many times and then tested the trained bees with the training star versus a series of similar stars of the same area that differed in one measurement: the length of edge. He found that the greater the difference, the better the discrimination. The score in the discrimination was related to the relative differences in the lengths of edge $(d_1 - d_2)/(d_1 + d_2)$—that is, an empirical law. It was not surprising, really, because the edge length was already known to be a parameter and it was the only variable that the bees could use. The results for four-pointed stars fitted a single curve but the curve for six-pointed stars was different. Later, the tests were repeated with other values of the diameter, the number of points, the angle at each point and relative length of the points (Schnetter 1972)—again, with no useful conclusion. There was no test of what the trained bees really detected or whether they could distinguish the stars from other patterns.

Schnetter's quantitative measurements served only to confuse the issue for decades. There was in fact no need for this cumbersome search for the best correlation. The relation between the length of edge and the number of correct choices is neither interesting nor informative. The effect of area was omitted and there should have been a logical identification of the cues by testing the trained bees.

Large patterns at Frankfurt

In 1958, von Frisch retired from Münich and, following custom, his staff and students moved to new positions and left space for Hans Autrum, his successor. Martin Lindauer became the professor at a new department in Frankfurt and soon found a student, Rudiger Wehner, who worked on pattern perception in honeybees.

I am sorry that, at this stage in the history, comprehending the next group of findings will need all your attention and constant referral to the illustrations. In their new effort in Frankfurt, Lindauer and Wehner presented the targets on a vertical plane, as in most of the prewar work. At first, they used two targets side by side, alternated in position to make the bees look and learn which target to visit. The patterns were centred and fixed and at first the criterion was landing on the reward hole, so the patterns subtended a large unknown angle at the point of decision. In Wehner and Lindauer (1966a), the bees were trained to discriminate between two patterns, in which case we now know that they ignored cues that were the same on the two targets. Wehner and Lindauer (1966b) trained with a single pattern versus a blank white target and later said that the results were similar, which showed that they were unaware of what the bees had learned.

Figure 4.2 a) Large crosses were discriminated when the criterion was landing on the target. b) Bees trained on (a) were able to discriminate (b) with the same score. c) Bees easily discriminated between two orthogonal bars with angular size of 45°. d) The same bees could not discriminate between the cross with angular size of 45° and the same rotated by 45°. e) Bees trained on the large grating (period 32°) were able to discriminate the two patterns of large squares (f), but it was not clear what they detected.

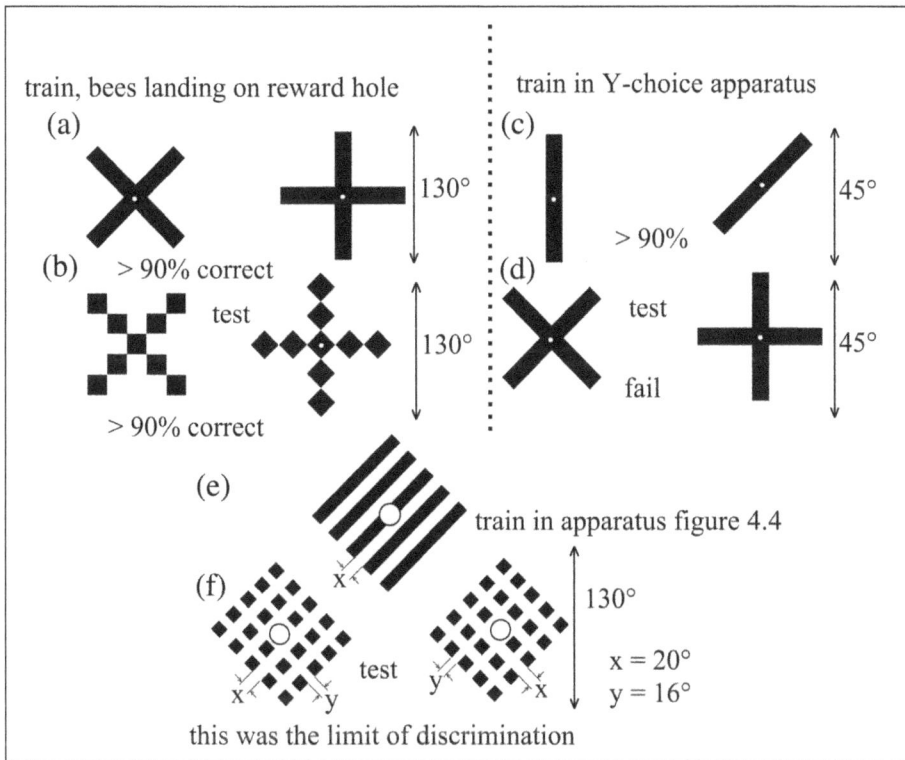

Source: (b) after Wehner (1967); (d) Srinivasan et al. (1994); (e) after Wehner (1971).

Bees trained on a square cross of two bars (18cm by 3.6cm), or to discriminate between two crosses at 45° to each other (Figure 4.2a), could discriminate crosses with edges that were stepped to destroy the edge orientation (Figure 4.2b), and did not distinguish between the plain cross and the same cross with stepped edges. So, '[t]he stripe contours may be dissected in a sawtooth-like pattern without affecting the [bees'] orientation to the inclination of the long axes of the black stripes' (Wehner and Lindauer 1966a). The edge orientation was clearly not a factor. They could discriminate a square cross of two bars (18cm by 2cm) versus the same cross rotated by 45° (Figure 4.2a) with a remarkable minimum difference in angle of only 4°, which was impossible for edge-orientation detectors, so it was finetuned to whatever feature was detected (the shift in black areas).

Figure 4.3 a) Bees were trained to discriminate the large crosses. b) They easily discriminated the peripheral parts but (c) not the central parts. d) Spots in the correct positions were adequate for the discrimination.

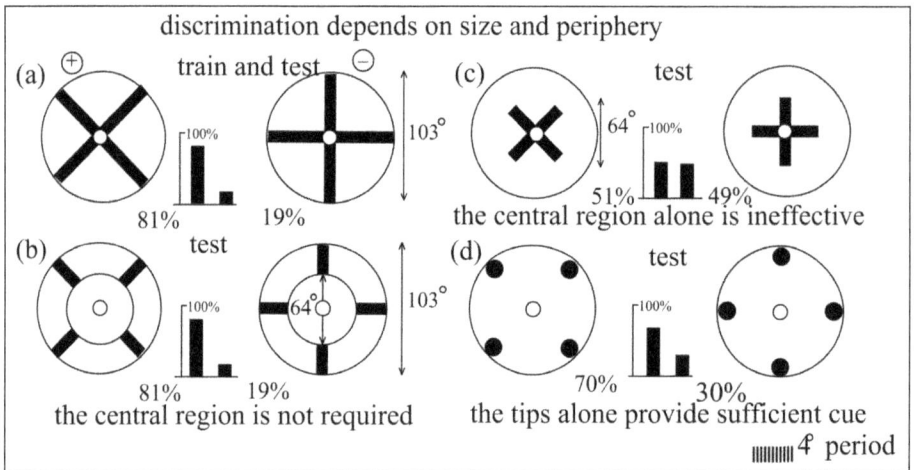

Source: After Horridge (1996c).

This result is of interest because later it was flatly contradicted. Using the Y-choice maze with smaller targets and simultaneous viewing of two patterns, Srinivasan et al. (1994) found that a square cross with angular size of 40° was not discriminated from the same cross rotated by 45° (Figures 4.2c and 4.2d). On this result, they based a notable theory that relatively insensitive orientation detectors were very coarsely tuned to edge orientation. Also, edges were summed in local fields within which orientation cues from neighbouring edges at right angles to each other were cancelled. In a reinvestigation of this disagreement, with simultaneous viewing and two targets, it was found that with very large bars the bees learned the positions of the black areas at the tips (Figure 4.3). The discrepancy between Wehner and Srinivasan was therefore due to the difference in the angular sub-tense of the training patterns (Horridge

1996b). The excellent resolution of the rotation of large patterns came from the sensitivity to the shift in position, not the edge orientation. Later, this result gave us the clue that adjacent regions of the eye remembered the positions of parts of large patterns separately and opened up the explanation of all the data collected when the criterion of success was the landing of the bee on the target.

In Wehner's other experiments, the training pattern was a large black bar, a square cross of two orthogonal bars or a grating of several parallel bars on a white background—all huge patterns. The criterion of success was landing on the reward hole. In tests, as the test pattern was rotated, the discrimination of it from the training pattern improved. The percentage of correct responses was plotted as a function of the angle of rotation. The empirical law relating the response to the angle of rotation was similar for a bar, cross and grating, although with a regularly spaced grating the position of black on the target did not change in the same way as the others. Wehner spotted this discrepancy and, having already excluded edge detectors by using stepped edges (but only for the cross), he postulated that the parameter 'is the orientation of the long axes of the black stripes, but not the direction of the black and white contours'. Although wrong, this was an interesting idea because it shifted the emphasis from edges to positions of areas in the pattern.

In the light of later work, the rotation of the regular grating must have been detected by the change in edge orientation, but Wehner could not accept that idea because he had already shown that edge orientation was not the cue for a rotated cross. He missed the point that there are two separate cues in parallel and the bees prefer to learn the position of black rather than the edge orientation.

To control the angular size of the training pattern during training, Wehner (1968) introduced a transparent screen 25cm in front of each training pattern and made the bees walk through this to the reward via a tube (Figure 4.4). Behind the screen, the pattern (with cues) was visible in a similar position on the bee's eye each time it arrived during the training. The patterns were huge (130° by 53°) as seen from the bees' point of decision, and were not viewed simultaneously, except perhaps from far away.

Shortly after, Wehner (1969) trained bees to come to a single, huge black bar, subtending 130° by 53° at a fixed range, versus a blank, and tested the trained bees with the same bar versus one that had been rotated through various angles about its centre—again looking for an empirical law. The percentages of correct responses at different angles fitted the idea that the bees distinguished better when there was less area of overlap (multiplied by a fudge factor) and more area of non-overlap (also multiplied by a fudge factor) between the positions of the training bar and the test bar (Figure 4.5). With the help of two arbitrary constants, the overlaps of training and test areas could easily be made to fit the data. The bee would then have a measure of the similarity of the training and test patterns. Apparently, it could not fail.

Figure 4.4 The apparatus for training bees with the target at a known range. Bees were trained on a single large bar of angular size 130° by 53° versus a white target. The trained bees were tested with the original bar versus a similar bar at various angles. Results in Figure 4.5.

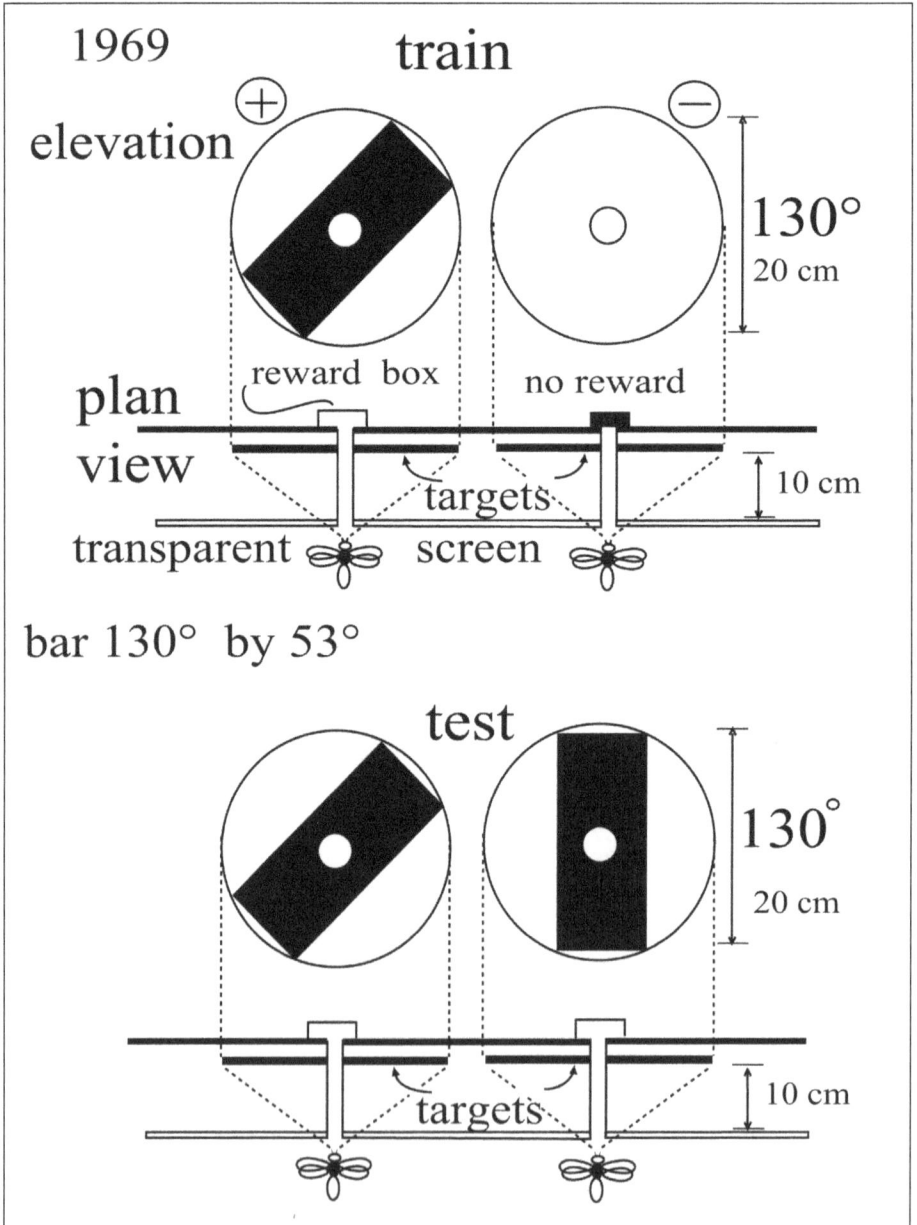

Figure 4.5 Data from the experiment in Figure 4.4. a) Calculated from the overlap only. b) Calculated from the overlap and non-overlap. d) The real shape of the bar.

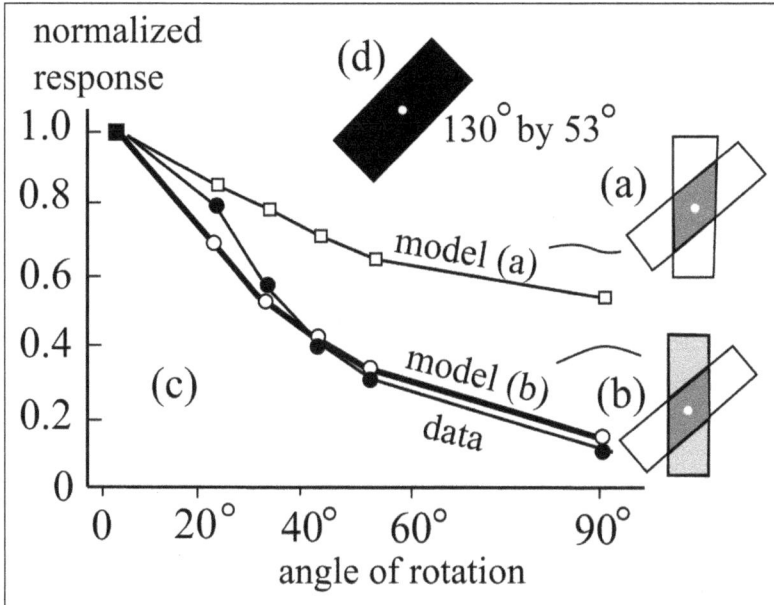

Source: (b) from Wehner (1969).

It was not mentioned that the measure of overlaps would be different for different initial orientations, but the bees would not know the angle of rotation. Therefore the overlaps were not useful as an indicator of shape. Also, the measure of the overlaps as a function of angle would be different for different shapes, so would be even less useful as an indicator of shape. It was not realised that the result applied only to targets that subtended very large angles. There were no scales on the illustrations.

This theory of matching positions of black areas was taken to imply that the bees remembered the *shape* of the training pattern and compared it with each test pattern, although that conclusion went far beyond what the data really showed. The retinotopic eidetic image in memory became accepted in the literature, but there are two twists in the story. First, in 2001, my bees trained on a small oblique black bar versus a blank white target learned neither the shape nor the orientation (Figures 9.3b and 9.3c), only the position and modulation (Figures 9.3d–f) because there was an order of preference for the different parameters. Second, in the discrimination of the rotation of a sector pattern, on which bees fixated, I found that they in fact used the position of the most horizontal sectors but failed to recognise the pattern (Figure 11.11). With very large targets, the bees detected the change in distribution of black over several local regions of the eye differentially (Figure 4.3). It was the parameters, not the pattern, that were retinotopic.

When trained with a single large target, the bees' preferred cue was always the position of areas of black (Figure 4.5). If the changes made in large targets to test the trained bees were too great, the bees refused to choose and simply went home. They had lost the black areas that were moved—that is, the expected landmark was missing.

In all the earlier work, from von Frisch (1914) onwards, in which bees were trained to land on vertical targets, orientation was scarcely mentioned. The reported parameters were the position, the area and length of edge, because the patterns subtended a large angle at the choice point and edge orientation was not preferred.

According to Schnetter (1972), 'Wehner succeeded in isolating the position of the black areas as the relevant parameter for direction information.' That was true, but by accident, because he chose to work with bars 130° long and 53° wide that overlapped several local eye regions. This story illustrates the belief that the bees see the patterns and how the unsuspected effect of pattern size has influenced the progress of the research. From this work, I drew a new conclusion. When the bees discriminated very large targets, they used the coincidences of features in well-separated parts of their eyes, as though they were viewing several landmarks in a scene.

An alternative theory

At the same time, Jander et al. (1970), working in the same institute at Frankfurt, found that wasps, trained to discriminate between a vertical bar and the same bar at 45°, were able to detect an oblique edge at 45° when tested with a black bar versus a black square of equal area, or even with white bars on a black background (Figures 4.6a–d). These striking results were interpreted in terms of strings of radially symmetrical local edge detectors, as found in vertebrate retina (centre-surround units). The authors pointed out three reasons why Wehner's theory of an eidetic image could not possibly be correct: it would require a different set of arbitrary constants for each shape, it could not apply to the discrimination of the orientation of a grating and Wehner (1968) had already shown that the response to rotation of a bar was relatively independent of the width of the bar. So, even as the new theory of overlaps emerged, it was attacked at the home base with arguments that still stand, based on a rival theory of feature detectors in parallel. At the time, there was a serious conflict but both ideas were in fact supported by data.

Figure 4.6 Contrasting results and conclusions. a–d) Wasps were trained with a single black bar versus a blank; the criterion of success was landing on the reward hole. b) The trained wasps discriminated the orientation, and (c) when black was exchanged for white. They failed to recognise the black bar when the orientation cue was the same on both targets, showing that they relied on the orientation cue. e–h) Bees were trained with a single large bar in the apparatus in Figure 4.4. f) The trained wasps discriminated the orientation. g) They discriminated weakly, when black was exchanged for white. They also recognised the black bar when the orientation cue was the same on both targets. They must have learned two cues: position and orientation.

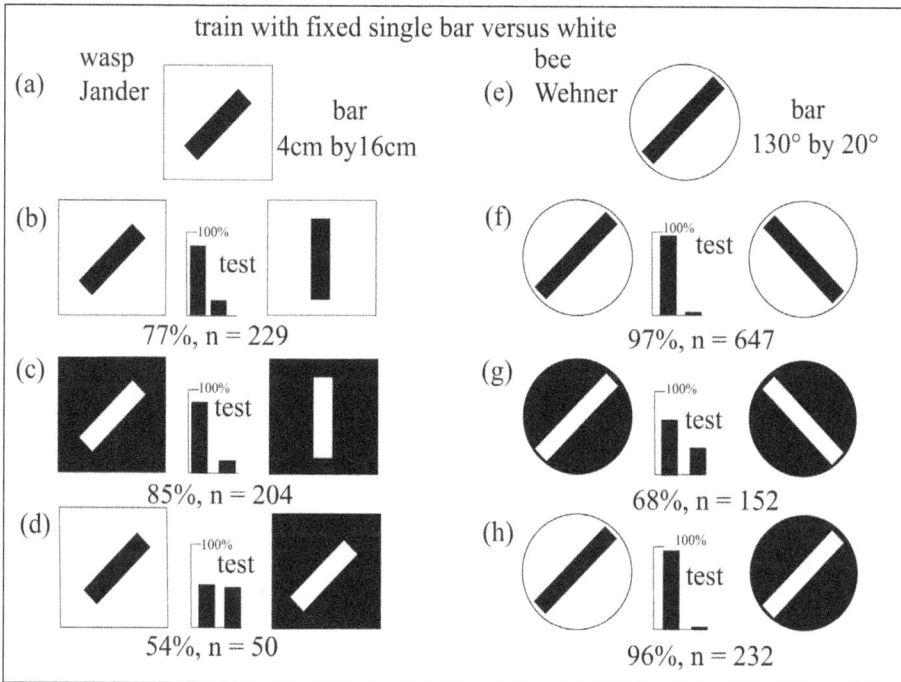

Source: Reassembled from the original papers (see text).

As though galvanised by this competition, Wehner (1971) trained with a single, large oblique black bar, at 45° to the vertical, subtending 130° by 50°, versus a blank target. The trained bees discriminated correctly between the training bar and a similar bar at right angles. They discriminated rather poorly when both test bars were white on a black background, but very well between the training black bar on white at 45° and a white bar on black, also at 45° (Figures 4.6e–h).

To explain these new data, 'analysers' of generalised orientation were intuitively inferred and 'an angle discrimination is more easily possible when the relevant analyser…is switched on by previous training' (Wehner 1971). The bees 'abstract from some special stimulus properties by generalizing the sensory input according to special cues, for example the direction of visual stimuli' and 'the information about the direction of the visual cue had been transferred to a new pattern configuration never seen by the bees during the training situation' (Wehner 1971). The words 'direction of the visual cue' are ambiguous and could mean the direction of black as seen from the point of choice or the angle of orientation on a vertical surface as an abstract cue irrespective of the real pattern. As expected from the training with a single fixed bar, most of the test results can still be explained by memory of the location of black, but both Jander and Wehner found that some discrimination of orientation persisted after the exchange of black and white (Figure 4.6).

Wehner (1971) also trained on a large, coarse grating (period 20°) versus a blank white target and tested on patterns of squares that were prepared by adding white bars of various width at right angles over the grating (Figure 4.2e–f). The limit for discrimination by the trained bees was a width of 16°. From the results, he inferred 'a real transfer of the information about the direction of stripes on a complex pattern'. Whether the bees really detected the orientation in these patterns of squares is open to question, because a separation of 16° approaches the separation of local regions of the eye and is greater than the resolution of position. It is now clear, 30 years later, that Jander's wasps and Wehner's bees had learned several cues—the position of black, the area, the modulation and some edge orientation—but not the complete pattern.

It was known at the time that in ants and the shore crab the edges of shapes were processed separately from the areas in the centres, but 40 years ago this was a new and counter-intuitive idea. For bees, it was clear that edges and positions of areas of black were separate parameters, but a model with several pathways in parallel was not conceived. Analysis was delayed by the use of very large patterns that overlapped several local eye regions and so allowed the bees to detect some configurational layout of the patterns.

In all this, the preconceived ideas of the principal antagonists were illustrated by their terminology. Jander used the word 'Kantenrichtung', meaning the 'orientation of the edge', but Wehner used words such as 'Winkelstellung' and 'Winkellage', meaning 'the place or position of the angle'. By 1971, however, Jander had left for the University of Kansas in the United States. For 20 years, there was no further mention of detectors that abstracted orientation irrespective of pattern, although the numerical parameters of total area and length of edge inferred by Hertz were redescribed several more times.

Up to this point, almost all of the work on training bees had been published in German; the theories were based on training performance and reached intuitively without extensive tests of the trained bees. There was no comprehensive review, so the details of the training, the results and the differing conclusions were scarcely known to English, French and especially American students.

Figure 4.7 With this apparatus, bees trained on a single sector pattern versus a white target were able to discriminate the training pattern from a similar one rotated by half a period, with the following results: two periods = 95 per cent; four periods = 90 per cent; eight periods (as illustrated) = 70 per cent; 16 periods failed.

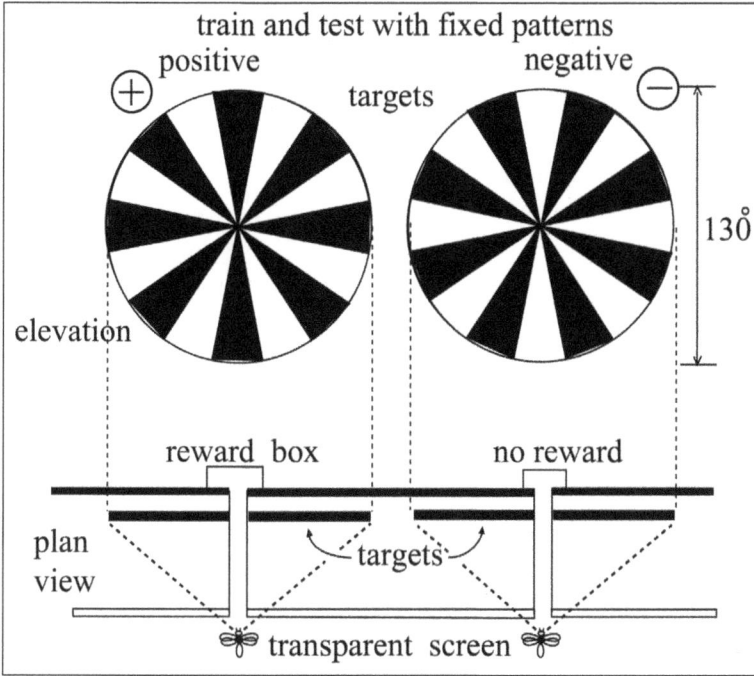

Source: Derived from Wehner (1981:Fig. 59).

In August 1972, at a symposium on insect vision at the International Congress of Entomology, Wehner proposed that the bee had sets of feature detectors that responded to relevant stimulus parameters. The theory of 'feature space' was popular at the time. He referred to the work of Jander and colleagues on feature detectors: 'Whether these "features" however are really used as parameters in an n-dimensional classification scheme, can only be decided by appropriate training tests.' Of course, there was a shortage of tests because the bees would not respond. He reproduced some of his data but was extremely cautious: 'All attempts to define the criteria of classification used by bees—or any other insect species—having failed, Cruse (1972a, b) returned to the method of calculating overlapping and non-overlapping areas between two figures by using a two-dimensional cross-correlation' Wehner (1975). But then, '[i]t is unlikely from another point of view that pattern detection can be done merely by cross-correlation methods'. It was impossible to fit the successes after reversal of contrast into any scheme with memory of overlaps. Within three years of its arrival on the scene, the theory of shape discrimination by matching area for area was considered unlikely by its originator, who had himself disproved it experimentally with gratings in 1967, yet later it turned out to be valid for very large patterns.

Wehner turned to work on the marvels of navigation by the desert ant *Cataglyphis*, but in his review (1981), he illustrated the discrimination of sector patterns from the same patterns that were rotated by half a period (Figure 4.7), with the comment: 'The only factor which can account for the bees' ability to discriminate between these gratings is the exact retinal position of black and white sectors.' This takes us back to square one, but is partially true. After 50 years of evidence of generalised parameters, however, it was a little confusing for the general reader, and it later turned out to be wrong anyway (Figure 11.11).

In this tangled period, the real data were excellent but clear conclusions failed to emerge because in the tests several parameters were changed at the same time, and few tests were done on the trained bees. The eidetic image was proposed but not proved. The fundamental difference between the parameters for areas and edges, and the order of preference for learning several parameters, were almost discovered. The topic was generally regarded as incomprehensible and intractable, as had been the case after the immense effort by Mathilde Hertz 40 years before. Even the large indiscriminate texts on comparative psychology and ethology omitted the known results.

What went wrong

In retrospect, the scientific method of the physical sciences of the time—varying the test pattern and plotting the responses in a search for an empirical law—was doomed to failure because several unknown parameters were changed together and the numerical relations said nothing about what the bees detected or the mechanism of vision. Not surprisingly, there was continual increase of correct choices as the test pattern differed more and more from the training pattern. The overlap of areas is a measure of similarity; the non-overlap is a measure of difference. The area of overlap (multiplied by a fudge factor) and the area of non-overlap (multiplied by a different fudge factor) were put into an equation that modelled the performance. For a single variable and one training pattern, this strategy could not fail.

For different kinds of patterns, however, it was necessary to have different constants in the equations and even different forms of the equations. Schnetter summed the edge lengths; Cruse added the two parameters together; Anderson multiplied them. It was all ad hoc curve fitting. There was no prediction of the performance from one pattern to another and few tests of trained bees on critical patterns. The broader criticism is that we are not very interested in the fraction of choices that are correct as a function of an arbitrary parameter or in an equation that fits the data, because there are many interacting cues.

What we need is an insight into the mechanism. These are perennial problems in biology, which needs to stick to what Forel called 'the intimate connections of the causality'.

The quantitative differences between scores depended on the length of training. In contrast, much of the recent work relied on yes/no decisions of whether or not the bees could learn to discriminate between two patterns that differed in a subtle way. If they could learn them, they were then given a large variety of yes/no tests.

That could not be the end of the eidetic image, because positions of natural landmarks were learned as positions on the eye. Also, there were still unexplained results when the criterion of success was landing on the pattern. Gould (1985), working at Princeton, showed that bees distinguished whether two flower-like patterns had different positions of coloured areas with a resolution down to about 10°. There was little popular understanding, however, despite 60 years of effort. There was no popular book, no signpost to the future or discussions of alternative theories. With good salesmanship, a final long review omitted any reference to Jander's work on orientation detection and concluded rather inadequately that 'some decisive pattern parameters such as contour density or area distribution have been unravelled' (Wehner 1981).

On second thoughts, none of the theories was satisfactory. A single summed parameter such as the total edge length is a stupid way to distinguish objects in the natural world, although it is used to classify fingerprints. Let me make an analogy. Suppose we remember the weights of 1000 people, then, when we meet someone, we identify them by their weight. We have introduced a totally unnecessary need for precision in measuring weights and also ambiguity because weights are not constant and many people have similar weights. A retinotopic memory is also not suitable for the detail in the natural panorama because the bee is always in motion, which is bad for collecting and comparing whole images. Changes in range, rotation or translation of the image relative to the head, or a view from a different direction, spoil the match. At any one time there is not one isolated shape in view. The idea of an eidetic image is also impossible to reconcile with the detection of a moving textured object on a textured background. Mechanisms of vision must avoid these problems.

A transatlantic revival

The long review by Wehner (1981) was in English, so it had the staying power to reach the New World, where it clearly inspired James L. Gould to experiment with bee pattern perception. Gould published an excellent book on ethology in 1982, with descriptions of his work on the dance language controversy and the sun compass of the bee, but unfortunately this was in press when Wehner's digest of postwar work appeared. Gould (1984) then published a review in the

Dahlem Conference report, with many references to bee pattern perception but no new findings, followed by a series of papers on the discrimination of coloured flower-like patterns, mirror images and inversions of the patterns. The conclusion was that the bees learned a retinotopic copy of the pattern, even if fuzzy, but in all of that work the criterion of success was the landing of the bees on the pattern, so the bees were able to detect parameters at very large angles to each other and obtain some information about the configurational layout of the patterns. Gould's student Adrian Dyer continues to use the same criterion to the present day.

Burps from an undigested topic

Despite all that, retinotopic matching became fixed into the literature on place recognition and fitted well into the intuitive idea that a small brain had vision like a camera. In the 1990s, the memory of a wasp for the nest site (Baerends 1941) and of a crab (Horridge 1966a) or a bee (Cartright and Collett 1987; Cheng et al. 1987) for the positions of surrounding landmarks was carried over to pattern discrimination. Here are four views. First, 'bees store an eidetic (i.e. photographic, or template-like) image of the pattern. This image is stored in the memory prior to landing during a fixation phase in which the bee hovers in front of the pattern' (Dafni et al. 1999). Second, 'the insect evaluates the ratio between the overlap of an actual retinal image with the stored template, and the total area of the actual retinal image' (Ronacher and Duft 1996). Third, with reference to sector patterns (Figure 4.7), '[t]he ability to distinguish between such patterns suggests that the pattern has been stored retinotopically, with the pattern only recognised when its elements fall in the same region of retina that viewed it during learning' (Collett and Zeil 1998). And again, 'the stored image and the actual image would have to be somehow superimposed. How are these images stored? Are they "real" images or parametrizations?' (Heisenberg 1995). Antagonists on one side or the other in these arguments quote any supporting statements they can find in the literature, however unsubstantiated by data.

A further muddle was introduced when Dill et al. (1993) and Dill and Heisenberg (1995) noticed that differences in the position of black accounted for discrimination by the fly *Drosophila*, and that discrimination failed if there was a shift of the pattern by 9° in the vertical direction on the screen. Recognition depended on the overlap as a fraction of the area, with no recognition of shape. The eidetic image brigade took this up as a validation of their theory, although it was exactly the same kind of evidence that had been known for 60 years (Friedlaender 1931). I told Martin Heisenberg, at a dinner in Cambridge for the Society of Experimental Biology in 1995, that the eidetic image was nonsense, but later discovered it was not entirely so. Meanwhile, unknown referees quoted this *fly* result at me and damned my papers on *bees* because I inferred parameters and feature detectors.

Fortunately, a new student refuted his professor (blessed be Würzburg for the action of one good student) and excluded the eidetic image by showing that the fly could learn to discriminate a few simple cues irrespective of the pattern, but not the overlaps and non-overlaps. The flies discriminated the height of the common centre of gravity of the black areas taken as a group, differences in total area, vertical and horizontal extent and the separation of pattern elements (Ernst and Heisenberg 1999).

A new effort to distinguish between the generalised parameters and the eidetic image began by training bees to discriminate between a triangle and a disc, either filled or in outline, in various combinations. The trained bees were tested with a variety of pairs of filled or empty triangles or discs in most of the possible combinations. The results could mostly be explained by postulating arbitrary parameters such as the triangular point at the top, or not. Unfortunately, the same results 'can also be explained by a template matching mechanism' (Ronacher and Duft 1996). Why do people do it? Or allow it to be published?

Efler and Ronacher (2000) therefore made another effort, training with pairs of black triangular shapes that differed in size or in disruption presented on a vertical surface. They tested with a variety of patterns that differed in the positions of black. They observed successful discriminations that should not be distinguished by a retinotopic memory and preferences for patterns that showed no overlap. They concluded that 'the bees must have used additional mechanisms and cues'. Yes, indeed they must have, but what were they, and why turn a blind eye to the literature?

Recently, Campan and Lehrer (2002) published a paper entitled 'Discrimination of closed shapes by two species of bee…' and certainly the bees discriminated between a square triangle, inverted triangle and square, and so on, presented fixed on a vertical surface. The bees explored the patterns closely and the criterion was landing on the reward hole, although previous work had shown that this technique yielded no conclusions. The authors suggested that the bees succeeded by scanning the patterns in flight and remembering the differences in the positions of the corners or edges (contours), not by memory of the shape. Again, this was an example of successful performance and the conclusion was an intuitive guess that did not contradict the data, but was not further tested.

I must mention at this point a strange lapse of the communal memory. After 1939 and until about 1995, the findings of Lubbock, Forel, Turner and others at the turn of the century, and also the works of von Frisch, Baumgärtner, Friedlaender and Wiechert on discrimination of vertical patterns, were almost erased from the record. Another memory lapse appears to have started with Cruse (1972), who actually says that, *in contrast with other authors*, Wehner (1968, 1969) trained with patterns presented in the vertical plane. In his long review, even Wehner (1981:533) states '[u]ntil then [referring to his own work] patterns had always been displayed on a horizontal screen'. The discussions and reviews from 1966 to 1990 show that they really believed this.

Retrospect

The paradigm for pattern vision changed from a qualitative description of performance in the first half of the century to one based on quantitative percentages of correct responses to selected parameters in the second half. The technique changed from groups of unmarked bees to individually marked bees, from training on a choice of patterns to training on a single pattern versus a blank, and from small patterns to very large ones. The most significant innovation was probably the apparatus that forced the bees to hover with the training or test pattern at a constant size in a constant position on the eye each time they inspected it (Figures 4.4 and 4.7), but the patterns still subtended very large angles. As time went on, however, it became obvious that the different techniques made little difference to the results with large patterns, from all of which the same four or five parameters were eventually inferred.

Another fundamental problem was caused by the use of targets that were fixed relative to the point of choice of the bees during the training. This detail did not change until 1990. The performance in the discrimination always increased as the mismatch between the training and test patterns was increased. This result gave the impression that the bees could really perceive the training pattern, implying the whole pattern. The retinotopic memory was a guess that was never corroborated, but there is an interesting twist to the story. Between 1996 and 2001, the retinotopic memory was disproved by direct experimental tests in a local region of the eye (as in Figures 2.2 and 2.3), but later it was realised that very large patterns would spread into two or more local regions of the eye and be detected as separate parts like separate landmarks. For 75 years, and even to the present day (Dyer et al. 2005), major advances were blocked by allowing the bees to examine fixed targets at a very large angular sub-tense, as though the target was a panorama.

The experimenter always imposed the training and test patterns, so the bees never had an opportunity to reveal their order of preference for different parameters. Almost every researcher started with a new set of patterns but found the same two or three parameters as the others. The misinterpretation of perfectly good experimental data was repeated over and over because some or all of the parameters in parallel were ignored. For the whole century, we have a collection of papers full of good data by excellent experimentalists who somehow did not do the right experiment and did not repeat the experiments of others or put their own theories to a critical test. Following the fashion, they refrained from open criticism and simply published and taught conclusions that differed from those of their predecessors, leaving a paper trail that no student was able to turn into a rational account. To each generation, for a whole century, the collected results on honeybee vision simply did not make sense.

Endnotes

1. This historical background might be better understood if read again after reading Chapters 9–12.

05 THE RETINA, SENSITIVITY AND RESOLUTION[1]

The combination of anatomy, optics and electrophysiology of the honeybee eye provides a splendid illustration of science in action and the way to figure out the mechanisms of processing in vision. It is a mature topic, with a wide variety of approaches—notably, optics, pigment movements, transduction, signal detection, successive arrays of nerve cells in parallel pathways, compromises between receptor sensitivity and resolution, distinctions between line-labelled channels and the interesting compromise between crowding-in more processing versus simplicity for speed of action.

The compound eye is composed of numerous simple eyes, called ommatidia, which are arranged side by side at a small angle to their neighbours. Together, the whole forms a diverging hexagonal array of visual axes that samples the visual world in angular coordinates (Figure 5.1a), in a different way to the vertebrate eye (Figure 5.1b), but with similar results. In directions where vision is most vital, the hexagonal pattern of facets is most regular. The wide field of view assists the detection of prey and predators and is essential for the recognition of a place by a relatively simple brain. The compound eyes of crustaceans and insects are similar in detail and in their functional arrangements, besides being controlled by similar genes, although there is no continuous series of fossil compound eyes between them.

Early observations

More than three centuries ago, it was proposed that the insect retina divided the image into small, separate receptor fields. In his book of 1665, the versatile English scientist Robert Hooke (1635–1703) inferred that each facet was a convex lens that formed a minute reversed image on a sensitive layer below. Hooke did not see the images but understood that they could not form a smooth composite image because they were reversed. Only the rays close to the optical axis could be effective. He was followed in 1695 by the father of biological microscopy, Antonie van Leeuwenhoek (1632–1723), who vividly described the reversed images in the flattened cornea of a fly. This led to a common view that the compound eye divided the panorama into an array of little pictures. The illustrations, however, were further confused because each facet showed the same image, making multiple views. The flattening of the cornea on the microscope slide caused this awkward artefact.

Figure 5.1 The compound eye (a) and the simple lens eye (b). In both, the panorama is projected to an array of receptors.

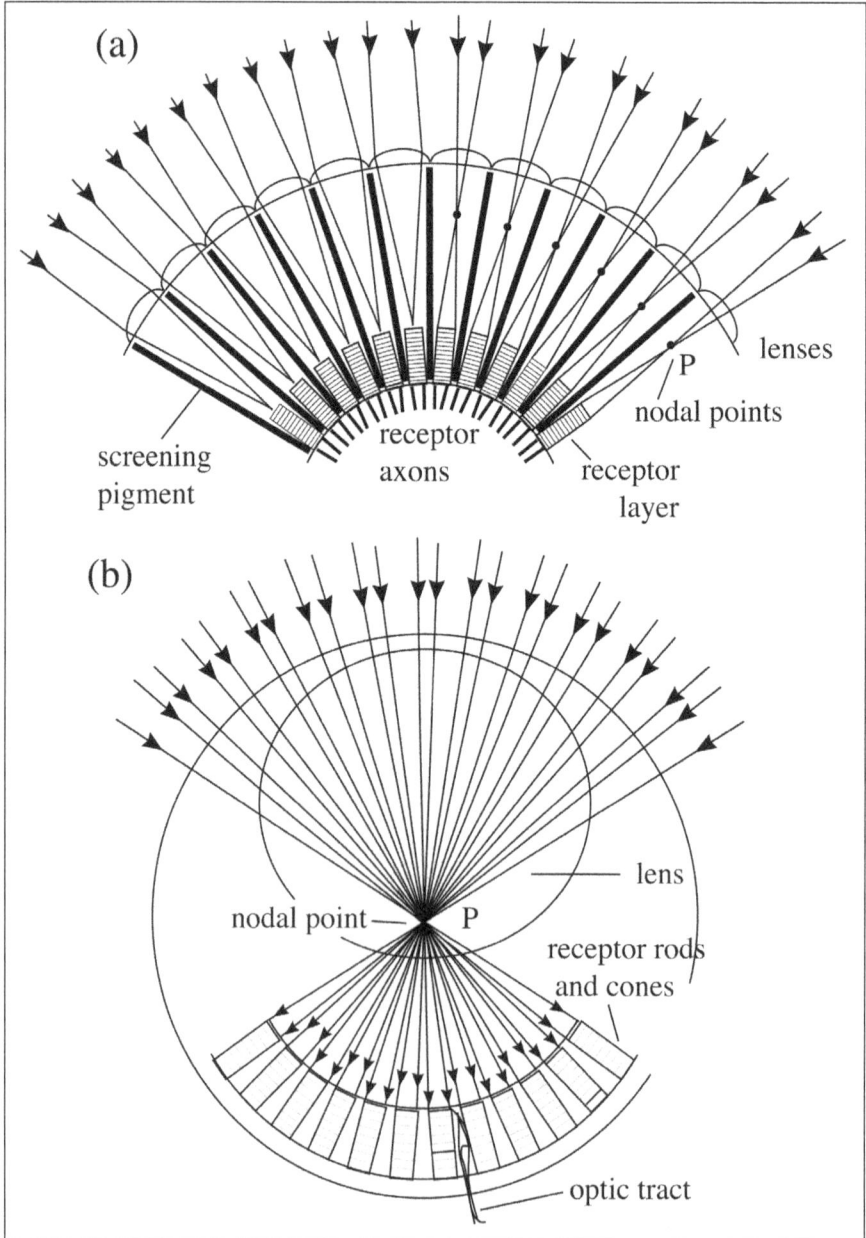

The little images were repeatedly mentioned in subsequent centuries, until Johannes Müller (1801–58), Professor of Physiology at the University of Berlin, simply bypassed them. Following Hooke, in his textbook of 1826, Müller assumed that the light passing through a single facet was concentrated to one receptor, so each facet must look in a single direction (Figure 5.1a, left side), but he also assumed that the panorama was divided without overlaps and without gaps (Figure 5.1a, right side). Both models were wrong in detail and Müller's theory was a simplification, but he carried sufficient authority to inhibit alternatives for 130 years or so.

As soon as new histological techniques were invented, very small, dense inclusions in each receptor cell, called rhabdomeres, were inferred to be the light-sensitive particles (Grenacher 1879). In many common large insects that fly by day—notably, bees, wasps, butterflies, locusts, crickets, mantids, dragonflies and the primitive wingless insect *Machilis*—the rhabdomeres were fused to a single rod, called a rhabdom, that extended inwards from the tip of the cone. As Exner noted later, Grenacher, like Müller, inferred that these eyes could have only one directional sensation for each ommatidium, but possibly more than one sensation of colour.

Many groups of insects—notably, bugs, flies, the primitive wingless insect *Lepisma* and many beetles and lower-order insects—however, had six, seven or eight separate rhabdomeres in the right place to receive the inverted image, so they could divide it into a few separate parts. Exner (1891), in his influential book, assumed Müller's theory and that the light was absorbed along the length of the rhabdom, like a single light guide. He examined only one species with separate (open) rhabdomeres—the drone fly *Eristalis*—but did not illustrate it, and subsequently the separate rhabdomeres were scarcely mentioned in twentieth-century texts, although they were the commonest type. The loss of a single gene, dubbed 'spacemaker', converts the open rhabdoms to the closed one.

Grenacher also inferred that broad rhabdomeres would function at lower light levels than narrow ones, that the field size would depend on the size of the rhabdomere and that ommatidia with large facets would be more sensitive than those with small facets. These relations between sensitivity and resolution were neglected until Kuno Kirschfeld rediscovered them in the late 1960s.

Advance was slow with sudden spurts. In a curious coincidence, Vigier (1907, 1909) in France, Cajal (1909) in Spain and Dietrich (1909) in Germany described the axons coming from the seven separate receptor cells of the fly ommatidium, and inferred that they looked in different directions. Each axon, however, meets with six others in the lamina layer. The effect of this intricate convergence of the axons is to sum together the parallel axial rays that enter the eye through each group of six facets, so making the optimum use of the little images by rotating and combining them. This amazing work was neglected because it appeared in journals that were little read and few were interested. The convergence was observed again in silver-stained sections, worked out in detail, and published by Valentino Braitenberg (1967) and Kuno Kirschfeld (1967). In all the sciences, we find this neglect of a topic for decades and then another sudden simultaneous interest—like goldminers rushing from one strike to another.

Figure 5.2 The structure of the retina of the worker bee. a) Vertical section; c = cone; p = principal pigment cell. b–f) Transverse sections at the different levels shown. g) and h) Details of the light path at the cone tip in the dark and light-adapted states, with exaggerated migration of the pigment grains.

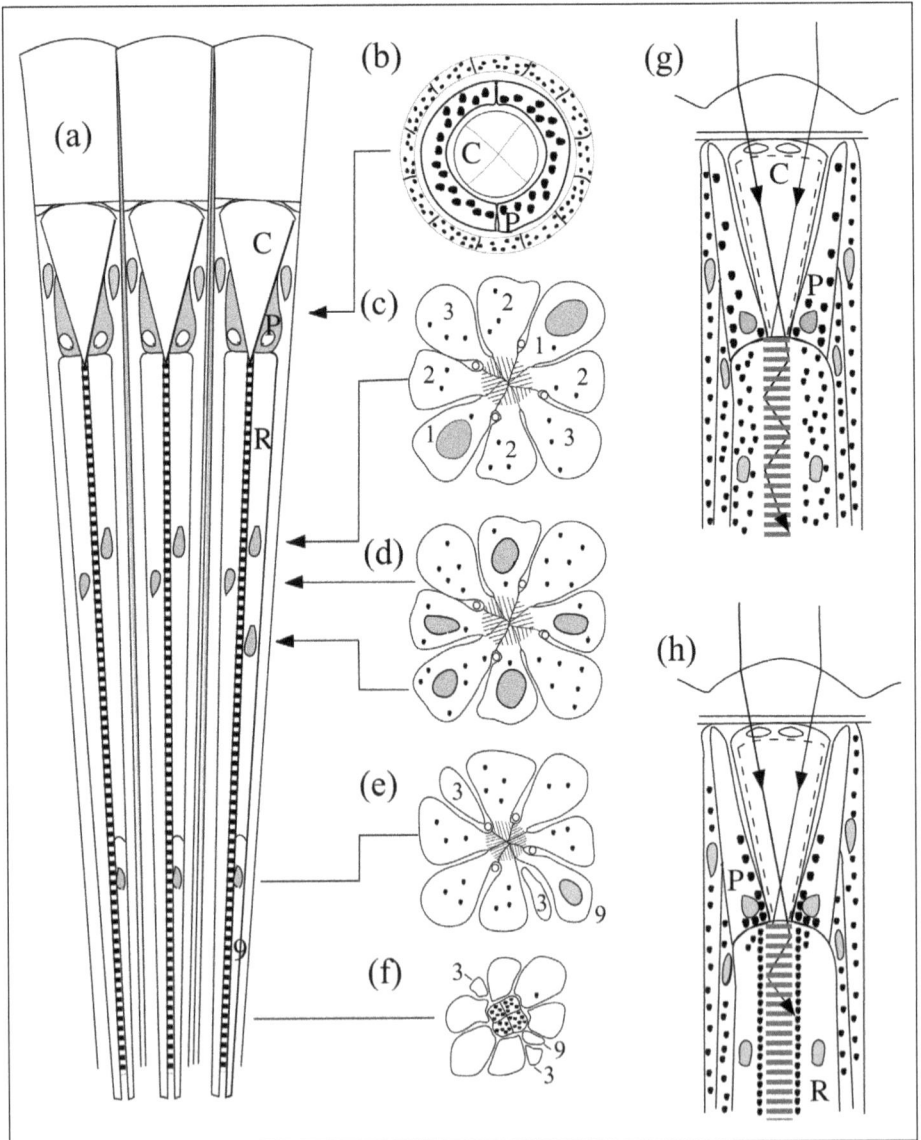

Functional anatomy of the bee ommatidium

The bee has a common type of compound eye in which each ommatidium has its own convex lens, formed by the cuticle of the cornea (Figure 5.2). Accessory pigment cells and trachea screen the ommatidia optically from each other. Light rays near to the axis pass through the lens then through a transparent region, called the cone, formed by four cells, and are focused on the distal tip of the rhabdom (Figures 5.2g and 5.2h). The principal pigment cells, identified by large pigment grains (Figure 5.2b), surround the cone and secrete the corneal lens in the embryo. In the bee, there are sensory hairs between the facets.

Figure 5.3 Absorption of light in the rhabdomere. a) A microvillus with oriented rhodopsin molecules and the preferred direction of absorption of the electric vector. b) The light path with internal reflection. c) A representation of light with electric and magnetic vectors in planes at right angles to each other. The polarisation plane is defined as that of the electric vector, e.

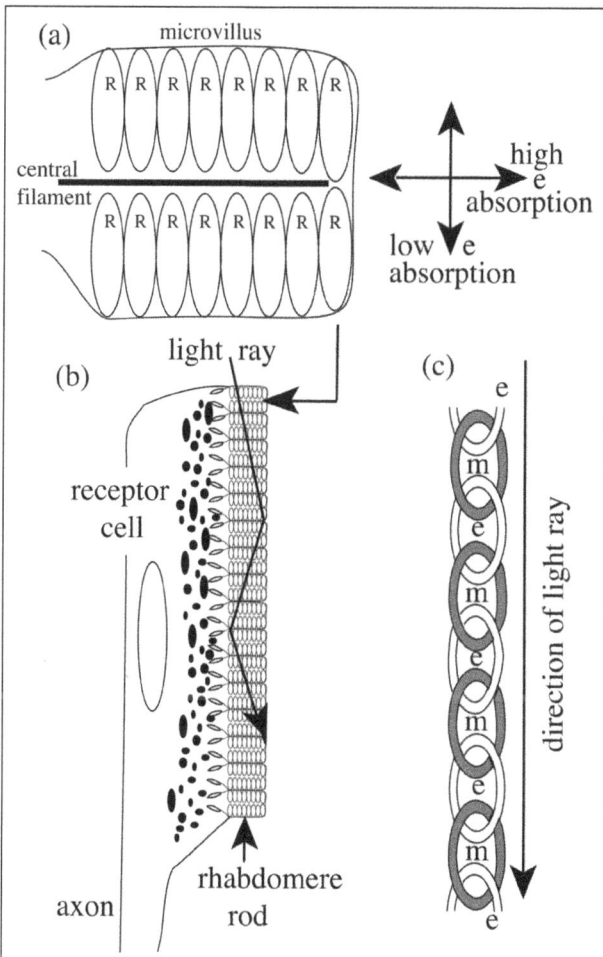

The retinula cells (usually eight, but nine in the bee) each secrete a rhabdomere, which is one sector of the long thin rhabdom that acts as an absorbing light guide down the middle of each ommatidium (Figures 5.2a–f) and have an axon at the base that runs to the next neuron layer, the lamina (Figures 6.3 and 6.5). Each receptor cell is therefore a sensory neuron with a receptor organelle. The rhabdomeres are composed of parallel tubules, called microvilli, that are each packed with about 1000 molecules of the visual pigment rhodopsin (Figure 5.3b), which progressively absorbs light that passes down the rhabdom rod. The rate of absorption is about 0.7 per cent per micron, so that at least two-thirds of the light is absorbed in the first 100 microns.

Retinula cells usually contain black, brown or red pigment grains, which can migrate close to the rhabdom by day and away from it at night (Figures 5.2g and 5.2h), so altering its absolute sensitivity and spectral sensitivity. In the bee, these changes are small. The ommatidia are not all the same; often the dorsal ones contain more blue and UV-sensitive cells while ventral and lateral ones are more green sensitive. Types of retinula cells differ according to spectral sensitivity and direction of their optimum sensitivity to polarised light. Some mainly aquatic insects that live in moist habitats have spectacular regular patterns of orientated rhabdomeres across the eye, as though they discriminate particular patterns of polarisation. Like the hand of a whale or the arm of a bird, the functions of the huge differences between ommatidia of different insects should be obvious, but most remain a puzzle. Several groups including bees have specialised UV-sensitive ommatidia with oriented microvilli along the dorsal rim of the eye for navigation using the polarisation pattern of the sky (Chapter 8).

Early theory

The convex corneal lens with the photoreceptor in its focal plane turns each ommatidium into an optical instrument like a camera with a single pixel. At the end of the nineteenth century, it was well known that resolution was limited by the aperture (D) and the wavelength of the light (λ). Incident parallel rays are concentrated to a 'blur circle' in the focal plane, not to a point. From the theory of diffraction, the minimum full width of the blur circle was calculated as $2.4\lambda/D$ radians or λ/D radians at the 50 per cent intensity contour (Figures 5.4 and 5.5). Mallock (1894) argued that the interommatidial angle ($\Delta\phi$) should be matched to the full width of the blur circle ($2.4\lambda/D$) so that the receptor would slide smoothly from one blur circle to the adjacent one as the eye moved. So, $\Delta\phi$ in radians equals $2.4\lambda/D$, and $D\Delta\phi$ equals 1.2μm for green light, with $\lambda = 0.5$μm.

Mallock's survey of 18 insects of different sizes gave a relation between $\Delta\phi$ and λ/D that was compatible with $D\Delta\phi = 1.2$μm, but he was ignored for 60 years. The theory assumed that there was a single rhabdom of negligible width on the axis, an inverse relation to the wavelength and contrast sensitivity similar to a human eye—none of which was actually realised, and there was no test of the resolution.

Figure 5.4 The two angles that define the compound eye, the field size of the ommatidium, $\Delta\rho$, and the interommatidial angle, $\Delta\phi$.

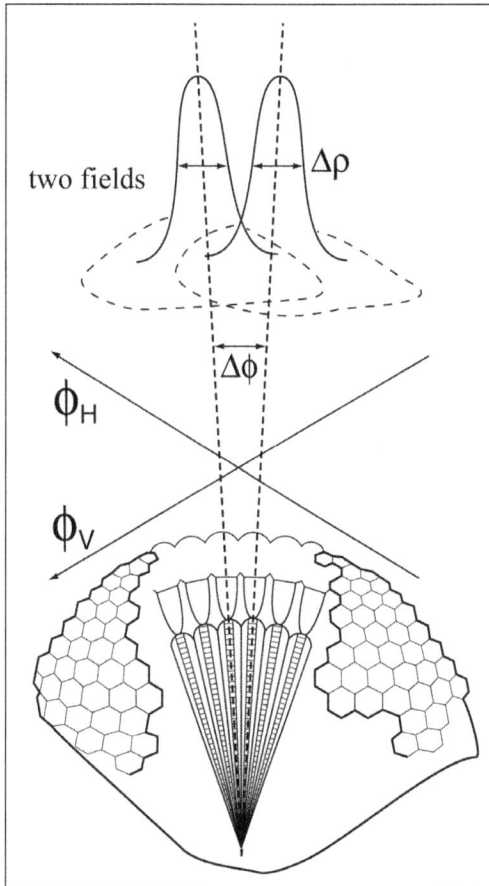

Mid-twentieth century

Unaware of Mallock's work, Barlow (1952) also assumed a match between the blur circle and the interommatidial angle. He considered three ommatidia in a row, with $\Delta\phi$, such that two distant point sources excited the two outer ommatidia sufficiently more than the central one. He predicted that $\Delta\phi$ should be less than λ/D but greater than about $0.5\lambda/D$. Again, there were only two factors—namely, the best focus and the optimum separation of neighbouring inputs—because he ignored the size of the rhabdom. Since, by geometry, $\Delta\phi = D/R$ radians, where R is the radius of the compound eye, $D\Delta\phi = D^2/R = 1.2\mu m$, so the radius should be proportional to the square of the facet diameter. Barlow was unaware of measurements of $\Delta\phi$ by Baumgärtner (1928) and assumed an isotropic eye for the bee. Although the data were incorrect, in a number of eyes of species of bees of different sizes, the average eye radius was proportional to D^2, which was compatible with the theory.

The anatomy of light capture

The width of the field of view of each receptor cell at the 50 per cent level of sensitivity is called the acceptance angle ($\Delta\rho$) and the angle between the axes of adjacent ommatidia is the interommatidial angle ($\Delta\phi$). These angles can be specified in the vertical and horizontal directions (Figure 5.4).

When there is one fused rhabdom, as in the bee, there is one optical axis in each ommatidium. In a single ommatidium, with a small circular lens, the spherical and chromatic aberrations are negligible and the distribution of intensity in the blur circle (sometimes called an Airy disc when well focused) can be approximated by a Gaussian function, which simplifies the theory but omits a weak halo of light around the edge. A convenient approximation is that the angular diameter of the blur circle at its 50 per cent intensity contour is $\Delta\alpha = \lambda/D$ radians (Figure 5.5a).

To achieve the best compromise, the distribution of photon absorption at the distal tip of the rhabdom must match the distribution of photons in the blur circle (Figure 5.5b). Larger receptors (Figure 5.5c) subtend a larger angle in the outside world and therefore catch more light, but waste the lens resolution— exactly as happens with large pixels in a cheap camera that make the image grainy however good the camera lens. Receptors subtending less than λ/D radians in diameter (Figure 5.5a) simply throw away sensitivity with no extra gain in resolution.

Because light is absorbed by the visual pigment at a rate of only about 0.7 per cent per micron along its length, the rhabdom is a long rod with the incoming light focused on its distal tip, like the rods in vertebrate eyes. To catch rays optimally, the rod points directly at the nodal point of its lens. When separated from each other by a medium of lower refractive index, rhabdoms or rhabdomeres act as separate light guides. When light guides are about 1–2μm, their capture cross-section for light can be approximated by a Gaussian distribution of diameter d at the 50 per cent level of sensitivity, where d is the diameter of the rhabdom. Therefore, to match the resolution of the lens to the capture cross-section of the receptor, we have Equation 5.1.

Equation 5.1

$\lambda/D = d/f$ radians

In Equation 5.1, f is the focal length, measured from the tip of the receptor (at the focal plane) to the nodal point of the lens (Figure 5.5b).

The nodal point is defined as the point through which rays pass as straight lines through the lens (Figures 5.1 and 5.5). The power of the lens of most non-aquatic insect eyes lies in the curvature of the outside surface—all the internal surfaces having much less power—so that the nodal point in an insect ommatidium usually lies near the centre of curvature of the surface of the cornea.

Figure 5.5 The effect of the rhabdom width d (μm) on the field of view of a single receptor. A point source in the outside world generates a blur circle, which subtends an angle of $\Delta\alpha$ at the posterior nodal point. The field of the receptor is generated as the blur circle moves over the receptor of angular width d/f radians, so the receptor field, $\Delta\rho$, is the convolution of the blur circle and the receptor absorption distribution. a) For narrow receptors, d_1, this reduces to $\Delta\rho_1 = \lambda/D$ when d_1/f is negligible and the sensitivity, s_1, is suboptimal. b) When diffraction and receptor width give an equal contribution, $\Delta\rho_2 \approx \sqrt{2}\, d_2/f = \sqrt{2}\,(\lambda/D)$. c) For wide receptors, $\Delta\rho_3 \approx d_3/f$ when $\Delta\alpha$ is negligible.

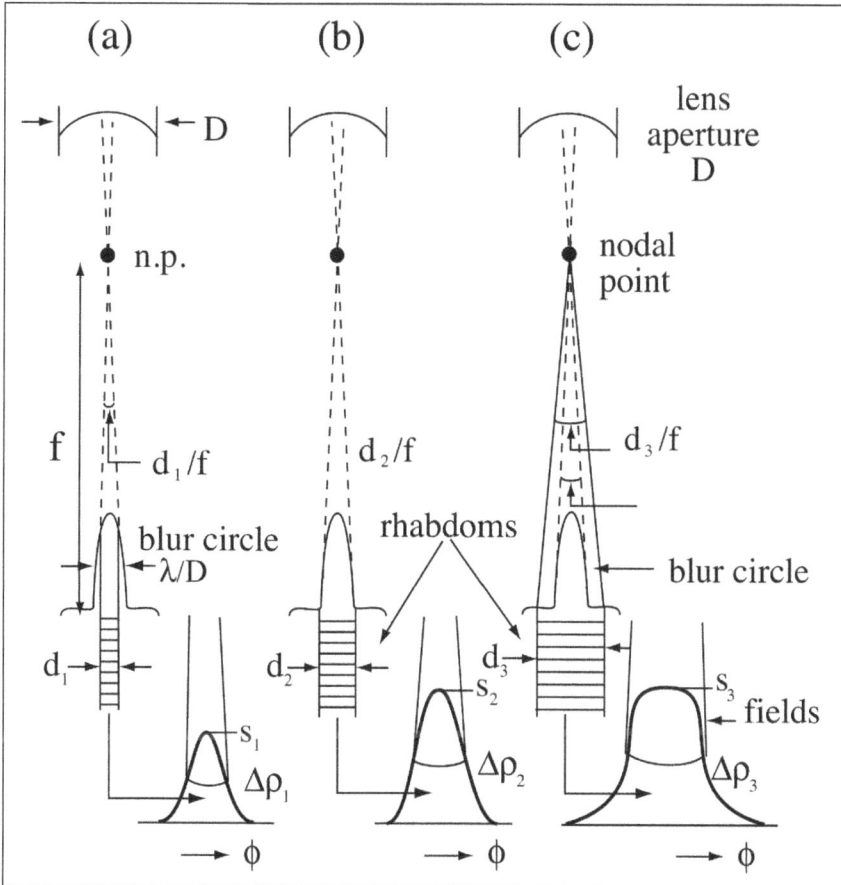

Rhabdom optics

A glass or transparent plastic fibre surrounded by a material with a lower refractive index transmits light along its length, and radar waves can be piped along a rod of polystyrene or wax. When the rod is thick, one can think of the process as total internal reflection of the waves, and classical optics based on ray tracing and absorption within the rod explains most of the properties of photoreceptors.

When the diameter of the light pipe approaches the wavelength of the light, the rays can fit into the pipe in only a limited number of ways, called modes, and the rod is properly called a wave guide. The number of modes carried in a wave guide is governed by the value of the mode parameter V, as in Equation 5.2.

Equation 5.2

$$V = \pi d/\lambda \, (n_1^{\,2} - n_2^{\,2})^{\frac{1}{2}}$$

In Equation 5.2, d is the rhabdom diameter, λ is the wavelength and n_1 and n_2 are the refractive indices inside and outside the light guide, respectively.

In fact, n_1 and n_2 are difficult to measure and the value of V is very sensitive to their difference. So, instead of calculating V from this equation, the first few modes have been observed in the light-adapted state in a few insects, from which the value of V has been inferred to lie between 1.5 and 4. Very approximately, the value of V and the number of modes in an insect rhabdom are equal to the diameter in microns, from which we can calculate that $n_1^{\,2} - n_2^2 = 0.05$, so the refractive index of the rhabdom, n_1, is approximately 1.39.

In butterflies, rays are reflected by the tapetum at the base of the retina and return through the rhabdom and out of the eye. When the cornea is optically neutralised with a little oil, the first few modes can be seen under a microscope. The effect of the increase in light intensity can be directly observed as a loss of the third and then the second modes as they are absorbed by pigment migration.

With thinner wave guides, a greater fraction of the energy lies outside, where it is absorbed by pigment grains within a range of 2µm. When the receptor rod is <1µm, light travels longitudinally along it without internal reflection—called the first mode of vibration. As the mode parameter V is increased by shortening the wavelength, more modes are accepted (Equation 5.2). A light guide carrying only the first mode accepts only photons that are near axial; the more modes, the greater is the angle of the accepted cone of light, up to the critical angle.

Mode theory applied to photoreceptors in the early 1970s led to the conclusion that the thinnest of the photoreceptor rods in small insects, about 1µm diameter, carried only the first mode for green light. More modes can be carried if the receptor is broader, but that throws away some lens resolution. The situation is quite different in a camera-type eye, where photoreceptors must also be as narrow as possible so that as many as possible can be packed in to optimise the spatial resolution.

A remarkable coincidence is revealed when we calculate the diameter of the receptor rod from two sets of assumptions. On the one hand, from the optics of the eye (Figure 5.5), d/f is approximately equal to λ/D, and f/D is the F number of the lens, so $d = F\lambda$. Since F ranges from 2 to 4 for insect eyes, and λ is 0.5μm for green light, the receptor diameter, d, is predicted to be between 1 and 2μm for all of the receptors that operate in bright light. So, a photoreceptor between about 1 and 2μm wide makes full use of the resolution of light. On the other hand, observations in butterflies, for example, reveal that a rhabdom between about 2 and 4μm wide is exactly the right size for the modes to be controlled by pigment migration outside, and the first and second modes carry most of the power. In fact, photoreceptor rods that operate in bright daylight are commonly about 2μm in diameter.

Absorption by the rhabdomere

The rhabdomere consists of numerous tubules (microvilli) of lipid-rich cell membrane, 80–90μm in diameter (Figure 5.3), on the inside walls of which are attached molecules of the visual pigment rhodopsin. The rhodopsin molecules, 12nm apart when packed in the microvilli, consist of a protein, called opsin, combined with a carotenoid related to vitamin A. There are about 1000 rhodopsin molecules per micron of microvillus. The alternating double bonds of the carotenoid capture passing photons, so the receptor acts as a photon counter and the light must be calibrated in photon flux rather than in energy units. The complexity of the vibration patterns of the rhodopsin molecule broadens the absorption spectrum—70 to 110nm wide at the 50 per cent level. This unusual property of the molecule is essential for its function in vision. The position of the peak of the spectral sensitivity depends on the particular opsin in the rhodopsin.

Perception of the plane of polarised light was popularised in the 1950s by publications from von Frisch on the bee, and simultaneously demonstrated by recordings from retinula cells (Autrum and von Zwehl 1962; Burkhardt and Streck 1965). The mechanism is that light in polarisation planes at right angles is absorbed differently by the rhabdomeres (Giulio 1963), called dichroism. It took another 25 years before it was shown that the polarisation pattern of the sky was detected by the line of specialised ommatidia along the dorsal rim of the bee eye, and not by ordinary ommatidia.

Polarisation sensitivity arises because the rhodopsin side chains are asymmetrical and elongated. The ratio of absorption along the preferred plane to that in the polarisation plane at right angles is up to 10:1 for each molecule. If the molecules lie at random in the plane of the walls of the microvilli, they will appear to be oriented because the incoming light strikes some of the membrane edge-on. This effect of the microvillus geometry alone yields an absorption ratio of 2:1 to the plane of polarisation. The electrical response of the receptor is not linearly related to the absorption of photons, so the ratio of the maximum to the minimum absorption of the receptor, called the dichroic ratio, cannot be calculated from the electrical responses, but must be measured optically or by electrical recording of individual photon captures (Doujak 1985). Even then, we do not record the value for the molecule.

The capture of a single photon by a single rhodopsin molecule causes a molecular change, which initiates a chain of amplifying reactions within the cell and finally results in the opening of channels in the receptor cell membrane, giving an electrical response. The initial unit of action appears to be the microvillus. One photon capture causes one positive-going miniature potential (called a bump) caused by the entry of a pulse of mainly Ca^{++} ions. Photon arrivals are distributed randomly in time, so when the light intensity is increased, the bumps come closer together in an irregular noisy summation and eventually fuse into a (still noisy) receptor potential. Bumps are very small in the bee, and are usually not seen in recordings.

The powerful amplification from a single photon to the electrical response of a capacitive membrane requires a great deal of energy, as indicated by the large number of mitochondria in the photoreceptor cells. This is another reason why eyes are small. Unlike vertebrate rhodopsins, insect rhodopsins, to save some energy, are bleached to metarhodopsin by the normal absorption of a photon. They are reconverted back to rhodopsin by photons of a different (usually longer) wavelength. This reaction, called photoregeneration, leads to an equilibrium concentration of available rhodopsin, depending on the wavelength content of the ambient light. One consequence is that screening pigments are often red or yellow and admit long-wavelength solar power for the regeneration of the rhodopsin without loss of resolution in vision.

The response of the photoreceptor cell

In some insects when dark adapted at night—notably, the locust, praying mantis and some beetles—single photon captures cause quite large depolarising responses, called 'bumps', so that it is possible to calibrate the transduction in the receptor cell by counting bumps and measuring the incident light intensity at the same time. Sensitivity can then be defined as the reciprocal of the number of photons required to generate a bump. It is possible to show that about 50 per cent of the incident photons are usefully captured, and also that an optical gain greater than 1 can be measured in some types of nocturnal eyes with optical overlap between ommatidia.

Figure 5.6 The method of measuring field sizes of single retinula (receptor) cells. A small source (the end of a quartz light guide) is moved around the eye on a calibrated cardan arm that measures to an accuracy of 0.1° in two coordinates. Flashes of the source at each angle cause responses that are picked up by a microelectrode from a single cell, amplified and recorded. Angular sensitivity is the reciprocal of the number of photons required to give a constant response at each angle on the eye. The light guide can be replaced by a lens to check from the shape and position of the pseudo-pupil that the optics of the eye are not damaged and to identify which eye region is stimulated.

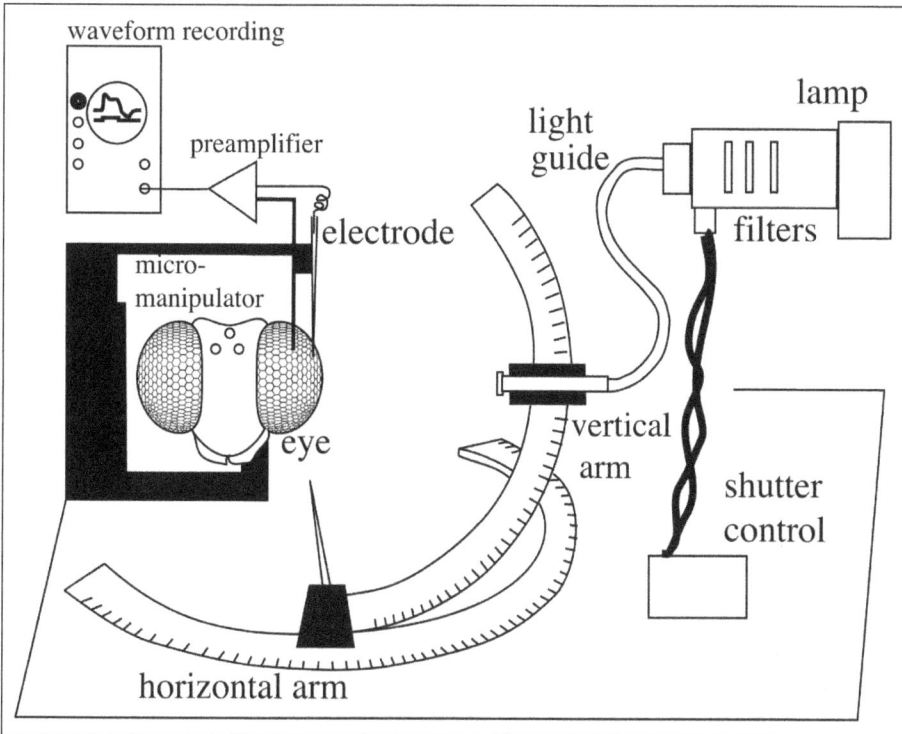

Rhodopsin molecules absorb photons as they pass down the rhabdom, causing a chain of amplification that eventually results in a depolarisation of the resting potential across the receptor cell membrane. The receptor potential is readily recorded with a microelectrode in many large insects (Figures 5.6 and 5.7a). Any measure of it on the oscilloscope screen is arbitrary; the real response is that measured by the next neurons downstream, and there are several of them responding in different ways. The receptor response is usually measured as the height of the peak to a brief flash or the initial peak at light 'on'. The shortest integration times over which light is summed linearly is about 10msec for common large diurnal fast-flying insects at 20°C. The peak depolarisation response increases with increasing light intensity in a smooth sigmoid curve

Figure 5.7 Responses of a receptor cell obtained with the apparatus in Figure 5.6. a) Responses to the point source on the axis as successive neutral density filters were removed from the light path. The logs of the filter densities are shown. b) Responses as the constant light source was moved in steps of $\frac{1}{3}°$ in the horizontal plane of the eye.

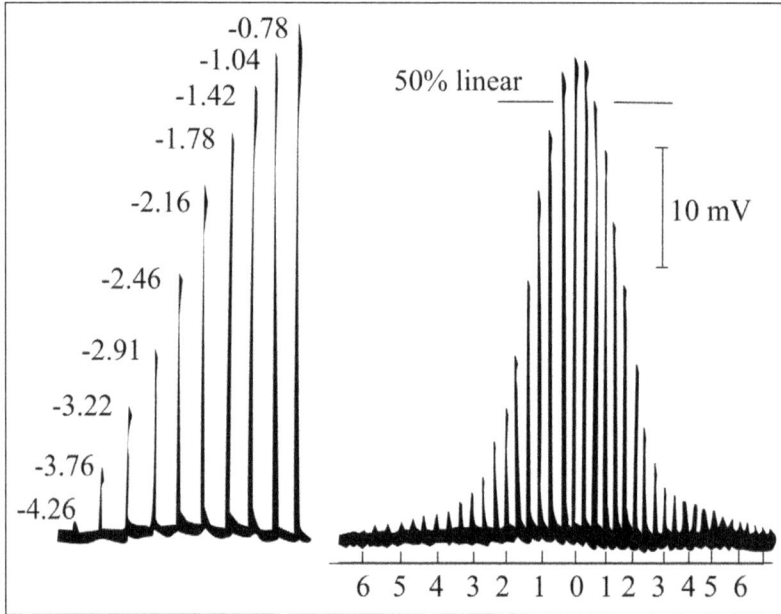

(called the V/log I curve), with a dynamic response over an intensity range of about 1000-fold. As in most other sense organs, in the eye, all response properties are relative and depend on the previous stimulation.

Adaptation to light moves the response curve to the right (Figure 6.5, Circles 1, 2 and 3), raises it upwards by the amount of the maintained response to background illumination and often makes it steeper. Repeated flashes cause a shortening of the response and a decline in the height. A sudden onset of a maintained light causes a rapid rise to a plateau that slowly declines. These changes are caused by a combination of several effects of pigment migrations, changes in membrane properties and probably extracellular potentials from the lamina that back off the steady-state response. Light-adapted eyes have higher flicker fusion frequency than the same eyes when dark adapted.

The responses of the receptors are graded—that is, without spikes. This makes transmission less noisy and faster over distances of less than a few millimetres. In evolution, the ganglia of the insect nervous system tend to fuse together and spike transmission is reduced.

One definition of sensitivity is the reciprocal of the number of photons that give a constant response—usually 50 per cent of peak. On this measure, the honeybee has a relatively insensitive eye by day, but behavioural experiments reveal a 1000-fold increase in sensitivity at night. Because there is no reliable relation between the intensity and the depolarisation of the receptors at any one time, one might suppose that this is a hindrance to accurate vision. Certainly, that would be the case in a camera. Visual systems, however, are interested in detecting features, not in measuring light intensity, or even relative intensity, except in colour vision.

Recent work has shown that the electrical properties of the cell membranes are tuned to the ecological requirements. Transduction is more sensitive and decays more slowly in slow eyes than in fast eyes. The retinula cell membrane acts as a low pass filter that smoothes the photon and transduction noise and increases sensitivity at the expense of speed. Fast photoreceptors have lower input resistances and voltage-sensitive potassium channels with delayed rectification that cut off the response with a large inward current and speed up the frequency response. This great expenditure of energy is necessary to charge the large membrane capacity of the rhabdom quickly. In slow eyes, these large currents do not occur (Laughlin and Weckström 1993).

The principle of univariance

The rhodopsin family of visual pigments has broad spectral sensitivity and, when the pigment molecule is excited by a photon, there is nothing in the response to indicate its wavelength, plane of polarisation or direction of origin. Effectively, the receptors are coarsely tuned photon counters with a smoothing filter. The response to flashes of increasing intensity is a monotonic graded increase in the peak and plateau (Figure 5.7). Responses to flashes of different wavelength, polarisation plane or direction on the eye can therefore be calibrated and related to each other in terms of the equivalent intensity on the optical axis that would give a criterion response (at a constant mix of wavelengths). In this way, the fields of the receptors can be plotted in terms of relative sensitivity for each independent variable and the resulting fields can be used to predict the relative receptor responses to other stimulus patterns.

Univariance means that the effect of intensity, angle, wavelength and polarisation can be calculated once the responses have been transformed to a sensitivity scale, because the variables are independent. Then, having measured sensitivity to various physical variables, we can discuss how the different compromises affect the receptors in terms of the laws of physics. Absolute sensitivity is best measured by counting bumps at calibrated low-light levels or by measurement of the photon flux required to yield a 50 per cent response in the V/log I curve to axial rays at the spectral peak of the receptor. Other measures of sensitivity are the slope of the V/log I curve and the noise amplitude as a fraction of the signal amplitude. All these measures are useful in their own way.

The field of the receptor

The concept of 'field' is fundamental to the analysis of nervous systems. The spatial field of a photoreceptor is defined as the angular distribution of sensitivity when a point source is moved outside the eye (Figure 5.6). The field of a neuron is the plot of the sensitivity to all its inputs, in all the dimensions in which they exist. In this case, sensitivity is defined as the reciprocal of the light intensity required to give a constant response. The field might thus depend on the choice of this arbitrary constant response and on other factors such as the polarisation plane or wavelength, so that even for a primary photoreceptor the field is dependent on the kind of stimulus. The optical axis of the receptor is defined as the axis of symmetry of the field. The responses also have important temporal properties. The field must be obtained by tedious exploration, which is why microelectrode recording is one bottleneck in the advance of knowledge of nervous systems.

Figure 5.8 The method for calculating the field size. a) Dark-adapted responses as in Figure 5.7b. b) The dark-adapted V/log I curve as in Figure 5.7a. c) Taking antilogs, one obtains the linear value in (d) at the corresponding angle in (a).

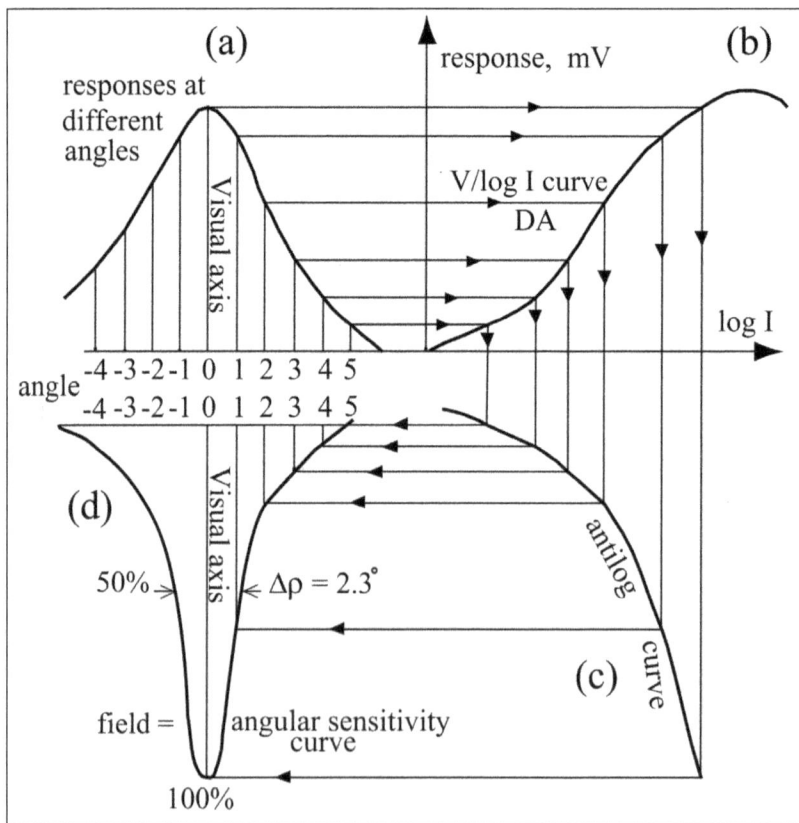

The spatial field is measured by moving a flashing point source at measured intervals in front of the eye (Figure 5.7b) while recording the heights of the graded electrical responses of a retinula cell with a microelectrode (Figure 5.7a). The point source is then kept stationary near the axis and its intensity at successive flashes is controlled with a series of neutral density filters. The graph of response versus intensity on the axis is called the V/log I curve and this curve acts as a calibration whereby the response measured at each angle is converted to a sensitivity relative to the maximum response on the axis (Figure 5.8). A more accurate but slower way to measure the field is to adjust the intensity of the point source at each angle relative to the axis until the same response is obtained at each angle, and then plot the reciprocals of these intensities. A problem can arise from the adaptation to the test light and the control of screening pigments can be via a separate pathway.

In a compound eye, there can be overlaps of the receptor fields of adjacent ommatidia, which increases the overall sensitivity, whereas in a lens-eye the receptors cannot be closer than side by side.

Convolution

We now come to a difficult but important concept that is not usually taught in school. When the field of an optical detector of any kind sweeps across a contrast in the panorama, the spatial distribution of the sensitivity of the field is multiplied point by point and moment by moment with the spatial distribution of the intensity in the image. So, if we plot the response to a contrast as a function of time, we see at first the background state in the receptor, then the initiation of a response that rises to a peak as the field moves on the contrast, and then decays as the field passes over. The process of continuous multiplication as the field passes is called convolution. If the field of the detector is very narrow, a reasonably faithful contour of the contrast (in one dimension) can be recorded. If the shape of the field and the shape of the contrast and their relative velocity and response times are known, the exact result of the convolution can be calculated, but the reverse is not true because many situations give rise to the same response.

In general, this is the situation in vision. The fields of the receptors and feature detectors are adapted to their tasks but the real shape of the image is not known. Therefore, the representation is never completely accurate. The fidelity of the response is also reduced by the adaptation to a repeated stimulus, the lack of control over the light intensity and the unknown detail of the motion of the eye—and we have a photographer's nightmare. Nevertheless, visual systems are remarkably effective because they are dedicated to the expected tasks and have many channels in parallel.

The theoretical size of the field of the receptors can be calculated approximately. The blur circle is nearly a Gaussian function of angular width $\Delta\alpha = \lambda/D$

radians (Figure 5.5a), and the optical absorption at the tip of the rhabdom is also approximately a Gaussian function of angular width d/f radians. The convolution of these two Gaussians is another Gaussian of angular width, $\sqrt{\{(\lambda/D)^2 + (d/f)^2\}}$. Therefore the field width at the 50 per cent sensitivity level—usually called the acceptance angle, $\Delta\rho$—can be calculated from the anatomy of the ommatidium. In some insects that fly in bright sunlight, $(\lambda/D) \approx (d/f)$, in which case $\Delta\rho = \sqrt{2}(d/f)$. In most insects, $(\lambda/D) < (d/f)$ because the rhabdomeres are larger than the blur circle, to help catch more light from diffuse sources, so $\Delta\rho = (d/f)$ radians.

Modulation at the photoreceptor

Contrasts always have a spatial as well as a temporal component that is generated by a moving eye. Spatial contrast is not detected directly, but simultaneous receptor responses can be correlated with each other at a deeper level in the visual system. A line, spot or edge modulates all the receptors as they pass, but each single receptor has no way to distinguish between them. The modulation frequency is a useful cue that depends on the density of edges in the pattern. It is the simplest and most frequent cue by which honeybees recognise a place.

Assuming that the receptive field is Gaussian in shape, from the work of Götz (1965), the relative modulation of light intensity caused by the movement of a spatial sine wave stimulus of period $\Delta\theta$ outside the eye (as in Figure 5.9) is given by Equation 5.3.

Equation 5.3

$$M = (I_{max} - I_{min})/(I_{.max} + I_{min}) = m.\ I.\ \exp\left[-3.56\ (\Delta\rho_{LA}/\Delta\theta)^2\right]$$

In Equation 5.3, m is the relative intensity modulation in the stimulus, I is a measure of the luminance of the stimulus and $\Delta\rho_{LA}$ is the width of the field of the receptor when it is light adapted to the mean intensity of the stimulus, not of the dark-adapted receptor, which is the measurement usually available. This equation is the result of a convolution of the sinusoidal input with the Gaussian receptor field, showing again that every operation in spatial vision involves convolution.

Fused rhabdomeres, as in the bee

In most insects, the commonest type of retinula cell has a peak that matches the most abundant environmental background colour. Frequently, as in the bee, four of the seven to nine retinula cells are green sensitive except in the dorsal part of the eye, where more are blue sensitive to match the sky.

If a photoreceptor absorbs all the light that falls on it, it will be black, so that discrimination of colour or polarisation will be impossible. To avoid this, vertebrate cones are short and absorb only a small fraction of the incident light, but sacrifice sensitivity. Some insects have rhabdomeres that are fused to form a long central rod along the axis of the ommatidium, as in the bee (Figure 5.2). Light passing down the composite rhabdom is absorbed approximately in proportion to the volume of each sector. The most abundant rhabdomeres, with peak absorption for green light, absorb less blue or ultraviolet. Each rhabdomere absorbs its own preferred kind of light, with the result that each cell retains its own sensitivity to colour and plane of polarisation, but all the light can be used. Theoretically, the spectral sensitivity curve of each receptor type is narrowed by the absorption of light in the others.

Figure 5.9 A regular striped pattern laid out in angular space in front of each facet can be represented by its sine-wave fundamental of period $\Delta\theta$. As the eye moves relative to the pattern, each receptor generates a modulated response, which is an oscillation *in time*. The graph shows the log modulation as a function of the pattern period for typical values of the acceptance angle, $\Delta\rho$. High sensitivity, as in dim light, requires large fields, and high resolution requires small fields, as shown in a different way in Figure 5.5.

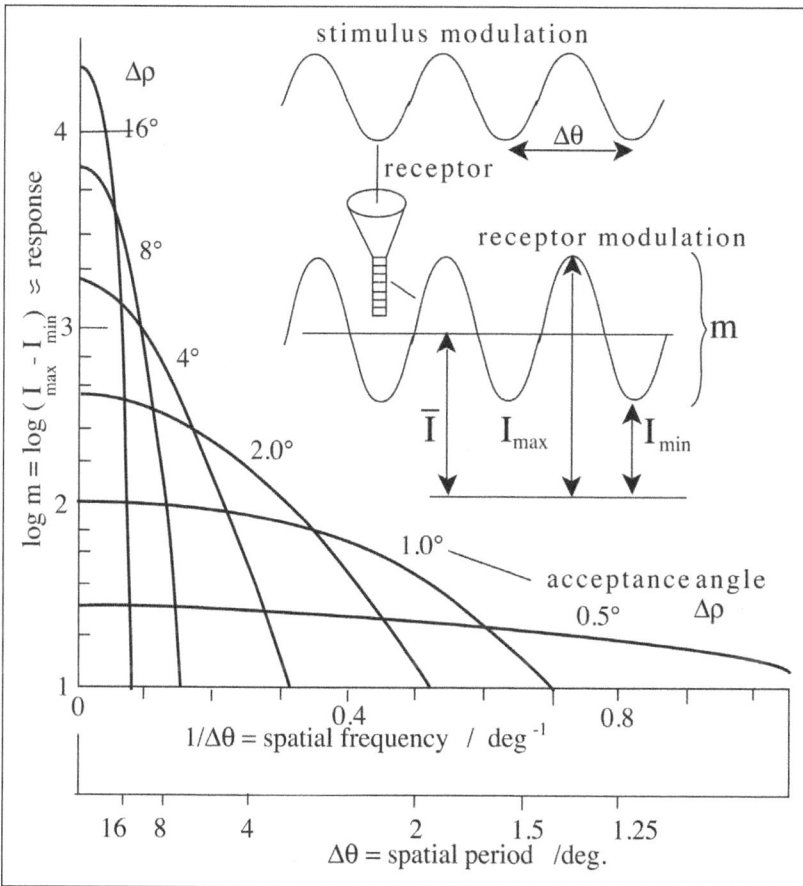

A different result follows when the retinula cells are tiered—that is, one before another along the light path, as in many insects. In the bee, one cell—number nine—at the base of the rhabdom, receives light that has passed through all the other rhabdomeres. The effect could be a marked sharpening of its spectral or polarisation sensitivity.

These effects of optical coupling are difficult to separate because there is also an unknown amount of electrical coupling between neighbouring retinula cells. The electrical coupling can be purely resistive leakage, which makes the sensitivity curves more similar, or it can be an antagonistic current flow, so that, for example, the activity of large numbers of neighbouring green-sensitive cells can hyperpolarise the blue-sensitive cells and modify their spectral and polarisation sensitivity curves.

The dimensions of the rhabdoms in the eyes of many large diurnal insects are exactly in the range where the light is controlled by very small radial movements of pigment grains. They act like a shutter around the outside of the rhabdom, with migration of pigment grains to within 1μm of the rhabdom surface in the light and away from the rhabdom in the dark (Figures 5.2g–h). They are able to absorb some of the light within because they reduce the internal reflection. Each cell acts independently, so their differences in spectral sensitivity can be detected histologically.

The principal pigment cells around the cone tip contain large black pigment grains that form a variable diaphragm controlling the entry of light to the rhabdom. When the corneal lens is neutralised with a little oil, the resulting closure of this diaphragm can in fact be seen from outside the eye by examination with an epi-illumination microscope.

One of the important lessons of the retinal array is that so few of the factors acting on the retinula cells are measurable that one has to be content with a list, such as the outline of the optics, the anatomy and the absorption curves of rhodopsins with different peaks. The consequences of all interactions on the cell are bundled together and conveyed as a single variable in the electrophysiological response, which after all is the output that acts on the next stage. In the nervous system in general, on account of unknown interactions, the redefining of the signal at each stage can be observed only by microelectrode recording, not by calculation. This is an important fundamental lesson for the study of any nervous system.

Sensitivity

The human eye, with an F number near 10, is not especially sensitive in dim light. Eyes or cameras that function in daylight usually have an F number between 2 and 16. Insect ommatidia are in this range. For example, if $\lambda = 0.5\mu m$ and $d = 2\mu m$, $d/\lambda = 4$, and, because $\lambda/d = D/f$, it follows that $F = f/D = 4$ (see

Figure 5.5). A rhabdom of diameter d = 2μm is exactly in the size range where the light passing down it is controlled by pigment grains that act as a sleeve diaphragm around the outside (Figures 5.2g and 5.2h).

When the honeybee eye is dark adapted at night, pigment grains within the retinula cells move away from the rhabdom (Figures 5.2g and 5.2h) and, at the same time, sensitivity to axial light is increased 1000-fold. The receptor field sizes are little changed but those measured behaviourally are greatly increased, so there is summation at a deeper level. Bees are not diffraction limited, they are not specially adapted for vision in bright sunlight and they commonly work in the shade. Rhabdoms of nocturnal bees are further enlarged. As the rhabdom width is increased, there is increasing sensitivity to a diffuse source, with sacrifice of resolution. Some insects that are active both by day and night—for example, locusts and mantids—increase the rhabdom diameter at night by a factor of at least 10, greatly increasing their sensitivity to diffuse sources but retaining lens resolution by day.

Spectral sensitivity

Colours are discriminated by collaboration between several receptor cells. In the bee, as in many insect orders, there are three types with spectral sensitivity peaks in the ultraviolet, blue and green (Figure 6.7b); the relative stimulation of these types gives the insect the opportunity to distinguish a range of colours. Bees have nine retinula cells in each ommatidium, one of which is basal and UV sensitive. Four of the other retinula cells have spectral sensitivity peaking near 540nm, two are ultraviolet (near 340nm) and two are blue sensitive (near 440nm). Most of the vision needed for mobility, obstacle avoidance, edge detection and so on, is colourblind and uses only the green receptor channel. For worker bees, black and white patterns on paper are just another set of colours, depending on their UV reflectance.

Polarisation

In the bee, electron microscopy shows that the microvilli of four of the retinula cells lie at right angles to the others in cross-sections, from which von Frisch et al. (1960) inferred that they detect the polarisation plane of the blue sky as part of the sun compass. This caused a lot of confusion because two directions of microvilli were not sufficient to detect all directions of polarisation without ambiguity.

Opinions differ about whether the rhabdom is twisted as a specialisation to prevent the colour vision being disturbed by polarisation of light that has been reflected at natural surfaces. The ommatidium is not twisted when carefully frozen before being fixed for sectioning. Electrophysiology shows that retinula cells of normal ommatidia are indeed sensitive to the plane of polarisation and that the polarisation sensitivity of single receptors is confounded with spectral sensitivity. The ninth (basal) retinula cell is also a puzzle.

Detection of the compass direction from the main axes of the polarisation pattern of the blue of the sky is done by specialised ommatidia along the dorsal edge of the eye, where the spatial resolution is poor but the microvilli are oriented in the pattern of a preset filter that matches the polarisation pattern around the position of the sun. By rotating itself, the bee receives a maximum or minimum stimulus from the sky (see Chapter 8).

The interommatidial angle ($\Delta\phi$)

When the eyes of many insects are examined with a lens, a small black spot appears to follow the movement of the observer. This is the place, called the pseudo-pupil, where light is not reflected back to the observer's eye because it is absorbed by the ommatidia. The angles between the ommatidial axes can therefore be measured by observing the eye on a goniometer stage (Figure 5.10). Many native bees have an obvious pseudo-pupil, but the honeybee eye has none, so its visual axes have been mapped with illumination from the back of the eye outwards through the optics. The result is a map (Figure 5.11) that provides fundamental data for any eye. Insects reveal their habits in their eye maps, especially in those that include facet size. This is a topic where the physics of the retina is related to ecology and behaviour (Figure 5.12), as documented in several earlier reviews related to spatial sampling and gradients of $\Delta\phi$ and sensitivity (Horridge 1978, 2005a).

Figure 5.10 Equipment for measuring D and $\Delta\phi$ to make a map of the eye.
The centre of the pseudo-pupil is the visual axis looking at the centre of the camera. Dust grains are used as markers on the surface of the eye, which is photographed every 5° or 10° around the eye in two dimensions. The angular coordinates of the pseudo-pupil are then marked on a linear map of the facets, from which a map of the visual axes is made in angular coordinates (as in Figure 5.11).

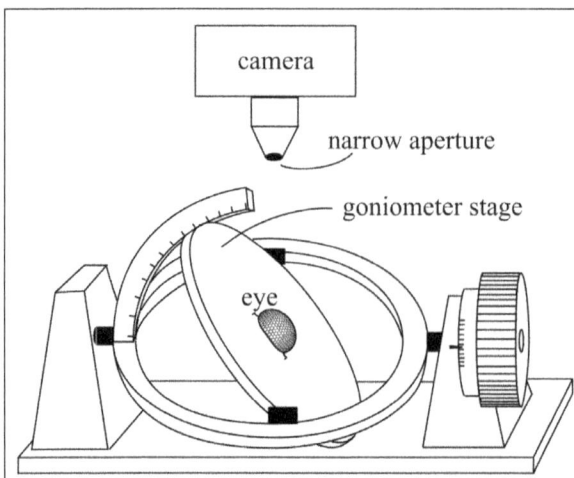

Figure 5.11 Map of the axes of the worker bee eye in isotropic angular coordinates as far as the mid-line (on the left). Inset: An enlarged map of a portion, also in isotropic angular coordinates, to show how the vertical compression of the facets produces a pattern of vertical rows.

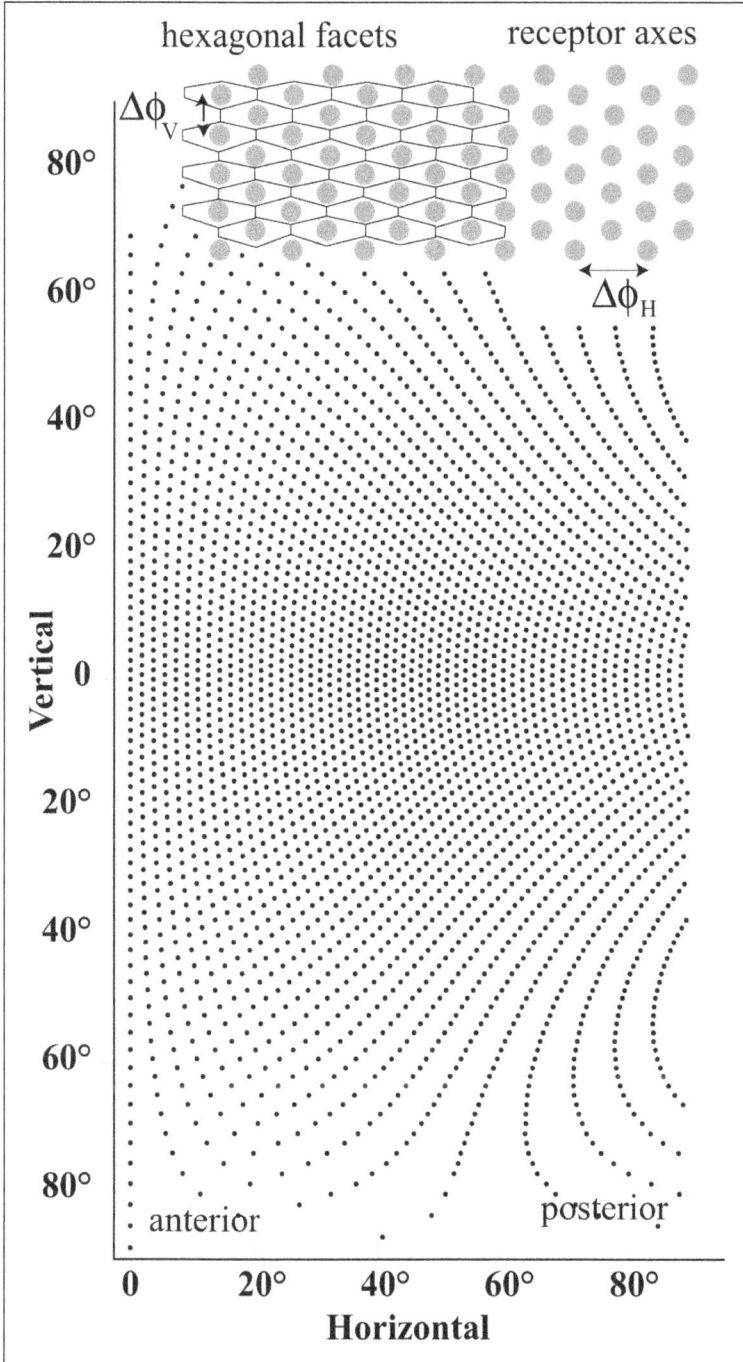

Most day-flying insects, including bees, have horizontal rows of facets at the front of the eye, but flies have vertical rows. Many authors use $\Delta\phi$ as the angle between nearest neighbours irrespective of the direction on the eye, which is convenient for eye maps (Figure 5.11). The convention usually followed is to use $\Delta\phi$ in this way in isotropic regions, but $\Delta\phi$ measured in the vertical and horizontal planes is commonly found in the literature. The use of the term interommatidial angle ($\Delta\phi$) must always be defined. An eye like that of the honeybee has very different values of $\Delta\phi_H$ and $\Delta\phi_V$ because the facets are hexagons.

Early measurements of $\Delta\phi$ in the honeybee eye were few and suspect, partly because the eye showed no pupil, except in the pupal stage. From sections, Baumgärtner (1928) had measured the interommatidial angles of the honeybee, ranging from $\Delta\phi = 2.4°$ in the horizontal direction at the front of the eye to $2°$ at the side, $4°$ at the back and $1°$ in the vertical direction. These measurements were not corrected in the literature for 70 years and led astray several researchers, including Srinivasan and Lehrer (1988) and Giurfa et al. (1997). Seidl (1982) surveyed the whole eye and located the real optical axis of each ommatidium (Figure 5.11), but Seidl's data remained unpublished until they were reworked by Andy Giger (1996) and published by Mike Land (1997a, 1997b).

The bee's isotropic pattern with vertical rows of axes is achieved by vertical compression, which is common in insects that fly by day. Bee eyes are not spherical and the radius of a horizontal row is different from that of the vertical row at the same place. Further, the optical axes are not perpendicular to the cornea. These complications reduce the validity of measurements of $\Delta\phi$ except by optical methods. In the worker honeybee, $\Delta\phi = 1.65–1.7°$ in all directions in the region around the centre of the eye (Figure 5.11). There is overlap of about $15°$ between the fields of the two bee eyes along the mid-line looking forward (Figure 1.3).

The effect of transduction noise

The theory so far might have conveyed the message that the lens resolution of individual ommatidia of insects, particularly those that fly in strong sunlight, can be limited by the diffraction of light, but this is not exactly so. A few large insects that chase prey in bright sunlight, with facet diameters less than 30μm, have acceptance angles near $1°$, implying that the rhabdom diameter contributes little to the field. On the other hand, most insects have sacrificed some spatial sampling for greater receptor size and have therefore sacrificed both receptor resolution and sampling resolution for increased sensitivity.

There is a variety of evidence suggesting that even the eyes of diurnal insects that function in bright sunlight are not at the diffraction limit. First, direct measurement on most insects reveals that the rhabdoms and the field sizes are larger than calculated from the aperture. In flies, the acceptance angle depends on rhabdom width, as direct measurements of fields demonstrate. Second, measurements show that interommatidial angles are often larger than expected from the apertures and behavioural resolutions are better than expected because of lateral inhibition in the lamina—that is, the eyes under-sample, meaning that there are gaps between the receptor fields. Third, many insects with apposition eyes are active in sunlight but have mechanisms of dark adaptation that remove screening pigment and increase sensitivity in dim light at the expense of lens resolution, but without change in the interommatidial angle. The bee is peculiar in that the eye is 1000 times more sensitive to modulation when dark adapted with little obvious change in the retina. The fields of the lamina cells are narrowed by strong lateral inhibition when in sunlight but perhaps rapid synaptic changes enable the bees to see in dim light.

Figure 5.12 Diversity of eye geometry for different ambient intensities and visual tasks. a) A diurnal eye that functions in bright light with small facets and high spatial sampling frequency. b) An eye for dim light has fewer, larger facets. c) An acute zone (a 'fovea') is made by local increases in eye radius, as in dragonflies. d) Coincidences of certain visual axes, as in the praying mantis, for prey capture at a fixed range. e) Coincidences of visual axes for discrimination of range along the head axis—almost always associated with the grabbing action of the mouthparts, as in dragonfly larvae.

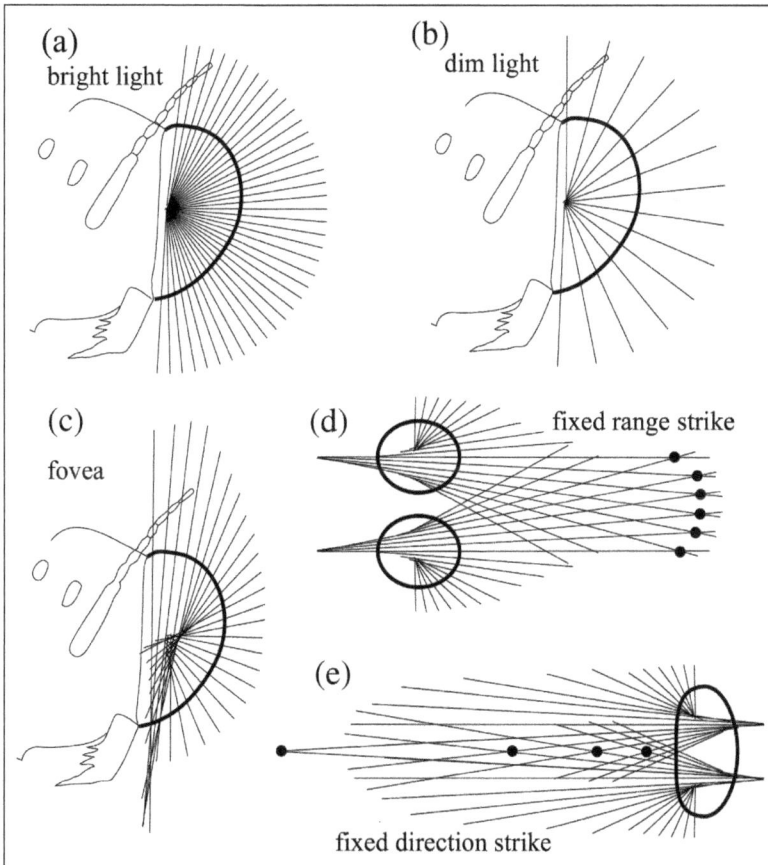

109

These observations are all supported by direct measurements of the noise. The arrivals of photons are Poisson distributed in time and therefore photon noise is proportional to the square root of the mean number of photons arriving in the sampling period. So, the signal/noise ratio is proportional to $N/\sqrt{N} = \sqrt{N}$. For the bee to see better in dim light, the signal must be increased relative to the noise, so this implies that the photon capture must be increased.

Each photon gives rise to a bump, and bumps fuse, so another way to reduce noise is to increase the integration time for the electrical response, as in 'slow' eyes, but this means a reduction in the speed of the response. The ways to gather more photons are to increase the aperture, D, the receptor volume, reduce the F number f/D or combine the signal from several ommatidia. The last possibility implies optical pooling with a superposition eye or neural pooling within the eye. Examples of all these ways occur in various insects, but the situation in ants and bees requires more work.

Direct measurements of the signal/noise ratio in retinula cells show that the diffraction-limited, noise-free eye is unattainable even in the brightest light. At low luminance, the noise comes mainly from the random arrival of photons, called shot noise, but at luminance of more than 10^4 cd m^{-2} (daylight), photon shot noise is negligible compared with the photoreceptor transduction noise that originates in the variety of sizes of potentials produced by single photon captures. So-called dark noise—the spontaneous activation of rhodopsin molecules in the dark—is negligible in insects.

The receptor capacity of the retinula cell to give a further increment of electrical response depends ultimately on the number of simultaneously active channels per receptor. In white-eyed *Drosophila*, the photon capture rate saturates at about 2.5×10^4 events, which overlap, which is about three orders of magnitude less than the number of rhodopsin molecules. Recordings from single locust receptors show that the signal/noise ratio follows the square root rule that is the consequence of shot noise up to moderate intensities, but then saturates at about S/N = 40. Saturation of the signal/noise ratio implies that Weber's Law holds ($\Delta I/I$ is a constant).

Eyes suited to low luminance are all dominated by the lower signal-to-noise ratio caused by more photon noise. Increase in facet size to increase sensitivity and resolution implies reducing the number of facets and therefore under-sampling. Compound eyes suited to bright environments have no way to increase sensitivity in the short term except by moving screening pigment, increasing receptor size or by having superposition optics of some kind. The bee is relatively insensitive when light adapted, but often flies in shade (Wolf and Zerrahn-Wolf 1935).

In the bee, the noise levels of the three receptor types together with their colour opponency predict very well the shape of the photopic spectral sensitivity curve measured behaviourally (Vorobyev and Osorio 1998). The noise in the receptor cells sets an absolute limit on the behavioural discrimination of different wavelengths (Vorobyev et al. 2001). Vorobyev et al. (2001) combined the skills of different specialists on bee training, physiology and computation as a foretaste of what must be done throughout the whole subject, including detection of motion and each of the feature detectors for edges. Interestingly, for a bee to discriminate a colour in a field of 60 ommatidia requires a photon flux per receptor cell about 1000 times that required by a fly to detect the direction of motion in a large field (Dubs et al. 1981), so the bee is insensitive by day.

Anomalous resolution

In the early 1960s, the resolution story took an extraordinary turn. From the work of Edmund Burtt and Bill Catton at Newcastle, it had been known for a decade that the locust was sensitive to the movement of a small spot or black/white edge by as little as 0.1°. At that time, there was still some confusion between the lens resolution, the spatial resolution of a grating or a spot and the least-detected motion.

A large neuron (the DCMD unit) of the locust ventral cord responded to a small movement of a grating of period 0.3°, which was 10 times smaller than the minimum calculated from the width of the blur circle (Burtt and Catton 1962, 1969; Catton 1998, 1999). Burtt and Catton inferred a larger aperture than a single facet and suggested that rays entering by several neighbouring facets summed behind the cornea to generate intensity patterns that improved the resolution.

Sections of the eye, however, show that the region behind the lens is packed with pigment grains so that each ommatidium is optically separated from its neighbours. John Palka (1965) and Horace Barlow (1965) published rebuttals that pointed out that the locusts had probably responded to a low harmonic of the period of the grating that moved behind a window, as one bar disappeared from one side of the grating and reappeared at the other side.

At the time, I had electrophysiological equipment, a few spare students and Vincent Wigglesworth had sent me some locusts. We placed the locust head at the centre hub of an arm on which was mounted a lamp behind a pinhole and recorded from the retinula cells while flashing the lamp. The angular sensitivity curve approximated to a Gaussian curve that had a width, $\Delta\rho = 4.4°$, in a region where $\Delta\phi$ was 2.4° (Tunstall and Horridge 1967). $\Delta\rho$ was the acceptance angle (or half-width) at the 50 per cent level of sensitivity. As shown later, this value was too large because the optics were damaged by the insertion of the electrode. Later, in 405 measurements of $\Delta\rho$ at the front of the locust eye, the average minimum $\Delta\rho$ was 1.16° in bright light in mid-afternoon and the maximum was 2.64° when dark adapted at night (work in Canberra published by Dr Wu in Chinese).

The whole matter was aired at an international symposium held in Stockholm in October 1965 (Bernhard 1966). Unfortunately, the revised edition of Wigglesworth's (1965) influential textbook on insect physiology carried the erroneous story, so a few generations of students worldwide were misled. That, however, was not the end of it. Claims of extremely high resolution of a small target persisted (Catton 1999). Further work popularised the adjustable strength of the lateral inhibition between second-order nerve cells in the insect lamina (Srinivasan et al. 1982). One of my students, John Scholes (1964, 1965), had shown that the night eye of the locust easily responded to single photon arrivals in single receptor cells, so that cells in a wide surround would deliver lateral inhibition. The peak at the centre of each field would be squeezed narrower, so improving the apparent resolution, but at the expense of an increase in noise and loss of absolute sensitivity. The earlier results with gratings were never explained.

As well as supposing that rays passing through several facets could sum together in the thick mass of dark pigment cells between the ommatidia, and the error of testing the eye with a grating of limited size with hidden harmonics, there was an error of understanding. Locusts are obviously not adapted to looking at gratings, so why the high resolution to them? The measured resolution of motion of a grating was the modulation summed over a large area, but in popular accounts, it was assumed to apply to any small object.

To estimate range before they jump, locusts move their head and measure the small relative movement of objects in front of them. Similarly, humans have evolved (sub-pixel) vernier acuity to infer the range of surrounding objects by measuring the parallax when the head is moved. These life-saving functions demand extreme sub-pixel resolution of a small movement of an edge, which is done by lateral inhibition and sophisticated summation along the edge, resulting in narrow edge-detector fields and even narrower modulation detectors. The feature detector for modulation is at the working edge of natural selection for sharp vision, as shown by the continual adjustment to light intensity to optimise the signal. In fact, for several vital reasons, such as having few facets or catching small prey, there are many insects in which $\Delta\rho$ is much smaller than $\Delta\phi$, and the field of the modulation detector could be narrower still.

Measurements of resolution in the bee

The resolution as measured in behavioural tests is related only indirectly to the interommatidial angle because the response depends on which feature detectors are in action. Baumgärtner measured the minimum angular sizes of blue and yellow rectangles that were detected or discriminated from a distance by flying bees and found that a coloured rectangle was detected more easily if the long side was vertical rather than horizontal. From his own measures of $\Delta\phi$, he inferred that the critical factor was the number of ommatidia involved. The minimum areas subtended about 8° at the eye, but for decades, Baumgärtner's measurements were misrepresented as evidence that the resolution was limited directly by $\Delta\phi$. Later, Gould (1985) found that bees discriminated flower patterns with two or three colours when the patches each subtended at least 10°, and he also commented that the resolution in memory was poorer than that of the retina. The resolution of a coloured patch and discrimination between two colours are both limited by the noise in the signal and, the lower the light intensity, the greater is the number of ommatidia involved.

The resolution of the bee eye was also measured by allowing each bee to walk freely on a glass plate beneath which a regular grating moved (Hecht and Wolf 1929). The bees turned against the direction of the motion, so this was directed locomotion, not an optomotor response. The minimum stripe period was near 2° in bright light, irrespective of the direction on the eye. Hecht and Wolf calculated from the optics that the minimum blur circle width was ≈ 1.14°. Referring to Baumgärtner, they saw that $\Delta\phi$ was not the limiting factor, which must be the modulation generated by the field size of the receptor, $\Delta\rho$. In dim light, the minimum period increased to 30°, so they postulated other receptors with wide fields and directional motion detectors with a wide span.

When trained honeybees were tested for discrimination of gratings against a plain grey target of the same average intensity, the minimum period in daylight was near 2.5° for horizontal and vertical gratings tested separately against grey (Srinivasan and Lehrer 1988). Referring to Baumgärtner again, the resolution was inferred to depend on the modulation—that is, $\Delta\rho$ and not $\Delta\phi$. When coloured gratings were used, with no contrast with the green receptors to eliminate motion detectors, the bees could still discriminate, although the resolution was not as good. Modulation was therefore detected by blue and green receptor channels. There was nothing to show whether the bees really detected the layout of the gratings.

Later, Giger and Srinivasan (1996) found that edge orientation was detected by modulation of the green receptors only. If, however, orthogonal gratings are oblique and without green contrast, they cannot be discriminated even when stationary (Horridge 2003c). Therefore, with vertical versus horizontal gratings with no green contrast, the cue must have been the difference in induced temporal modulation of blue receptors, irrespective of measurements of $\Delta\phi$.

Do the bees see the grating?

When no other cue is available, bees trained on a checkerboard versus grey, or with alternating vertical and horizontal gratings versus grey, learn only the modulation cue and the lower limit of resolution is 2.5° irrespective of the type of pattern (Horridge 2003c). As it flies, a bee scans in the horizontal (yaw) plane so a grating with vertical bars generates a lot of flicker (modulation) at the eye, but a horizontal grating generates much less, so the bees could rely on the difference in modulation, which is the preferred cue anyway. When trained on these gratings, however, they also learn the edge orientations.

When bees were trained on a single black-and-white grating versus white paper, they responded almost as well to a pattern of black spots versus white paper as they did to the grating, so they cared little for the pattern (Horridge 2006b). Tests showed that they had learned the modulation and the orientation, as well as to go to anything black.

Bees also learned to discriminate between an oblique grating at 45° from a similar grating at 135°, with no difference in the modulation caused by scanning in flight, so the edge orientation alone was the cue. When the resolution tests were repeated on these trained bees with oblique gratings of various periods, the limit was at a period of about 3.5° for orientation, not 2.5° for modulation. As the training patterns were rotated, the bees switched from the modulation to edge orientation as the cue.

More convincingly, when trained on a grating at 45° versus the same grating at 135°, with no contrast to the green receptors, the bees could not detect the orientation cue, and no other difference was available, so they failed to learn. Clearly, they did not remember the stripes.

The gratings provide another example showing that bees detect and learn cues, not patterns, and the results expand the concept of resolution of the bee eye. From now on, we must think of resolution of edges in terms of the feature detectors, including modulation. When the positions of areas of colour or black are learned, larger regions of the eye are involved (Chapter 10).

Measurement of sensitivity and optical gain

Educated guesses from models are instructive but a measurement by microelectrode recording is definitive. Sensitivity can be measured as the photon flux per facet required to give a threshold or a 50 per cent of maximum response, or as the slope of the curve of the response in mV plotted against intensity.

In the dark-adapted locust, the retinula cells can also be calibrated by counting the effective photon captures (bumps) by intracellular recording at fluxes less than 10 per receptor per second. It is then observed that a reasonable signal/noise ratio is reached at a flux of approximately 100 photons per receptor per second, which is approximately 1000 photons per facet per second or 10 photons per facet per 10ms period. At these light levels, individual photon captures are seen as bumps of about 1mV in the recording. Lillywhite (1977) showed that 50 per cent of the axial photons arriving on a facet were captured by the rhabdom. In the gyrinid water beetle *Macrogyrus*, which has a superposition eye, there are two photon captures for every photon (of green light at 552nm) that falls on the facet belonging to the receptor that is recorded from (Horridge et al. 1983).

For a facet of $500\mu m^2$, full sunlight provides 5×10^5 photons per facet per 10ms period, which is more than the transduction can use. In shadow, when intensities are down by a factor of 100, an eye will cope quite well, but in deeper shadow or at sunset, we reach a flux of 10^8 useful photons/cm/s at the cornea, which is approximately 5 photons per receptor per 10ms period—that is, the lowest limit for useful vision. Plenty of insects, however, find it necessary to fly in luminance lower than moonlight, which is a factor of 10^7 less than sunlight. The increased sensitivity is found in two ways.

First, parallel rays passing through several facets are deflected across a clear zone by the optics and converge on the layer of rhabdoms below, forming a superposition eye, as in skipper butterflies and some moths, *Neuroptera*, many night-flying beetles and a few others. Alternatively, they are absorbed in adjacent ommatidia and the convergence is done by convergent axons ending on lamina ganglion cells, as in flies and maybe all others with separated rhabdomeres. Second, there might be a large rhabdom that during the day is screened by pigment cells or pulled down deep into the retinula cell, as in many eyes with open rhabdomeres. Alternatively, the rhabdom diameter could increase tenfold, as in the mantis and locust.

The temporal resolution of compound eyes usually decreases in dimmer light. Long ago, Autrum distinguished 'fast' and 'slow' eyes—the latter characteristic of species that were active in dim light. The temporal properties of the membranes and synapses of the receptors and lamina cells (like the whole nervous system) depend on the mix of ion channels of different types in the membranes (Laughlin and Weckström 1993). This control of integration times at the front end, like all the other physical mechanisms in the retina, is felt through the whole visual system.

Comment

The analysis of the retina illustrates what happens when serious scientists get their hands on a versatile subject that yields hard results and promotes worthy discussions. In fact, the supporting philosophy of this science was not fundamentally different from the 400-year-old use of data to calculate the tracks of the planets or the use of costs and prices to run a business. There is similar convergence of concepts and experiments, essential training, the expert use of

complex equipment and familiarity with a large reservoir of previous studies—
however full of errors they might be—but observation, imagination and logic
are still the main foundations.

Endnote

1. The study of the retina does not tell us what the bees see, but it is an excellent example of
    how a variety of techniques, combined with a lot of hard work, expose the optimisation of the
    mechanisms by long evolution.

06 PROCESSING AND COLOUR VISION[1]

As we look deeper in the optic lobes, we find progressively altered maps of the visual panorama. Behind the retina there are three successive neural regions, crowded with dendrites and synapses, called respectively the lamina, medulla and lobula (Figures 6.1–3). The neuron cell bodies, with no electrical activity, form a thick coat around them. The regions are separated by tracts of axons and are the result of the growth of the eye at the edges, so that groups of local circuits are reduplicated side by side in columns to form successive arrays of dendrites in layers. From the retina at least as far as the lobula, each successive layer is a retinotopic array with a different function in processing the parallel inputs, so the layers are successive stages of processing. Tracts from the lobula continue to the optic tubercle, the calyces of the mushroom bodies (Figure 6.4) and to motor centres of the neck, and finally in descending tracts to the segmented groups of neurons and motor centres of the ventral ganglia.

The basic unit of the inward pathways is the column of neurons corresponding with each ommatidium, with at least 10 synaptic relays between the visual input and the motor output. We now have neuron recordings at about eight different levels, plus some associated visual behaviour, though not all in the same system or in any one species. Functionally separate types of neurons are found side by side at every level, but few of these relate to clearly distinguishable behaviour patterns. Latencies and temporal properties are also important aspects of visual processing. The mechanisms and neurons that code decisions, long-term behaviour and learning are still obscure.

Processing in the lamina

The lamina of the bee is essentially the pre-processing neuron layer immediately below the retina. The exact 1:1 projection from each ommatidium to each cartridge of the lamina is continued through the lamina to each column of the medulla. The lamina cartridges are packed side by side. Each consists of eight to

10 neurons of which about five are lamina monopolar cells (LMCs), at least three of which have no spikes. Each LMC has a distal cell body, local dendrites in the lamina and an axon crossing in the first chiasma to the medulla (Figure 6.3).

The insect lamina is an excellent example of reasonably complex neural processing that is understood from many points of view, and is best known in the locust, fly and dragonfly. It illustrates how to investigate the central nervous system and what kind of conclusions we are likely to find with existing techniques, but analysis is difficult and our knowledge is incomplete.

Figure 6.1 General arrangement of the parts of the optic lobes and other nervous system of the bee.

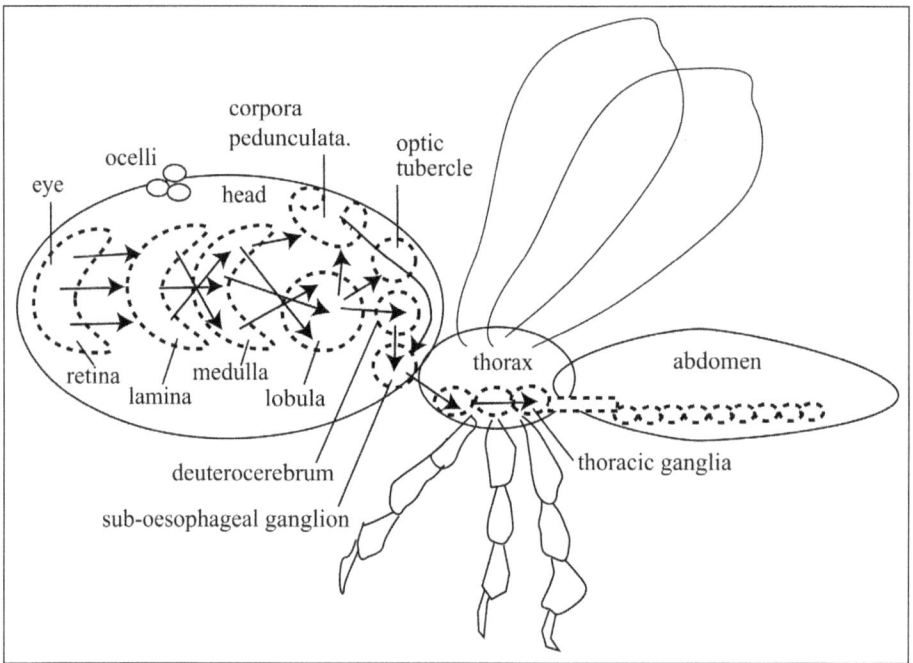

Figure 6.2 General arrangement of the successive regions of the optic lobes, leading into the brain, ventral ganglia and muscles, with the visual feedback loop.

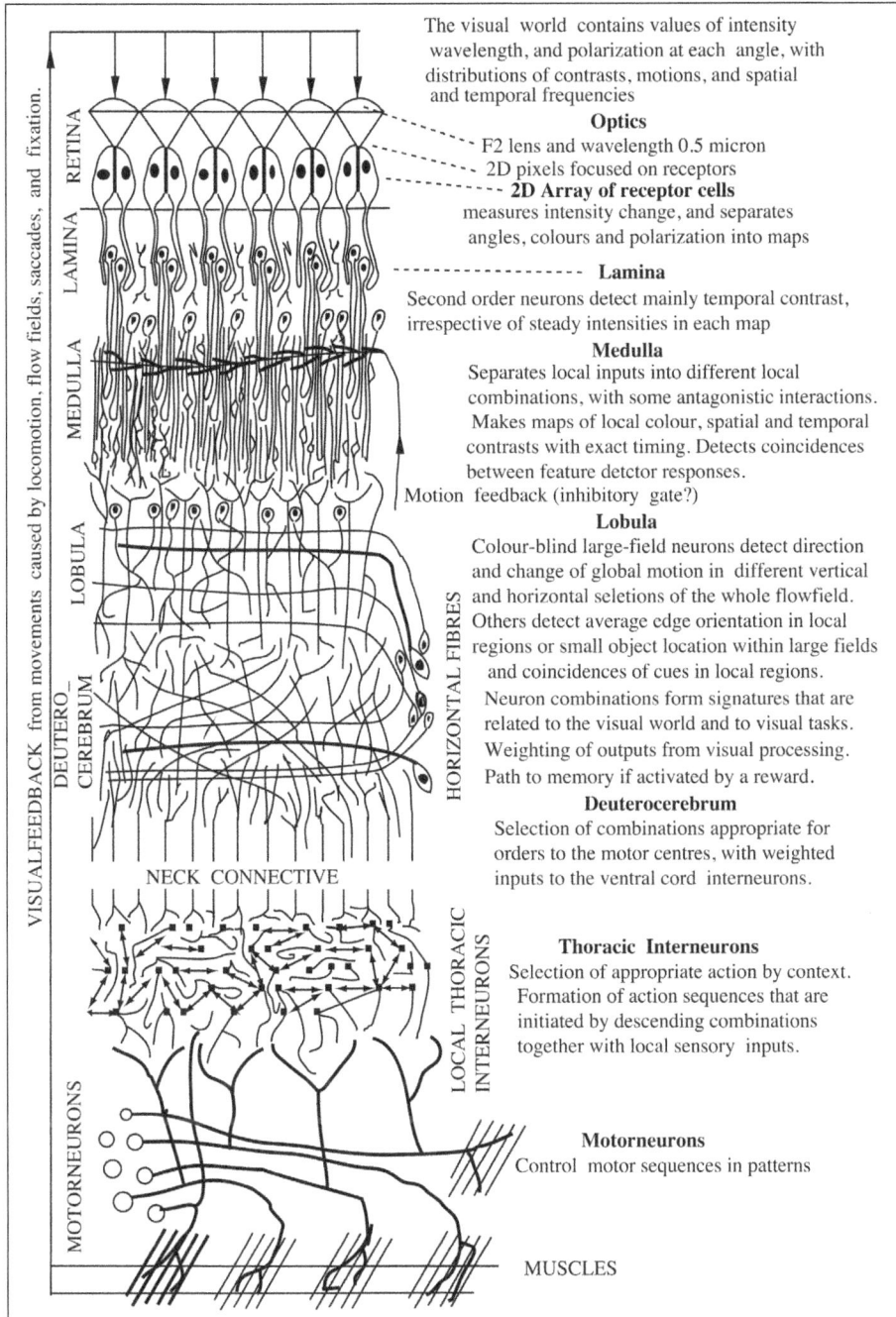

VISUAL FEEDBACK from movements caused by locomotion, flow fields, saccades, and fixation.

RETINA

LAMINA

MEDULLA

LOBULA

DEUTERO-CEREBRUM

MOTORNEURONS

NECK CONNECTIVE

HORIZONTAL FIBRES

LOCAL THORACIC INTERNEURONS

The visual world contains values of intensity wavelength, and polarization at each angle, with distributions of contrasts, motions, and spatial and temporal frequencies

Optics
F2 lens and wavelength 0.5 micron
2D pixels focused on receptors
2D Array of receptor cells
measures intensity change, and separates angles, colours and polarization into maps

Lamina
Second order neurons detect mainly temporal contrast, irrespective of steady intensities in each map

Medulla
Separates local inputs into different local combinations, with some antagonistic interactions. Makes maps of local colour, spatial and temporal contrasts with exact timing. Detects coincidences between feature detctor responses.
Motion feedback (inhibitory gate?)

Lobula
Colour-blind large-field neurons detect direction and change of global motion in different vertical and horizontal seletions of the whole flowfield. Others detect average edge orientation in local regions or small object location within large fields and coincidences of cues in local regions.
Neuron combinations form signatures that are related to the visual world and to visual tasks. Weighting of outputs from visual processing. Path to memory if activated by a reward.

Deuterocerebrum
Selection of combinations appropriate for orders to the motor centres, with weighted inputs to the ventral cord interneurons.

Thoracic Interneurons
Selection of appropriate action by context. Formation of action sequences that are initiated by descending combinations together with local sensory inputs.

Motorneurons
Control motor sequences in patterns

MUSCLES

119

Figure 6.3 A small selection of the neurons forming the columns in the optic lobes of the bee, as described by Cajal and Sanchez (1915).

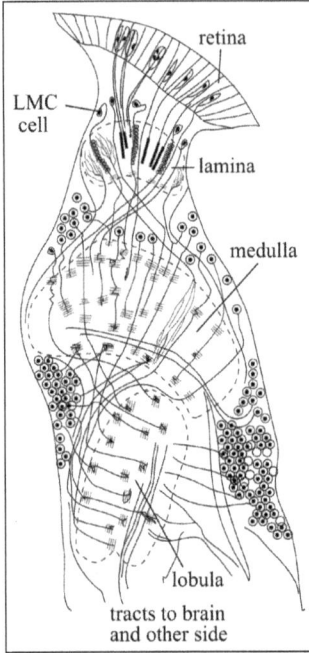

Figure 6.4 Section through the head of the bee to show the layout of the main lobes of the brain and the tracts of ('horizontal') neurons from the optic lobes and the antennal lobes to the mushroom bodies and to the opposite side of the brain.

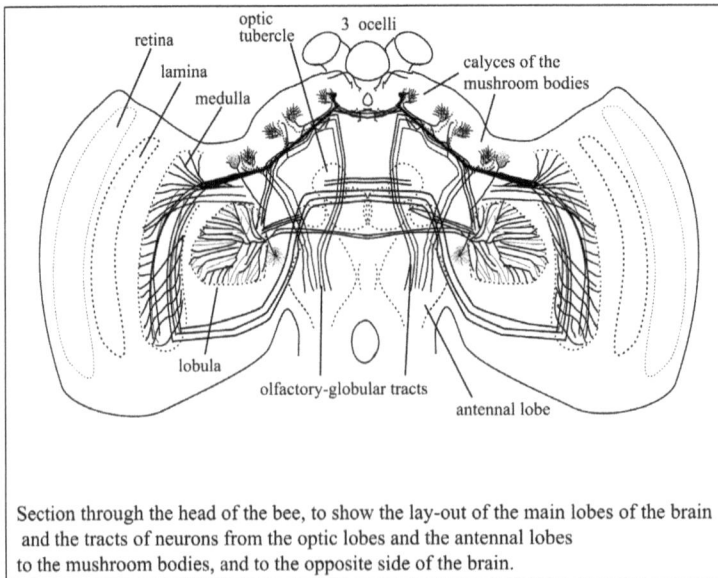

Section through the head of the bee, to show the lay-out of the main lobes of the brain and the tracts of neurons from the optic lobes and the antennal lobes to the mushroom bodies, and to the opposite side of the brain.

Source: After Jawlowski (1958) and Hertel et al. (1987).

Figure 6.5 The transformation at the first synaptic layer, the lamina: data for the dragonfly. On the left are the retinula cells and the corresponding lamina ganglion cells, L1 and L2. On the right are the stimulus/response curves of the ret and lam cells at three different states of adaptation. The three background intensities are shown by vertical arrows. The points measured are indicated by the horizontal arrows, with the character of the curve. Absolute values of photon flux (524nm equivalent) are plotted against the membrane potential in millivolts. The curves are numbered with the corresponding background intensity.

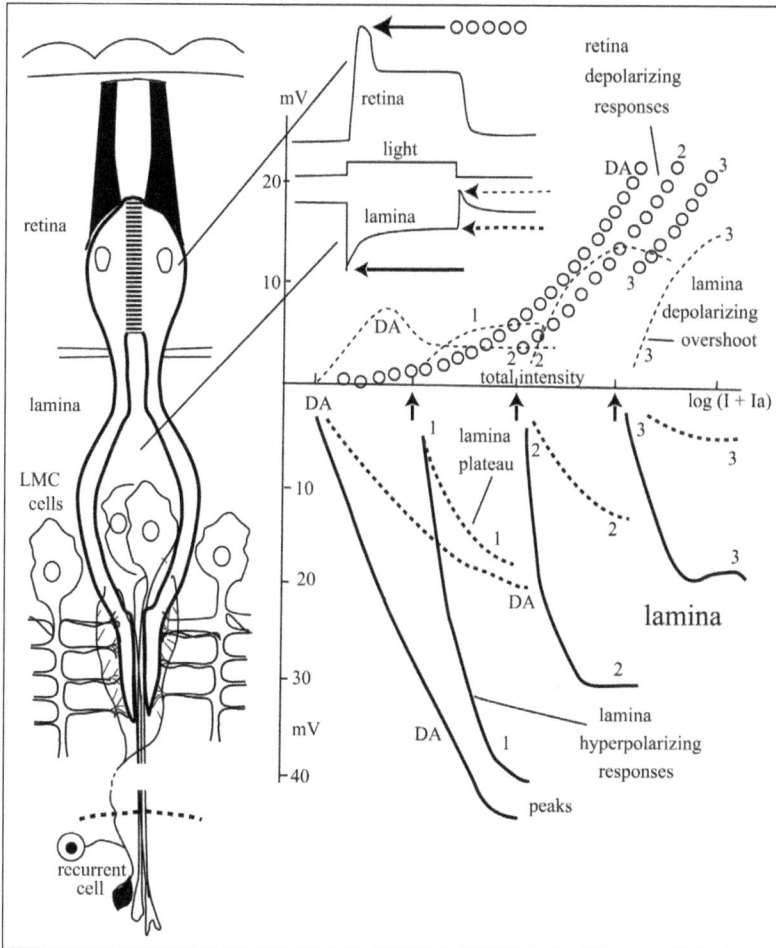

Source: After Laughlin (1975).

Responses of large second-order cells (LMC 1–3)

The lamina cells detect spatial contrasts. As a bright spot passes, the retinula cells give a depolarising response (Figure 6.5 ret) and the large lamina neurons give a hyperpolarising graded response to an increase in intensity, but ignore the background intensity (Figure 6.5 lam). The height of the peak of the lamina response and its duration depend on the intensity in an approximately logarithmic relation over about a tenfold range. In contrast, the retinal cells track intensity or photon flux and respond with a graded depolarisation in an approximately logarithmic relation that spreads across about a 1000-fold range of intensity (Figure 6.5). The lamina response is therefore more sensitive. The appropriate measure of sensitivity is the slope of the relation between the intensity of a flash and the voltage response, because at this level the system is interested in measuring the rate of change of the stimulus.

The multiple synapses in parallel that connect each retinula axon terminal to its corresponding LMC (about 200 synapses for each axon in the fly) illustrate several principles. The transmitter is histamine, which opens numerous post-synaptic chloride channels (Hardie 1987, 1988a). There is a very high gain around the midpoint in the response amplitude caused by requiring about three histamine molecules per chloride channel. Apparently, the LMC membrane is a passive integrator with no local voltage-sensitive channels and the post-synaptic gain is independent of light adaptation. The hyperpolarisation of the LMC is opposed by the large inward currents, presumably of potassium ions, which cause the overshoot at 'off'. The high bandwidth of the propagation of graded potentials is achieved by abnormally high LMC membrane resistance.

There is a substantial body of information about the anatomy and physiology of the lamina—notably, the types of neurons in flies and bees, the electron microscopy of synaptic connections and the fields of the large lamina monopolar neurons LMC1 and 2, but the functions of the smaller neurons are obscure. Most of the neurons, but not all, are monopolar with axons to the distal layers of the medulla (Figure 6.3). In insects with colour vision, there is a well-developed colour coding among the lamina ganglion cells, sometimes with antagonistic effects of different wavelengths. Where it has been measured (in dragonflies and butterflies), light adaptation alters the LMC spectral sensitivities. Except for colour processing, the lamina in the light-adapted state is mainly a high-pass, high-gain filter.

When the responses to changes of intensity on a background level of illumination are plotted for retina receptors and their corresponding LMCs (Figure 6.5), we notice several important differences. The 'on' responses in millivolts plotted against the log of the light intensity to base 10 (called V/log I curves) for the lamina are steeper (higher gain) than for the retina, but their slopes change less with adaptation. Simon Laughlin proposed that the shape of the V/log I curve of the large LMCs was the optimum to transmit the typical distribution of contrasts in the visual world, transmitting best the commonest contrasts between 30 per cent and 70 per cent, but transmitting very small and very large contrasts less precisely. The standing potentials of the receptors are largely ignored in the lamina cells (Figure 6.5 inset). The depolarising response of LMCs to 'light off' could also be an effective signal downstream. The LMC responses to common contrasts are rather sharp because the response is brief and low frequencies are cut out. The thick LMC axons and rapid graded responses are suited to the accurate timing of responses, as required for the processing of motion.

The shapes of the LMC fields change with background intensity, from a narrow deep surround in bright light to a broader centre and shallow surround in dim light. Together with the floating zero, the fast response would optimise the number of contrasts at borders that could be detected in a natural scene. Insects seem to ignore real contrast amplitudes, however, and look for contrast frequency in their visual behaviour. Bees use a measure of the modulation in local areas as a cue for the recognition of a place—for example, in discrimination between a vertical and a horizontal grating. Colour vision appears to depend more on the photon flux in local areas than on chromatic contrast at edges.

The large LMCs behave as though their visual world consists of rapidly repeated moving contrasts. Their small circular fields provide the best compromise of high spatial and high temporal resolution for detecting moving sharp edges when light adapted and contrasting blobs when dark adapted. They teach us that for early visual neurons, the main requirement is spatial and temporal resolution in the detection and timing of contrasts on each visual axis. Their axons end in terminal bulbs at different specific layers of the next neuropile, the medulla. Their outputs are seen downstream in the medulla transient cells.

Other lamina neurons

Small spiking units were recorded below the lamina of the fly, with a centre 'on' and an inhibitory zone on either side (Arnett 1972), suggesting an early stage of motion detection there. Recently, lamina tangential dendrites sensitive to flicker but not motion were found, while another efferent neuron, C2, was sensitive to motion in either horizontal direction (Douglass and Strausfeld 1995–96). Small-field retinotopic directional responses have not been found peripheral to the lowest layer of the medulla.

As well as these and the LMCs, there are several other neuron types in the lamina: a) local amacrine cells with no axon; b) presumed efferent neurons with an axon from the medulla and arborisations in the form of a basket around one or several columns of the lamina (Figure 6.5); and c) presumed efferent neurons with widespread arborisations, which stain with antibodies for polypeptides and are probably neuroendocrine neurons. Apart from the largest LMC cells, and their responses, and colour coding in dragonflies and a caterpillar, little is known about the physiology of lamina neurons.

Compression of the image in the lamina neurons

The array of retinal receptors captures the image but it contains far more information than the nervous system is able to process in real time. For a start, the light at each receptor covers a range of about 10^{14} in intensity and varies rapidly in colour and angular velocity. The motion of the image can range from zero to several thousand degrees per second in any direction in fast turns. The compression of the image into a smaller number of neurons is one of the best examples we have of processing in neuron pathways—due mainly to the work of Laughlin and his colleagues. At least five separate principles are shared between retina and lamina; let us list them.

1. Intensity compression

First, the huge range of intensities is reduced by two factors in the retina: the approximately logarithmic response of the receptors over eight orders of magnitude, and their changes in sensitivity by four to six orders of magnitude as they adapt to light by movement of absorbing pigments and other effects. As a result of backing off the background intensity, and because natural contrasts are relatively small, the responses of the receptors are up to 15mV or so, riding on a background potential that depends on the scene. As a result of the adaptation, the calibration of intensity is lost and LMC cells operate with contrast.

2. Line labelling

Second, single receptors respond to changes in colour, polarisation plane and angle of incidence of the incident light without distinction, but, by having differently tuned receptors in each ommatidium, these aspects of the stimulus are separated into different lines in parallel. Having several receptor types means more division of the signal, so there must be an economy of types. Usually in each ommatidium of large day-flying insects there are three or four types of colour receptors and three preferred orientations of polarisation plane among the cells with peak in the green. The green receptors feed the perception of motion, which is therefore colourblind. There is commonly a basal or distal cell with a small rhabdom that must be less sensitive, providing for a higher range of intensity but with different colour processing. Commonly, receptors for ultraviolet mediate a colour-specific escape response. In flies, six of the receptor cells in each ommatidium are green sensitive and sum on the second-order cells, so polarisation sensitivity is reduced in the motion-detection system. Colour and possibly polarisation are detected in flies by comparisons between receptor cells seven and eight and the other six. The way the different aspects of the stimulus are line labelled potentially tells us the priorities, but we are only just beginning to understand the details.

3. Noise limitation

Third, noise must be minimised relative to the signal by an assortment of optical tricks that increase the signal, including long rhabdoms that catch a large proportion of the incident light. The signal is amplified as early as possible in the transmission line before synaptic noise is added to the photon and transduction noise. Between each receptor terminal and its large second-order neurons, LMC 1 and 2, there are many synapses in parallel that smooth out the transmission, but even so the synaptic noise is about equal to the receptor noise. Additional noise caused by conversion to impulses is avoided by using graded potentials as far as the third-order neurons. Effort is concentrated on the high frequencies where the signal is weak relative to the noise, because where the signal/noise ratio is high, extra signal produces little extra information.

4. Redundancy removal

The fourth principle is the neglect of the redundant part of the signal. One kind of redundancy is when two parts of the signal amount to the same thing. Having started at the receptor level to respond to change of intensity rather than intensity itself, at the second-order cells the background intensity has even less effect. The large lamina ganglion cells respond almost as though to the temporal derivative of the intensity on their visual axis, with maximum slope of the response curve at the most common level of contrast. This self-inhibition removes the temporal correlation introduced by the duration and shape of the impulse response of the photoreceptor.

In the spatial domain, the resemblance between neighbouring points in an image is a form of redundancy or predictability that is reduced by lateral inhibition, which has the effect of amplifying edges and spots relative to areas of constant intensity. Both of these effects partially compensate for the smoothing of the response as a result of convolution of the signal with a filter as the eye moves. The retina introduces much of the redundancy in the lamina. Everywhere in the nervous system, filters are adapted to the predictable pass bands of previous stages, rather than to the signal, which is less predictable.

5. Under-sampling

As illustrated in Chapter 3, the blur circles behind adjacent facets never overlap sufficiently for complete sampling of the outside world, mainly because there is not room on the eye for enough facets. Reverse-motion perception due to aliasing (Figures 3.3d and 3.3e) seems never to be a problem, perhaps because adjacent ommatidia collaborate in motion perception and regular gratings are unnatural. The under-sampling improves the efficiency of the eye because it removes some of the redundancy in the adjacent regions of the image. In retinas with open rhabdomeres, under-sampling is a consequence of the separation that prevents cross-talk between rhabdomeres.

6. Speed of response

Increasing the bandwidth is the most effective way to increase information carrying capacity. It used to be thought that the response must be rapid and must decay rapidly to be useful when images move. Flying insects respond to contrasts moving at angular velocities up to 2000° per second and some detect flicker up to 200Hz when warm. Speed, however, is costly in energy terms and often not necessary. Because it takes time to collect photons in small receptors, a high speed of response is not compatible with high sensitivity and there are many examples of slow sensitive vision restricted to large objects in dim light.

7. Information transfer

In a splendid fit of collaboration, Snyder et al. (1977) showed that the design of the retina yielded the maximum transfer of information at the ambient intensity in which the insect was active. By including the distribution of spatial frequencies in the visual world, and the expected angular velocities, van Hateren (1992) carried these ideas into the temporal domain and found that the required filter at the front end of the visual system agreed with the known properties of the LMC cells—that is, temporally low passes in dim light to make use of the high power in low spatial frequencies, but faster and biphasic in bright light to increase the bandwidth, with reduced gain at low frequencies where noise was relatively small. In contrast with the theories of de-blurring, also based on lateral inhibition, predictive coding gave a similar picture and predicted that LMC cells were adapted to the detection of blobs and edges (Srinivasan et al. 1982).

8. Resolution

As will be seen, bees detect features, not the image on the retina, and therefore vision is adapted to detect features optimally. The simplest feature—contrast modulation and its position on the retina—can be the response of a single receptor, narrowed by lateral inhibition (Figure 9.2e) and is used, for example, for the detection of a small moving prey or queen bee in flight. Detection of edge orientation requires simultaneous responses of at least seven ommatidia (Figure 9.2) and motion detection requires successive responses of at least two, so the minimum detectable signal is also related to the precision in timing and the interommatidial angle. Colour, polarisation and position of black are detected by larger groups of neighbouring ommatidia, so resolution is poorer. In each case, the post-lamina processing must be related to its corresponding feature detector, as well as to the maximum signal/noise ratio.

Colour in the lamina

Discovering the above principles has dominated the study of the lamina, but the processing of colour might be quite different in that spatial and temporal resolution matter less. To discriminate colour requires that the receptor types have parallel, or at least predictable, intensity/response curves. The first stage of colour processing always involves convergence of two colour types with an antagonistic interaction on a post-synaptic neuron. Except in *Diptera*, this process starts in the lamina and continues with greater numbers of neurons in the medulla.

In the lamina of the bee, recording is difficult, but there is some evidence of spectral opponent cells, UV-sensitive cells, depolarising cells and spiking cells, while most recorded cells are green sensitive.

The dragonfly *Hemicordulia* has five types of retinula cell with spectral peaks near 330, 430, 490, 520 and 620nm. From each ommatidium there are six retinula cells ending in the lamina, two with peaks at 520nm, two at 490nm and one each at 620 and 330nm. Five types of hyperpolarising monopolar lamina cells have been found (Yang and Osorio 1996), three of them driven directly by short retinula axons and one by collaterals of a long visual fibre. The first (m1) sums several receptor types. The second (m2) has a peak in the green and also in the ultraviolet and a surround that peaks at 360nm. Adaptation by green light reduces its sensitivity in the green. The third (m3) has a peak similar to that of the 430nm receptor; m4 has a peak in the green and also in the ultraviolet. Adaptation by ultraviolet enhances the UV sensitivity and by green abolishes UV sensitivity. In both cases, the sensitivity to green is unchanged. Finally, m5 has a peak similar to the 525nm receptor. Adaptation to 430nm narrows the spectral sensitivity. At least two other colour types run directly to the medulla. Since we do not know whether dragonflies have colour discrimination, only colour-specific responses or both, sorting out the destinations and functions of these neurons will be a difficult task.

In all insects studied, the usual responses in motion perception, edge detection and several behavioural responses, except colour discrimination itself, all turn out to be colourblind, with inputs only from green receptors. So far, the neural processing of colour has not been correlated with behaviour.

Processing in the medulla

The medulla, as shown by painstaking work in a few preparations, is where the first real work of vision is done, making combinations of inputs and detecting their coincidences. There is a column of small neurons corresponding exactly with each ommatidial axis, with a sudden and early expansion of the numbers of neurons in each column (Figures 6.2 and 6.3). The principle is that the map of the visual world on the retina is reduplicated at an early stage into numerous and different successive maps, each of which is composed of its own type of feature detector. There are about 50 types of diverse small-field feature detectors in each column, even in a small fly. These arrays receive the same inputs from the retina but process them differently, in distinct layers in some cases. All complex visual systems have such stages, after which they introduce large-field collector and feedback neurons. Each of the medulla columns has several projections to the next neuropile, the lobula.

Six or more lamina neuron types, and the terminals of receptor cells seven and eight, project directly to the medulla. The neurons of the medulla columns are mainly:

1. Narrowly arborising, 'on–off' or sustaining units as described below, but superimposed on these maps are at least three other systems.
2. Layers or strata of horizontal axons with widespread arborisations at right angles to the columns (Figure 6.4). At least some of these are whole-eye motion detectors, which are efferent to the medulla from the deeper optic lobe— some ipsilateral, some contralateral. They could be rapidly acting gates, for example, for cutting out background motion to show up relative motion.
3. Efferent neurons to the lamina.
4. Numerous types of medium-field neurons, only a few of which have been described.

The neurons of the medulla are mainly small, with numerous similar anatomical and physiological types, which are difficult to record from and characterise in terms of appropriate stimuli. Only two species have yielded useful data: the larval eye of the butterfly *Papilio* and the locust. The locust medulla has been studied with a stimulus that is sufficiently sophisticated to classify the spatio-temporal fields of the neurons in a non-arbitrary way.

Locust medulla neurons

Analysis of locust medulla neurons by James and Osorio (1996) omitted colour, which the locust appears not to use much, but concentrated on the spatio-temporal properties of the column neurons measured by a rapid method. The local neurons are diversified in their polarity, latency, time course, adaptation properties and sensitivity to motion. There is a huge diversification of small and medium-sized fields providing the higher-order neurons with plenty of combinations of inputs. This conclusion presumably holds for all insect groups.

To identify the unknown properties of its field quickly while the neuron was held on a microelectrode, a method of white noise analysis was introduced. As a first step, an oscilloscope screen is divided into 64 squares in an 8 x 8 array. Each square can be bright or dark in random sequence (or a calibrated grey level if required for greater accuracy). Care must be taken to get the right spatio-temporal scales, as seen by the insect eye. This two-dimensional randomly flickering distributed stimulus is moved to cover the field of a newly penetrated small neuron and the graded responses of the neuron are correlated online by computer with the exact previous occurrences of the stimulus in two spatial and one time dimensions. Another way to say this is that the summed response seen in the neuron is divided into the constituent parts arising from each of the 8 x 8 squares on the screen for all previous combinations in space and all latency periods. The result is that the multidimensional spatio-temporal field of the neuron can then be examined in any of a number of ways, called kernels. For example, the spatial field can be plotted for any time after the onset of a flash or the response can be plotted as a function of time for any point in the field. The first-order kernels show some linear temporal or spatial responses to intensity changes or modulation. These can be subtracted from the total to give the non-linear parts. The second-order cross-kernel shows the responses plotted against various delays in one spatial direction and against delays in the other direction. It indicates directional motion irrespective of edge polarity, like a motion detector. The third-order kernel indicates contrast gain control and the fourth-order kernel indicates the non-directional motion, like an edge detector.

The square array of flickering stimuli classifies the field but is not appropriate for many neurons. At one extreme, the spatio-temporal responses of photoreceptors, and even of the lamina monopolar cells, are too simple to justify an extensive white noise analysis. At the other extreme, dedicated neurons respond to a very small part of the white noise mixture. They might respond to bars of certain orientation or small spots, so that their responses are lost in the noise caused by the large stimulus array. Those with peculiar spatial/colour antagonism are not suitable because they are not excited. Another problem is to avoid trends caused by adaptation to intensity, flicker or motion.

To find the fields of complex dedicated neurons, a combination of a dedicated stimulus and white noise is used. If the neuron responds to the motion of a bar, random bars can be presented in different orientations in random parts of the visual field and moved at random speed for a random distance. By correlating the responses with the various inputs, the whole field can be plotted in the selected ways. Another method for a neuron that responds to bar motion is to project two bars alternately on the screen, separated by about the angle between visual axes, and modulate them by two independent white noise signals. The bars appear to move in either direction at a range of speeds and the responses are correlated with the stimulus changes over all latencies. For a neuron with a spatial field with a centre and surround, a spot on the centre and a large patch covering the surround are similarly modulated at random, so that all possible interactions are measured as a function of time. As recorded by these methods, the common small-field neurons in the locust medulla are of three main types.

- Linear sustaining cells that respond with graded potentials in opposite directions to 'on' and 'off', like LMC cells; they could be the inputs to many later systems. Some show spectral or spatial opponency. They encode local intensity with various adaptation rates so that some are effectively transient, but still give opposite responses to 'on' and 'off' and show spatial summation within small fields. They are commonly tuned to flicker frequencies about 15Hz in the dark-adapted locust and do not respond to motion within their field.

- Non-linear transient cells that respond with spikes equally to 'on' and 'off' and are sensitive to motion in any direction rather than flicker. They have a long but remarkably constant latency as though important in timing events. Some have fields corresponding with a single ommatidium.

- Directional motion-detector neurons with a variety of temporal and spatial constants are revealed by second-order kernels. Some are sensitive to either black/white or white/black edges (as shown by their first-order linear kernels), but only one with a small field has been recorded in the medulla (or anywhere else) in the locust.

Figure 6.6 A tentative circuit for part of the motion-detection pathway in the medulla and lobula columns of the fly, based on recording from neurons dyed through the electrode. Note the feedback to the lamina and to the medulla (arrows). The labelled cells have the following properties: ndir = directional; pdir = partly directional; sdir = strongly directional; wdir = weakly directional. The outputs to the brain are all of horizontal fibres.

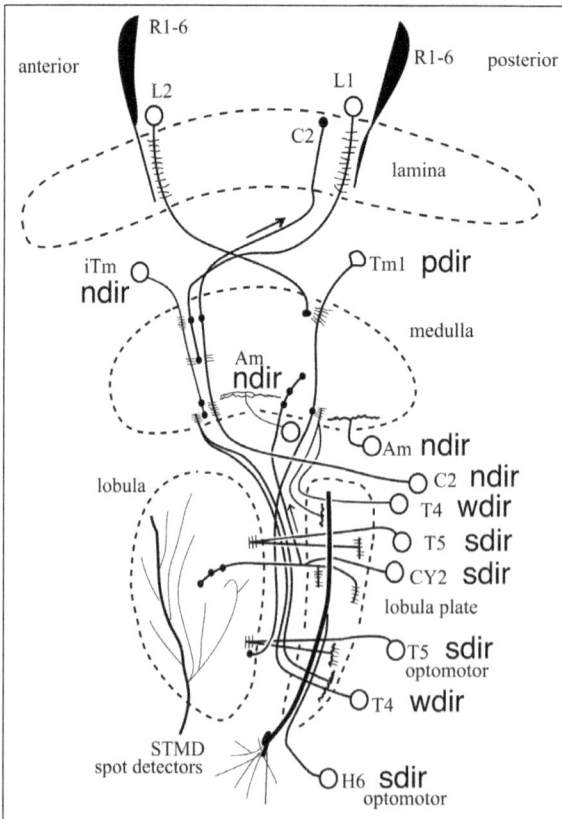

Source: Composed from work by Douglass and Strausfeld (1995–96).

Medulla transient cells

These abundant cells fire with one or two spikes only, with a remarkably constant latency of 25–35ms in response to intensity increment or decrement. This is rather slow for inputs to a motion-detecting mechanism. Some have fields of 2–5°, others 7–20°, corresponding with their dendritic spread. The motion of a contrast within the field in any direction causes a steady stream of spikes by successive stimulation in an array of presynaptic small cells of the medulla columns (Figure 6.6). Successive stimuli, even at a single visual axis, cause strong localised self-inhibition, which decays after about 100ms, but stimuli of one polarity do not suppress responses to the opposite polarity, suggesting separate inputs for 'on' and 'off'. The interesting feature is that the response returns when the contrast is increased after adaptation to contrast. An increase of 20 per cent contrast is sufficient to bring back a full response. The result is a rapid habituation to the background contrast level, so that the visual system could ignore a constant modulation at every point, but respond to a novel contrast. Their non-directional response to motion is at the highest spatio-temporal resolution achievable by the retina, so the inputs are single columns. They have fields of different sizes with a range of time constants. Therefore, at all points in the visual field, the medulla contains a large number of non-directional motion detectors forming overlapping maps with different scales in space and time (compare Am cells in Figure 6.6).

In the natural visual world, large areas of background can have a texture at about the same contrast level. As the eye scans across this background, numerous transient cells respond only when they come to a boundary at a greater contrast. A near object will often have a sharp boundary when seen against a distant background, even if only because of different illumination and shadow; and this detection of an outstanding boundary by medulla transient cells is independent of contrast polarity and average background contrast. The precise timing of their spike responses suggests that they participate in coincidences with others. Also, to distinguish a spot requires a precise timing to discount coincidences of responses from a moving background.

At several levels in the insect medulla there are strata of horizontal fibres at right angles to the columns. One type of transient cell of the locust, called the tangential medulla amacrines, with fields of about 20°, spreads over a wider region of the medulla, synapsing on many medulla columns. In the butterfly, most of the large medulla neurons appear to be wide-field directional detectors, similar to the detectors of flow-field patterns in the lobula plate of flies. Some of these come from the lobula, and all eventually adapt to a continued motion, so that they detect velocity changes, not steady motion. Some cross from the other eye, others terminate with the directional lobula neurons in the deuterocerebrum and make synapses with descending interneurons to the motor centres of the thorax.

Neurons with fields of 10–20° are possibly the regional detectors of the cues that are inferred in trained bees (see Figure 10.8). In the bee, there are tracts from the medulla and the lobula to the calyces of the mushroom bodies (Figure 6.4).

Directional motion detection

Elementary motion detectors

The elementary motion detector (EMD) is a convenient hypothetical circuit, a kind of reduplicated miniature black box, first introduced by Reichardt about 1956 as the mechanism of the optomotor response, and subsequently of the large-field directional motion-detector neurons that fit so well with the optomotor responses. They measure the rate of passing of edges—that is, the temporal frequency, not the angular velocity—and are very slow, with peak response near 10Hz, but the motion detection occurs at the maximum spatial resolution of the eye, down to very low contrasts and low intensities, with a low ratio of noise to signal. The maximum response occurs when the jump of an edge is the angle between adjacent visual axes, falling off rapidly over a few visual axes. This local limitation of the interaction was confirmed by stimulating single photoreceptors of the fly retina. For some, the Reichardt EMD is the behavioural unit of motion detection, but any algorithm that extracts the direction of a shift in the square of the contrast is sensitive to directional motion.

The detection of directional motion at first assumed that the lateral interaction was between adjacent ommatidia (Hassenstein 1951; Götz 1965) and the true situation was never properly published (see Figure 3.4). Experimental analysis eventually showed that explanations of the response were complicated by receptor and regional diversity, by sub-adjacent interactions and by pooling of channels in low light. Moreover, under-sampling is the rule, so the true direction of motion is not detected in the finest patterns that the insects can detect by temporal modulation.

Directional motion perception of all kinds at neuronal or behavioural level is consistently colourblind in insects and does not measure angular velocity. The mechanism saturates at low contrast and low velocity and adapts, so that the feature that is detected consistently is the local direction of velocity change over a huge range of light intensity. At the lowest intensity threshold for motion detection in the fly each photoreceptor averages 6 photons (two to three bumps) per second.

Despite extensive work by Buchner and his colleagues showing that the metabolism of desoxy-glucose during motion perception implicates some layers of the medulla but not others, we have no firm anatomical or electrophysiological indication of where the directional motion detectors are. The medulla is certainly not stuffed with EMD neurons.

Mechanisms of motion detection

Electrophysiology with one electrode cannot find the locus of motion detection because there is no suitable place to probe and motion detection is not a single process. Douglass and Strausfeld (1995–96, 2005) recorded from a few types of small-field neurons in the fly and proposed a circuit for motion detection, based partly on latency measurements and anatomy (Figure 6.6). Axons L1 and L2 connect to short iTm fibres of the medulla that are sensitive to non-directional motion. The iTm fibres connect with T4 cells of the lobula that are weakly directional. Long Tm1 fibres also connect with T5 fibres of the lobula that are strongly directional. T4 and T5 fibres connect with the large directional lobula neurons that run to motor centres of the neck muscles and the thorax. In later work, they also found differences between the angular velocities at which these and other neurons gave peak responses. The conclusion was that local directional motion detectors had inputs from non-directional ones. These circuits alone, however, would not provide a suitable substrate for the measurement of angular velocity and control of flight speed by non-directional inputs in bees flying in a tunnel (Chapter 7).

The bee probably resembles the locust rather than the fly. Directional motion detectors with retinotopic medium-sized fields, but not small fields, are abundant in the medulla of the locust. The directional mechanism appears to be a directional delayed lateral inhibition with the same time constants as the self-inhibition of the transient cells, acting over 30–100ms, with inputs from many columns.

Some details fail to agree with the rest of the account. Curiously, direct stimulation of the two large lamina monopolar cells, L1 and L2, in the fly failed to excite directional motion detectors. The directional response to a single 1° jump of a single edge and the memory of a stationary pattern both suggest that a DC component is essential as one or both of the inputs to a directional motion detector, but evidence that L1 or L2 can carry a DC signal for long is doubtful. In bees, dragonflies and butterflies, the motion perception has inputs only from green receptors, but L1 and L2 are not so restricted in these insects.

In conclusion, the medulla is an expansion of the visual input into local combinations of features in space, time and colour, with some wide-field neural input from other visual centres. The medulla contains many maps of local features and passes on coincidences of combinations of the inputs. There is range fractionation (that is, division of the total range into parts) for motion detection, angular velocity, spectral sensitivity, edge orientation, latency, direction of motion, adaptation rate and more, so there could easily be 50 types of neuron in each medullary column.

The lobula and lobula plate

The third optic neuropile, the lobula, contains retinotopic projections of small fibres but is usually considered as the terminus where the spatial representation is converted to local sums and coincidences of feature detector responses in large fields.

Large directional motion detectors

Large neurons that detect directional motion were first discovered in the lobula in the 1960s. They have large fields in different parts of the eye with different optimum directions of motion (Figure 3.5). The directional sensitivity curves are very wide, with angular widths near 90° at the 50 per cent level. As described in Chapter 3, their properties match the optomotor response, but it is not yet clear how they act during normal flight because they adapt to the visual feedback from the moving surroundings.

In the fly, butterfly, dragonfly and locust, there are at least two types of direction-selective motion-detector neurons—one slow and matching the optomotor response in time constants, the other fast, with responses increasing up to angular velocities of 1000°/s or 100Hz or more. Both respond to contrast frequency irrespective of velocity. They terminate lateral to the oesophageal canal. In the bee, the fast fibres respond more constantly to velocity irrespective of contrast frequency and so are of interest for possible relations to the flight speed, either as sensory input or feedback control. As in many arthropods, the bee has tonic and phasic muscle fibres, corresponding motor neurons and probably slow and fast pre-motor interneurons of the ventral cord.

Other movement and small-object detectors of the lobula

Abundant behavioural studies have shown that dragonflies, pond skaters, hoverflies, mantids and others detect small moving spots against a moving cluttered background. In the lobula of hoverflies, Barnett, Guerten, Nordström, O'Carrol and co-workers in various combinations find large numbers of interesting neurons lateral to the oesophageal canal, which are thought to be detectors of small moving objects or complex features. The field sizes range from 10° to 30° at the front of the eye, which is all that is required to home in on a prey or potential mate, or even 40–50° in female hoverflies, which is sufficient for escape. The range of sensitivity to angular velocity corresponds with detection of a similar insect flying at a typical speed at a range of 1 metre. Some neurons detect a small object moving over a patterned background. When presented with several small targets simultaneously, the whole animal is confused and two small targets close together inhibit the response (for example, Meyer 1974). The neurons behave similarly (Guerten et al. 2007). Recordings from the dragonfly lobula reveal large neurons that detect small black or bright spots down to the minimum allowed by the width of the acceptance angle as sharpened by lateral inhibition—about 1°. These detectors of small spots are usually non-directional and are not restricted to green-sensitive receptors. The response soon fades on repetition.

Similar neurons have been found in the lobula of the bumblebee, which seems more suitable for recording than the honeybee (Paulk et al. 2008). Responses of different types are segregated into at least six different layers—some are precise in the timing of the first spike; others are more erratic. Obviously, there is much more to come from this animal.

Neurons that detect non-directional motion and directional motion in various proportions can be found running between the lobula and the midbrain in the bee. Similar neurons in the locust measure angular velocity irrespective of the stimulating pattern.

Neurons that are sensitive to the edge orientation have been recorded in worker bee lobula after a long search (Yang and Maddess 1997). The plot of the response against the angle of a thin bar has two opposite lobes because the feature detector is symmetrical about one axis (Figure 9.2). They could be non-directional motion detectors. The required length of edge is small and the orientation tuning is very coarse, mirroring the behavioural tests. They run from the lobula to the deuterocerebrum on the opposite midbrain, where they connect with descending neurons of the ventral cord. A tract of axons from the lobula to the calyces of the mushroom bodies of the bee deserves examination.

A large neuron in the lobula of the locust, the DCMD, sensitive to ultrasound and very small non-directional movements of edges and spots but not large targets, was originally thought to be an alert mechanism, but turned out later to be a detector of impending collision. Again, it soon adapts on repetition and the first neuron to fire on the left or right side inhibits the other. Analysis suggests that the inputs to this neuron are the small-field transient cells of the medulla.

All who have recorded from the lobula of large insects remark on the great diversity of responses that are not understood, and it is quite likely that some of these neurons will reveal to be large-field summation of feature detectors of edges, spots or orientation, or even parallax, when the proper tests are made. For example, there are different motion detectors that show range fractionation of contrast frequency, but it is not known whether these relate to the measurement of optic flow.

For some reason, the mechanisms of learning and the location of memories have not appeared among the neuron studies, yet presumably they are initiated in the optic lobes. The traditional sites—the four mushroom bodies at the top of the brain—are essential for learning odours but extensive work shows that they are not required for tactile, visual or motor learning in *Drosophila* (Wolf et al. 1998). It is still possible that the layout of the sensory input is somehow coded in the calyces of the mushroom bodies.

It is pertinent to note, however, that there is no sign among the neuron responses of the rapid learning of muscle control and flight posture that is demonstrated in flies when the visual feedback loop is reversed (see Chapter 7). There is also no sign of a separate pre-motor control from non-directional motion detectors, which might control the flight speed or measure the distance flown. Just asking these questions is a reminder that, despite 30 years of effort, theories that are based on curve fitting to whole-animal or single-neuron performance tell us little about the underlying interactions, and they avoid the hard work of unravelling the circuitry. It is of some consolation that in the vertebrate retina, which has been explored far more intensively, we still lack all the neuronal mechanisms.

Colour in insect vision

Colour vision has to be recognised by a test and therefore is defined by performance. Possession of two or more receptor types with different spectral peaks is suggestive, but insufficient. Some responses, such as food appraisal, might show colour vision while others, such as motion vision, are colourblind in the same animal. Third-class colour vision passes the test of discriminating at least one wavelength from all shades of grey or separating two colours irrespective of intensity. Second-class colour vision discriminates different colours from one another irrespective of saturation and intensity. First-class colour vision recognises a colour even when the wavelength mixture in the illumination changes, irrespective of saturation or intensity, as in humans and bees, but confuses some mixtures of wavelengths that combine to make the same colour, just as green for humans can be made from various mixtures of blue and yellow. Few animals have been tested for colour vision. Insects also have receptor-specific vision, in which the outputs of different colour-coded receptors are used separately for different responses. All of these kinds of colour vision can exist side by side, together with colourblindness of the same responses in dim light. The performance gives us little information about the number or types of receptors or the neuronal mechanisms of processing—or vice versa. Even if we understood our own colour vision, it would not be a model for insects.

Colour cues

Confusion surrounds the question of specificity of insects' visual responses to colour because there are numerous observations but very few critical tests. A sharp distinction between colour-specific responses and colour vision based on inadequate data probably adds to the confusion. Certainly, there are responses that are tied to one type of retinula cell, so the response is monochromatic but responds to quite a broad band of wavelength. For example, all insects tested detect motion via the receptors with a peak in the green and some make their escape towards UV light or detect the polarisation plane with a single type of UV receptor.

Colour cues suggest colour vision, but each must be analysed and usually the colour is effective only in context. Receptors with a peak in the ultraviolet occur in the eyes of all insects—male and female—and there is sufficient ultraviolet in daylight for it to be a useful colour. Some insect sexes and white flowers are discriminated by their UV reflectivity. There are many examples where trigger cues depend on special receptors, but the question of colour vision remains open. Butterflies are at the centre of the confusion. For example, the white butterfly *Pieris* has at least four receptor types with peaks at 360, 450, 540 and 620nm and several behavioural responses that depend on the colour and intensity of the illumination. Feeding is triggered by red (600nm) and especially by blue (447nm), egg laying by discrimination of green (542nm) and yellow induces a tactile test with the feet. The same butterfly is attracted to blue or yellow flowers and ignores green when feeding. Male butterflies in flight will turn and chase another suitably coloured butterfly as a potential female or territorial intruder. An unexpected finding is that *Pieris* females reflect more of the ultraviolet than the males, but in lycaenid butterflies it is the other way round. In each case, the males easily discriminate the sexes visually. Some butterflies—for example, the satyrid *Pararge*—distinguish some colours from all shades of grey. The diurnal hawkmoth *Macroglossum* prefers blue flowers but is easily trained to reverse its preference. The butterfly *Papilio demoleus* prefers blue flowers with few petals, and models that resemble them, rather than other colours or models with more rays.

Red is little used by insects except by dragonflies, some wasps and butterflies with red markings. Butterflies of the family Papilionidae, when hungry, select red from all shades of grey in mistake for flowers. The large black *Papilio aegeus* has red spots on the wings and both sexes have red receptors in the eyes— presumably to recognise their own species.

Some insects, such as the cockroach, have only two types of receptors, with peaks in the ultraviolet and green. Whether they have some form of dichromatic colour vision is unknown. Some dragonflies, butterflies and wasps have four types among the six large retinula cells, with peaks in the ultraviolet, blue, green and red, together with one or two types of small cells, one at least of which is UV sensitive. Each receptor type could be the input for a monochromatic response and two or more of them might collaborate in behaviour that passes as one of the forms of colour vision. Narrow-band wavelength-specific responses are rare. A well-known example is the bee, where the dorsal light response, escape to the light and the sky compass depend only on UV receptors; landing, scanning, landmarks and behaviour that depends on motion, only on the green receptors; dim light vision is colourblind, with ultraviolet, blue and green simply summed together; and finally all three receptor types collaborate in antagonistic interactions in first-class colour vision. In the dragonfly, ultraviolet participates in the antagonistic responses of lamina monopolar cells, and in the bee the UV cells of the retina have long axons to the medulla, with side branches in the lamina, perhaps to allow them to participate in two different activities.

Electrophysiology of the optic lobes reveals far more neuron types than needed for a minimal model of colour vision. The neurons, with multiple inputs and alternative outputs, deserve careful analysis because in fact they, not the model, are the mechanism.

Figure 6.7 Physical background to colour vision of the bee. a) Photon flux of sunlight at different wavelengths. b) Spectral sensitivity curves of the three types of retinula cells of the worker bee. c) The threshold minimum difference in wavelength that is discriminated by the bee at different wavelengths.

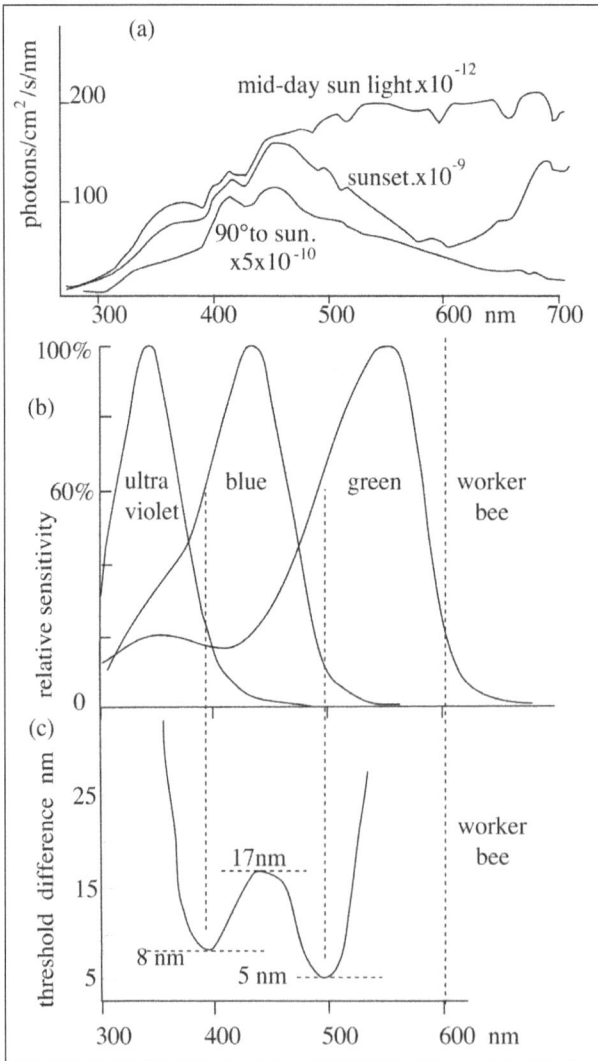

Sources: (b) from Autrum and von Zwehl (1962); (c) from von Helversen (1972).

Colour discrimination

An enormous literature on colour vision in bees tells us much about performance but nothing about mechanisms of discrimination. Bees have three types of large retinula cells with broad spectral peaks in the green, blue and ultraviolet, with uneven distribution over the eye. In the honeybee, unlike other insects, colour vision has been studied in sufficient detail to show that bees discriminate colour from all shades of grey in their selection of food targets and at the hive entrance. Certain mixtures of colours interact together to produce 'bee white', which is close to human green. Colours are discriminated irrespective of the amount of bee white (called saturation) in the mixture. Data on trained bees strongly suggest that they detect the relative positions of at least two neighbouring patches of different colours.

Over the region of the spectrum where the spectral sensitivities of the receptors overlap (Figure 6.7b), the honeybee can discriminate differences of about 20nm in wavelength irrespective of intensity (Figure 6.7c). In dim light, the spectral types are apparently added together, so increasing the sensitivity, but bees are poor at discrimination of brightness differences, which suggests that the colour discrimination system has only opponent neurons. Neurons with opponent wavelengths in both centre and surround, as in primates, have not been found in insects.

Electrophysiology of the bee's optic lobe (Yang et al. 2004) revealed at least 10 colour types of neuron without antagonistic responses and eight types with antagonistic responses—for example, excited by blue and inhibited by green and ultraviolet, as in the caterpillar medulla (Figure 6.9). A centre-surround organisation, as in primates, was not found. Presumably, there are many more types. Antagonistic responses help explain why the discrimination of a small change in wavelength is optimal near the wavelengths where the spectral sensitivity curves are steepest and cross (Figure 6.7b). Most importantly, the invariance of colour vision—such as independence from intensity, object size and repetition rate—is explained in principle by the opponency of neurons. In the bee, as in humans, colour discrimination is recalibrated according to the colour of the illuminating light, as though the weighting functions of the three receptors are modified so that known objects such as clouds or leaves are detected as expected. Beyond the medulla, large-field neurons involved in motion perception tend to be green sensitive and colourblind and those sensitive to spots tend to have various spectral sensitivities.

The various distributions of spectral receptor peaks, distributions of the various colour types of receptors, responses to colour and numerous details of bee responses to coloured targets are available in a large but well-reviewed literature, showing endless adaptations to the world of colour. It seems certain that, before flowers evolved, insects had receptors with a variety of spectral peaks and colour-specific responses. A recent finding is that flower colours have spread themselves out across the range of colour vision of the pollinating insects, just as they are spread out in time of day or in different seasons. Smaller steps in flower colour are observed at the wavelengths where pollinators discriminate smaller colour differences. Conversely, the spectral peaks of the photoreceptors are near the theoretical optimum wavelengths to discriminate the observed range of flower colours.

A superficial model of colour vision

A useful model of colour vision that necessarily fits a system of receptor types with broadly overlapping spectral sensitivities is a colour triangle for three receptor types (Figure 6.8) or a tetrahedron for four types. The equilateral triangle represents a plane that cuts across the corner of three Cartesian coordinates. Three lines drawn perpendicular to each side from any point in the triangle represent the normalised responses of the three receptor types as a fraction of the sum of all three responses. Each point in the triangle then represents the relative responses of all three types irrespective of total intensity. When the responses of the receptors to two different colours are calculated from the spectral sensitivity curves, the colours are hard to discriminate if the two points obtained lie close together in the triangle. Each pure wavelength has a position in the triangle according to its relative stimulating effect on three receptor types (Figure 6.8c). The line of the spectrum curves from the corner of the short wavelength receptor towards the corner of the middle wavelength receptor and terminates near the corner of the long wavelength receptor, but never reaches the corner of the middle wavelength because this receptor is never stimulated alone. A point on the plane near the middle is indistinguishable from grey or white. Points that are more separated on the plane (in stimulus space) are more easily discriminated. Each point on the plane is really a small patch, the size of which is a measure of the noise level (Figure 6.8e). Because there are three receptor types, the data from discriminations of colours can be related uniquely to their measured spectral sensitivities. Any point within the spectral line can be reached by a very large variety of combinations of wavelengths, so the exact mixture of wavelengths is not recoverable—for example, there are many mixtures of various blues and yellows that will match any green.

With this model, it is impossible to see two colours simultaneously at the same place because there is only one output. We know, however, from the spectral sensitivities of different responses, even in the honeybee, that receptor types can be used independently of each other. We do this ourselves with our ears when we hear separately the notes in a chord, because our cochlear is able to detect two or more simultaneous notes. Even more, we detect constituents of odours and flavours, but we cannot see two colours at the same place.

Figure 6.8 The classical representation of the colour triangle for the bee and for humans. a) Three vectors representing the normalised responses of the three colour types of receptor. b) The fractional contribution of each receptor is represented on the triangle. c) and d) The position in the triangle for each pure wavelength for the bee and humans. e) The stimulus/response curve of a retinula cell, showing how the increased noise at low levels limits the possibility of discrimination of intensity and therefore of contrast and colour.

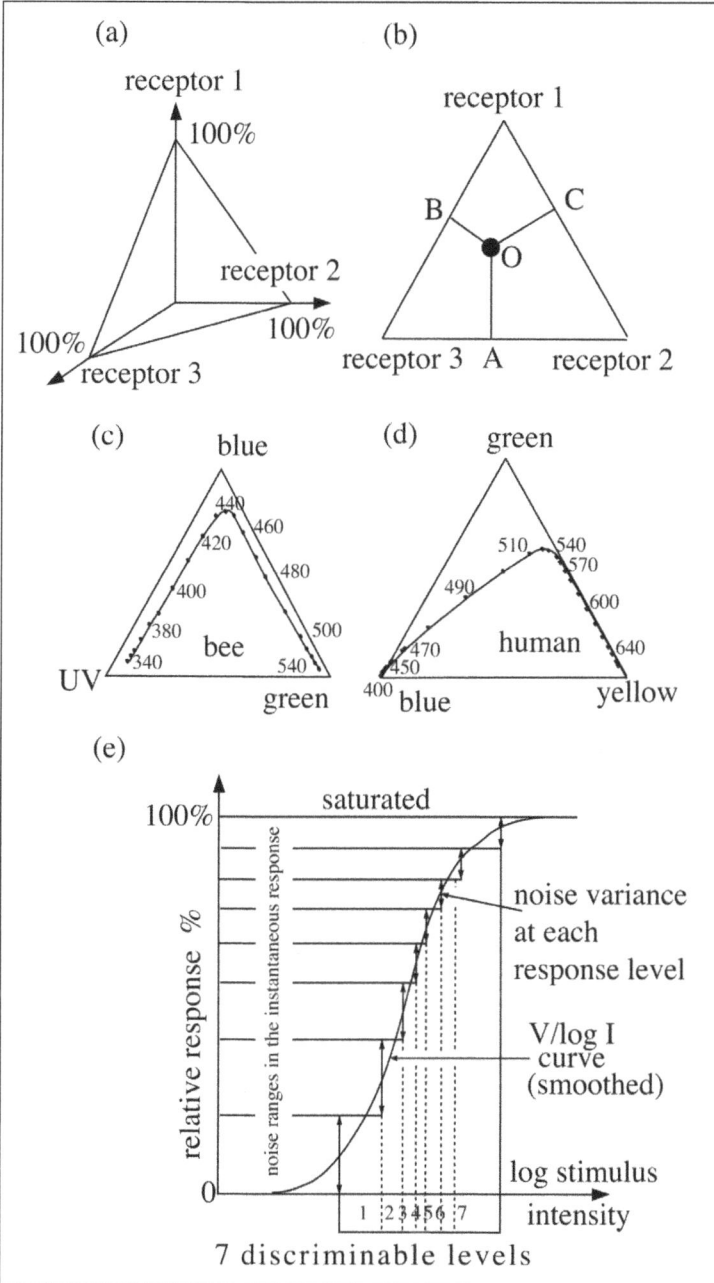

This model is much simplified for many reasons. It is really just a geometrical way of representing a summary of one part of the data. The behavioural responses are influenced by effects of summation of inputs, adaptation over time, regeneration of photo-pigments by long wavelengths, screening pigment, and so on. It illustrates how three input filters with broad overlapping tuning curves result in a great many discriminations, but is simply a consequence of the properties of the three types of receptors. There is a version of the model for the bee (Backhaus 1991) that postulates central antagonistic interactions, based on the responses to various wavelengths by just two medulla neurons, but this does an injustice to the variety of colour types of neurons that are really present.

Figure 6.9 Receptor, lamina and medulla neurons of the caterpillar of *Papilio*.
a) Each eye is a single ommatidium; lamina cells are filled black; medulla cells are white on black, with a tract to the corpora pedunculata. b) The six eyes look out in different directions on each side of the head. c) Spectral sensitivity curves of the three types of retinula cells.

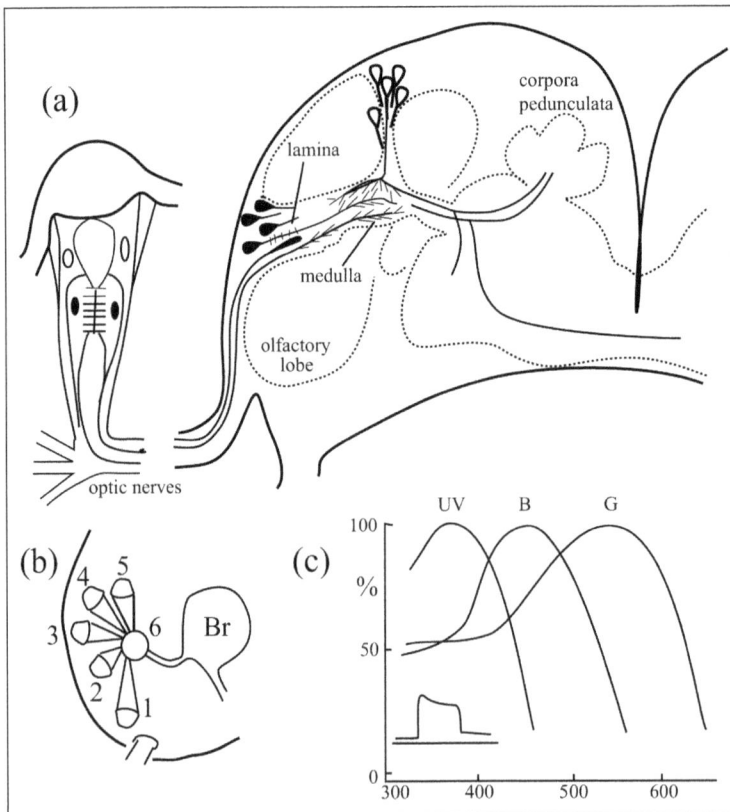

Source: Based on works by Ichikawa and Tateda.

Colour processing in the medulla

One would not have thought that the humble caterpillar would help us elucidate colour mechanisms, but it is a simplified version of the adult at the right level of complexity to show what we can expect in the adult insect. The medulla of the larva of *Papilio* has a wide variety of colour-coded neurons, which can be recorded with microelectrodes. There are only six separate simple eyes (stemmata), each like one ommatidium of a compound eye, each with seven photoreceptor cells of three spectral types beneath a small lens. The neuroanatomy is similar to the adult, but much simpler (Figure 6.9).

The receptors each have one spectral peak near 380, 450 or 540nm, which are typical values for many insects. The photoreceptor axons are of two types, long and short, of which the latter terminate on a monopolar cell of the lamina. The total of 24 LMC axons project to the medulla, where most of the neurons have graded potentials with small superimposed spikes. In the medulla are numerous interneurons with antagonistic interactions in all ratios that generate a large variety of neuron types in parallel (Figure 6.10). This is an example of range fractionation that is common in sensory systems.

One group of 50 or so intrinsic medulla neuron types responds to different combinations of the six stemmata (Figure 6.10). Eleven of them receive different types of *spatially opponent* colour inputs from a few stemmata, mostly all different from each other, but looking forwards. In their responses, they look like colour-pattern detectors, with areas separated from edges. Seven neurons show a relatively homogeneous spectral sensitivity over their whole receptive field and have inputs from two or three dorsally directed stemmata; three of these neurons are tonic and three phasic. Another seven neurons are spectrally homogeneous with large receptive fields covering four to six stemmata. Some of them show spatial summation, others spatial antagonism within their receptive fields. In their responses, they look as though they serve phototaxis.

The 50 or so neurons receiving inputs from single stemmata have been examined with reference to chromatic and neutral backgrounds. They look like local colour discriminators. Eight types are from stemmata with only blue and green receptors. Of these, six have specific colour opponency on different backgrounds; one has strong colour opponency on a neutral background and the other five have colour opponency only with coloured backgrounds. The most complex has excitatory responses on a black background, inhibitory responses on a white background and various colour opponency on coloured backgrounds. Two other types show simple summation of blue and green receptors. Eight types with inputs from trichromatic stemmata (ultraviolet, blue and green) have colour opponency in different combinations—some of them depending on the background colour. The neurons with inputs from several stemmata are not summations of the inputs of constituent stemmata or post-synaptic to single-stemmata neurons: they have different colour combinations but from different stemmata. There is little sign of centre/surround organisation of fields, unlike in vertebrates. All of these medulla neurons could be third order on the visual pathway. They clearly generate many of the total possible combinations of broadly tuned inputs in colour and direction (Figure 6.10), but have not been related to visual tasks.

Figure 6.10 The distribution of 25 medulla cell colour types among the six eyes of the caterpillar *Papilio*. Capital letters mean a large contribution, small letters a small contribution. Note the phasic units for edges and the tonic units for areas, and the resemblance to the coding in olfactory systems by coincidences in diverse populations of neurons.

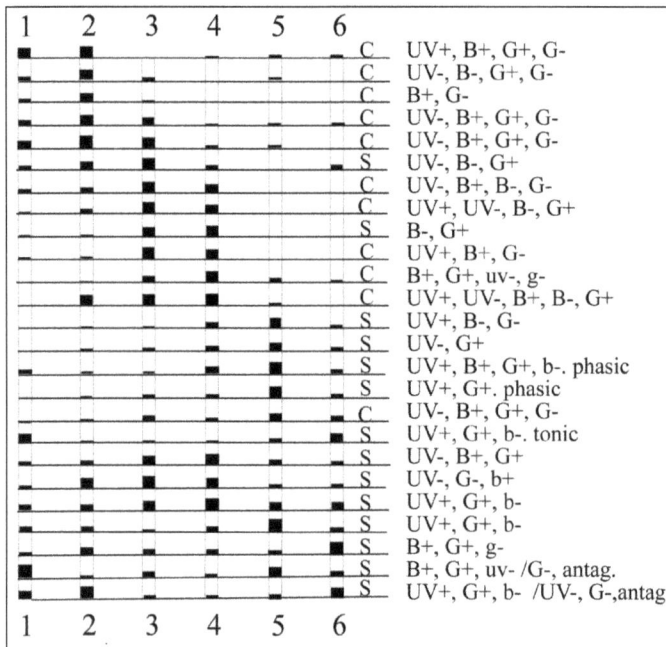

1	2	3	4	5	6		
						C	UV+, B+, G+, G-
						C	UV-, B-, G+, G-
						C	B+, G-
						C	UV-, B+, G+, G-
						C	UV-, B+, G+, G-
						S	UV-, B-, G+
						C	UV-, B+, B-, G-
						C	UV+, UV-, B-, G+
						S	B-, G+
						C	UV+, B+, G-
						C	B+, G+, uv-, g-
						C	UV+, UV-, B+, B-, G+
						S	UV+, B-, G-
						S	UV-, G+
						S	UV+, B+, G+, b-. phasic
						S	UV+, G+. phasic
						C	UV-, B+, G+, G-
						S	UV+, G+, b-. tonic
						S	UV-, B+, G+
						S	UV-, G-, b+
						S	UV+, G+, b-
						S	UV+, G+, b-
						S	B+, G+, g-
						S	B+, G+, uv- /G-, antag.
						S	UV+, G+, b- /UV-, G-,antag
1	2	3	4	5	6		

Sources: Based on Ichikawa (1990, 1991).

It is remarkable that such a small ganglion in such a simple visual system contains neurons with so many combinations of inputs. We must infer that the caterpillar uses colour as a source of information about its surrounding leaves, but with so many neuron types in parallel, any model relating outputs to inputs would be hard to prove uniquely.

There are strong indications that the same system occurs in other insects, and perhaps in primates. In the bee, Yang et al. (2004) recently recorded non-opponent and also opponent cells, some broadband and others fed from one type of receptor. Combinations such as (UV + B − G −), (UV − B + G +) and (UV − B + G −) were recorded—in all, 50 types—but no spatial opponency. In dim light, the opponency disappears at the same level as colour vision is lost. Spatial fields were huge: >50°. As in the locust medulla, edges and motion are coded by varieties of related neurons. The point is that the coding of small-column neurons of the medulla is of very simple local features that are close together in feature space and that are combined in further selections by collector neurons downstream.

Each colour is represented by its own pattern of activity in many neurons: edges by phasic neurons, areas by tonic ones. Colour discrimination then involves the detection of familiar coincidences of neuron activity. Similarly, the colour system of the primates contains a great variety of colour-coded neurons that cannot be correlated individually with the behavioural data.

Conclusions

Electrophysiology of identified neurons reveals what is going on, cell by cell, and leads to many conclusions about information flow, but fails to explain or predict behaviour because that depends on the combined action and coincidences of many neurons. Sensory input is the easy part. Discovering the feature detectors is the next step. Some feature detectors are inferred from behaviour, others from the responses of the optic lobe neurons, and a few from both. One example where the feature detector has been isolated is the modulation detector, which is a lamina ganglion cell that is sharpened by lateral inhibition. Another example is the orientation detector, which has a maximum and minimum size of three facets long and three wide. A third is the elementary motion detector, where stimulation of a single optical axis with a flash, followed by the same at a neighbouring axis, shows that two adjacent or two sub-adjacent axes are effective, but we know that they are not as small or as homogeneous as is usually proposed (Figures 3.4 and 6.6).

In general, we can infer from the electrophysiology that there is a high-speed inflow of sensory information, then an expansion into combinations of inputs in the medulla and lobula, followed by a rapid integration and reduction to a relatively small number of neurons in the tracts to the brain. This reveals nothing about discrimination or about the coordination of inputs of different kinds, except that processing must depend on coincidences between different neuron responses at every level. Long-lasting modulating transmitters and neuron hormones are known to exist, but how they participate is largely a mystery. The analysis of a nervous system is a difficult, tedious and never-ending task.

Endnotes

1. There are two impediments to the understanding of neural processing. First, the work has been done in a small number of laboratories, where the techniques are nourished for decades and different conclusions emerge. Second, the study of one neuron can continue for many years and each new researcher entering the field tends to introduce a different preparation.

07

PILOTING: THE VISUAL CONTROL OF FLIGHT[1]

What a delight it is to watch insects go about their daily life on a summer day. Most obviously, a large male butterfly flutters by on regular patrol around its territory and suddenly it recognises a female of its own species. A group of hoverflies hovers in separate stations in a shady place between trees; bees move from one flower to another of the same kind; a large fly weaves from side to side as it dashes past; and along the footpath a dragonfly hunts for mosquitoes. They all appear to see quite well and, after centuries of discussion about insects' perception of their visual world, it is now possible to outline some of the mechanisms that coordinate their manoeuvres in flight. We can distinguish more than a dozen kinds of generalised visual tasks in relation to their three-dimensional world and can offer details of some of them. Bees control their flight manoeuvres visually and by a variety of receptors of the joints, hairs and muscles. All of these mechanisms are strongly modified by learning. Here, we will stick to the visual control of flight.

When we examine the nineteenth-century works of Fabre, Lubbock, Forel and others, we find descriptions of performance but no clear ideas about mechanisms of control. The twentieth century brought some experiment but much of it beside the point. The works of Loeb, Crozier, Mast and others, summarised in the textbooks by von Buddenbrock and Wigglesworth, contain numerous accounts of varied responses to light but few that can be related to mechanisms of normal behaviour. There were discussions of reflexes versus central nervous pattern generators (from Sherrington to von Holst) and classifications of responses (from Kühn to Fraenkel and Gunn)—now mostly forgotten. In mid-century, the new techniques of electrophysiology (Autrum, Burkhardt, Burtt and Catton, Pumphrey, Pringle and Roeder) were diverting but revealed little more about visual control of motion than did the behaviour itself (Autrum 1979–81).

From the 1950s, motion perception became the basis of insect vision. A part of that story was the huge diversion of resources to the optomotor response—that is, when an insect responds to the rotation of a drum around it by turning in the same direction. More recently, we have clear-cut examples of analysis of visual behaviour in free flight. The main reason why there has been so much discussion, even acrimony, about how insects in nature perceive the three-dimensional world is that several mechanisms always operate in parallel and it is hard to demonstrate the relevance of a response to a laboratory stimulus. One reason for the slow progress is that for obvious reasons insect vision is narrowly dedicated to the real tasks required in the ecological context and we are usually ignorant of these.

Responses to light

In the classification of movements that bring insects to their preferred places (Kühn, Fraenkel and Gunn), the responses affected by the direction of the light are called taxes and those that are undirected are called kineses. Orthokinesis is a dependency of locomotion speed on stimulus intensity, so the insect finally stops in the dark or light, and stays there. Klinokinesis is when the frequency of turning depends on the intensity. Klinotaxis is when the insect compares the stimulus on two sides by making successive movements to left and right. Tropotaxis is when the comparison is simultaneous. Telotaxis is fixation on one goal at a time. The old term 'tropism' vaguely covers all these and is sometimes more convenient because the human categorisation is not sharp.

As a student, I had to learn this classification, but it did not help in understanding insect vision. The experiments scarcely approximated the natural situations and the results were described in behavioural terms that failed to connect to the anatomy, physiology or life history of the same animal. In brief, there was too little thoughtful experimental analysis and too much naming and categorising of the performance.

Freely flying vehicles need a cue for staying the right way up. Having little gravity sense in flight, many insects use the dorsal light response and the general brightness of the sky as a sign of 'up'. Some make use of the direction of the light beams rather than the intensity of the sky above. Insects that swim upside down have the reflex reversed. Some insects can fly in complete darkness, but most, like locusts and dragonflies, cannot.

The UV light of the sky is poorly reflected from natural objects, so it comes almost entirely from above. Most insects studied have more blue and UV receptors on the dorsal part of the compound eye than elsewhere. A disturbed bee flies towards the brightest UV part of the sky to escape and uses the ultraviolet of the sky to help it stay the right way up in flight. Bees turn a forward somersault in flight if they fly over a mirror that reflects ultraviolet upwards.

Many insects have three small eyes called ocelli at the top of the front of the head. In the dragonfly, they are detectors of the average position of the sky and they stabilise flight in dim light. In some species, they are partially focused laterally, in the expected direction of the horizon. The large apertures of the ocelli and the summation of the receptors on the neurons below account for the extraordinary sensitivity to the position of the sky, earth and horizon. In the locust, the ocelli detect deviations from the direction of the horizon even in starlight, but in the bee, their functions and interactions with the compound eyes are not yet clear.

Binocular vision

Because insects have two eyes does not mean that they have the vertebrate binocular mechanisms to measure range. Insects have no accommodation of the lens, convergence of eye movements or receptors at different focal planes. Cats and primates have an array of binocular neurons in the visual cortex with a variety of angular offsets in the visual axes of the two eyes—so-called disparity units. Combinations of these disparity neurons can measure range even off the midline. Insects do not have this mechanism.

The most skilful fast-flying insects that catch prey in the air commonly have some binocular overlap of axes at the front and top of the eyes but little separation between them, and they measure range by moving in flight. At the opposite extreme are the mantids, dragonfly larvae and a number of predatory insects that have two widely separated eyes. They slowly assess the situation for grabbing prey by turning the head and use triangulation for range estimation (Figures 5.12, 5.12d, 5.12e). As far as we know, the triangulation is done by coincidences of corresponding visual outputs on motor centres, not by a congruent mapping of the spatial array of one eye into the opposite optic lobe. The direction of the mantid's leg extension is controlled partly by hairs at the neck. Finally, there remain a vast number of insects with a little binocular overlap, but they can triangulate over short distances for the operation of their own mouthparts (Figure 1.3).

Saccades

Saccades are spontaneous jerks of the eye at intervals of up to a few seconds—so small as to be scarcely noticeable, so they were ignored until recently. The word 'saccade', from old French, meaning 'twitch', is not in my edition of the *Oxford English Dictionary*. In 1963, David Sandeman noticed that in crabs a saccade was initiated by bringing a contrasting object into view. Land described them in the fly in 1975 (Figure 3.5), and their role in the vision of *Drosophila* was analysed in detail by Heisenberg and Wolf in the 1980s. They serve three essential purposes in active vision.

First, saccades give the fly a way to calibrate the motion detectors by making a voluntary motion of a standard size. Second, a saccade activates a synchronous input along all visual axes that look at contrasts in the visual world, so that the retinotopic location of every contrast is renewed. The saccade is the engine that arouses the feature detectors when insects take a snapshot of the angular distribution of landmarks.

Third, in the flies *Drosophila* and *Musca*, a voluntary turn is initiated by making a fast saccade towards the direction of the intended turn, overriding the optomotor system's visual control of the head position, and the body then follows. In this action, the saccade is the unit of self-directed turning (see Figure 7.7). In the flying locust, active turning of the body begins at the same time as the saccade of the head. No doubt, voluntary turns will be found that are not accompanied by a saccade, as in the smooth tracking by the head of a mantis that follows a moving fly on a featureless background. On the other hand, when tracking a prey against a patterned background the mantis is obliged to make a series of saccades (Rossel 1979). Spiders move the retina inside the head to make saccades and to fixate on prey. In most insects, the saccades have not been studied.

The optomotor response

The optomotor response is limited to the visual stabilisation of the head position when there is an unexpected displacement. Traditionally, it keeps a moving insect on a straight course and a hovering or floating insect on station in spite of wind or water currents. The classical systems analysis (Figures 3.1c and 3.2.a) was in terms of the perceived angular velocity at the eye although at the 'time it was also inferred that the motion detectors responded to contrast frequency, not velocity. How this was resolved was not explained. The optomotor response is rather delayed, with a latency of 40–50ms in the fly, and is also tuned to low temporal frequencies, rising rapidly to a peak near 1Hz, then falling off to zero near 10Hz, so any particular response can arise from two different stimulus situations. A fly hovering in a drum will not oscillate faster than 0.1Hz, which is useless for the control of flight. The response is to the passing of edges, so, within limits, it is similar if the number of edges is halved but their velocity is doubled. It adapts to a steady motion and then responds afresh at each unexpected change in frequency of the passing of edges. The optomotor response signals direction but not angular velocity and is dependent on a residual slip motion, so that it alone never completely compensates for a deflection from a straight track and never recovers the original direction of heading.

For all these reasons, the optomotor response cannot account for the way insects fly in a constant direction at a preferred speed over the ground or how they total their successive turns. Finally, as described below, the optomotor response is learned, like all short-term postural control of the legs, head or antennae. It is a stabilising mechanism with directionally sensitive motion detectors and a learned control of the steering muscles (see Chapter 3).

Figure 7.1 Situations encountered during flight. a) The optic flow in forward flight. b) The relation between the forward air speed, V, the angular velocity, dΦ/dt, and range, X, of a nearby object. c) Opening and closing parallax when seeing an object against a background.

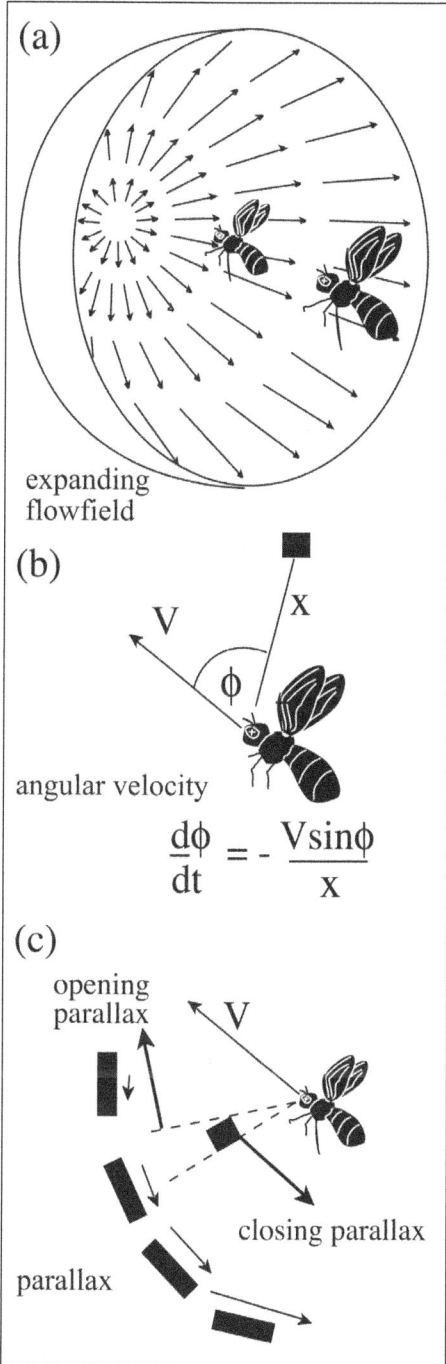

(a)

expanding
flowfield

(b)

V

X

φ

angular velocity

$$\frac{d\phi}{dt} = -\frac{V\sin\phi}{X}$$

(c)

opening
parallax   V

closing parallax

parallax

Optic flow and the avoidance and centring responses

In the past decade, the central factor in flight control has turned out to be the optic flow (Figure 7.1). The first indication was an accidental discovery of the fact that bees refused to approach rotating sector wheels, but were not concerned about the flicker of a light at the same temporal frequency without motion. The avoidance response was not at first linked to the perception of the three-dimensional world, but the general idea now is that insects interpret relative motion as the nearness of something and their three-dimensional world is composed of the ranges of different contrasts in different retinotopic directions around the head, irrespective of the real patterns.

In 1950, J. J. Gibson published a groundbreaking book suggesting that we recognised the three-dimensional arrangement of our surroundings directly from the gradient of velocities in our visual field as we moved about. The null point in the flow field is the place we aim for and the induced angular velocity at any point in the visual field is a measure of the nearness or inverse of the range (Figures 7.1 and 7.3). Gibson, who at the time was working for the US Air Force, took these ideas from classified work at Farnborough by G. C. Grindley, who died in 1976 leaving no published record of his work (Mollon 1997).

To fly in a cluttered, unpredictable three-dimensional world, insects must have mechanisms in different parts of the eyes to measure perceived angular velocity irrespective of pattern, so that they can control speed, range and steering. To control the visual input and also record the flight position and speed, the first analysis was made with bees trained to fly along a tunnel in which the walls on either side could be moved in either direction separately (Figure 7.2). Later, large television screens replaced the walls so that the patterns could also be moved up or down. The bees flew as though they equalised the angular velocities on their two sides, irrespective of the pattern or the direction of motion that they saw. Somehow, with contrast frequencies up to 150Hz, they measured angular velocity but ignored its direction. The mechanism is therefore faster than the optomotor response, but both responses are colourblind and based on green receptors. The avoidance response enables insects to fly rapidly between objects without risk of collision. In addition, anything coming into view causes a sudden turn. Self-guided mobile robots that negotiate between obstacles and walls were made with this design in the early 1990's.

Figure 7.2 The apparatus for experiments with the angular velocities on the two sides of the flying bee. The bees learn to enter on the left and fly along the tunnel to obtain the reward in the box, R, on the right. They then return by the same route. The side walls can be moved either way at a controlled speed and the patterns on them can be changed. A video camera tracks the bees from above. The average track, T, of the bees is such that the angular velocities are equal on the two sides, irrespective of the pattern.

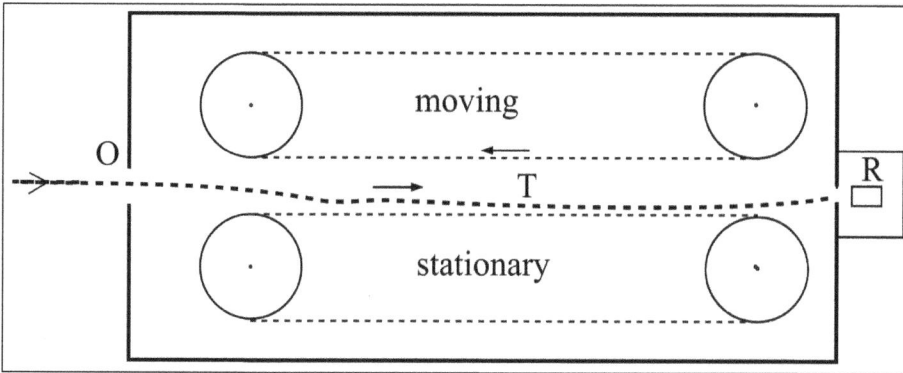

Source: After Srinivasan et al. (1991).

The analysis of the visual feedback

A mechanism based on saccades has been worked out for the fly *Drosophila* (see Chapter 3), but saccades give information about location and angular spatial frequency of surrounding contrasts, not about range or flight speed.

Forwards or sideways motion of an insect in flight, if stabilised against rotation, causes an induced relative angular velocity of surrounding objects that is inversely proportional to range (Figure 7.3a); so, distant objects appear to move little and slowly. Like a one-eyed man who estimates the range of objects of unknown size dead ahead, the insect makes a lateral movement by a known or predetermined amount (Figure 7.3b). When the flying insect sees a regular striped pattern of constant stripe period at the side, the perceived angular velocity is inversely proportional to the range, but the contrast frequency—that is, the rate of passing of stripes—is independent of range (Figure 7.3c).

The analysis of the ways that bees use the perceived angular velocity began with the observation that they fly through a gap between obstacles, along the side of a wall without collision or without turning into the wall, irrespective of the pattern on either side, and they can be trained to come to a target at a particular range. To measure the range of an arbitrary pattern, it is essential to move the eye without rotating it and measure the resulting apparent angular velocity of the target. For an object in front, at the side or below, one way to do this is to use the fact that the range is inversely proportional to the perceived angular velocity (Figure 7.3). This calculation requires a measure of the forward speed in relation to the ground.

To measure the range at an angle to the direction of motion there is a sine or cos function to consider (Figure 7.3). The range to an object dead ahead can also be measured from the perceived rate of expansion on the eye, which is another simple relation (Equation 7.1).

Equation 7.1

Range R = (forward velocity)/(angular velocity of the edge)
Or: (time to contact) = 1/(angular rate of expansion)

In these relations, flight speed over the ground is measured in metres per second, angular velocity in radians per second and range in metres.

Air speed is measured by sensitive mechano-receptors that detect the bending of the first joint at the base of the antenna, due to the air pressure in flight. Similarly, hairs on the head detect air movement, but these cannot report speed over the ground.

A problem arises from the fact that the most easily observed motion detectors, in the optomotor system, are sensitive to the rate of the passing of contrasts (called the contrast frequency) and also the direction of passing, but not to angular velocity. For an eye looking at a textured background at the side or directly downwards, Equation 7.2 applies.

Equation 7.2

Angular velocity = (contrast frequency)/(angular spatial frequency)

For regularly repeated edges, as in a grating, the spatial frequency is the reciprocal of the period. This relation implies that the absolute value of the average period in the pattern passing by is part of the calculation, but the bee in the air does not have a measure of it. The bee does, however, measure its perceived velocity over the ground, as shown by many experiments. It must learn to use a preferred angular velocity that suits its preferred or real flight speed.

How is the angular velocity measured? The bees clearly measure their speed over the ground by non-directional detectors and they learn the speed at which they negotiate a familiar route. Several authors have recorded from deep optic neurons that show an increasing response to increasing angular velocity, irrespective of the spatial frequency, but so far all the neurons sensitive to high velocities are directional. A possible mechanism can be designed in several ways—for example, by measuring the delay as a contrast passes from one visual receptor to the next, or the next but one (called the stopwatch method), but there is no evidence that this happens or that arbitrary times are measured.

Figure 7.3 Relations between flight speed, V, range, R, and the induced angular velocity for different directions relative to the direction of flight. If rotation is eliminated, (a) forward flight or (b) sideways scanning causes an induced relative angular motion of surrounding objects. This visual feedback is inversely proportional to range. c) The frequency of passing stripes (the induced contrast frequency) is proportional to forward velocity, V, and inversely proportional to the pattern period, P, but independent of range, R. Angles are measured in radians.

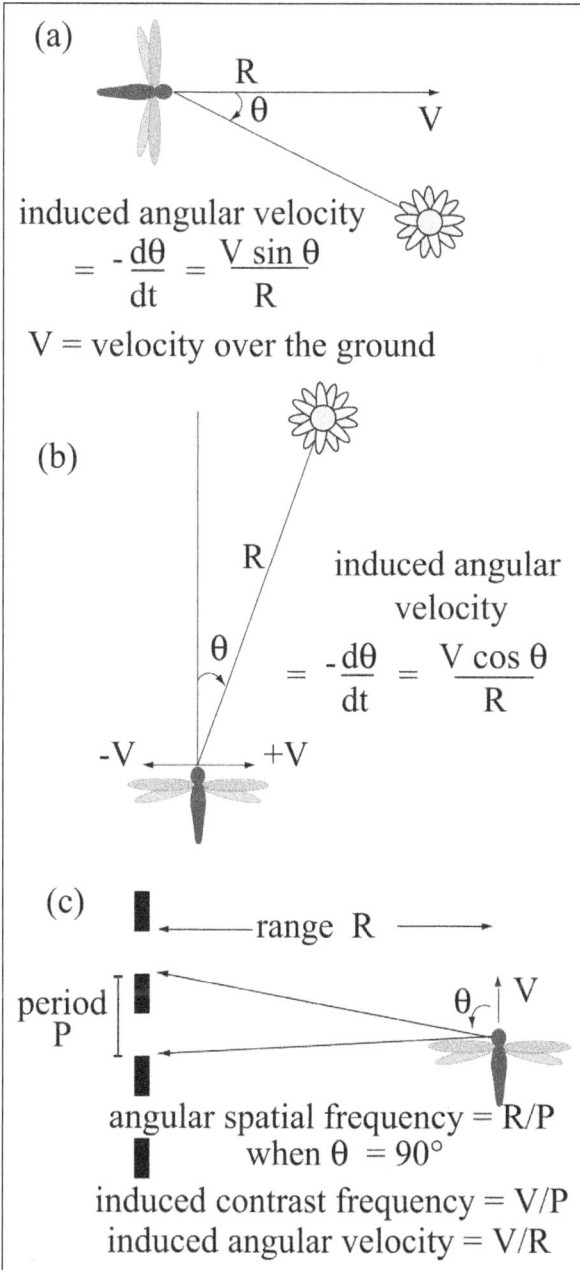

(a)

R

θ

V

induced angular velocity

$$= -\frac{d\theta}{dt} = \frac{V \sin \theta}{R}$$

V = velocity over the ground

(b)

R

θ

induced angular velocity

$$= -\frac{d\theta}{dt} = \frac{V \cos \theta}{R}$$

-V   +V

(c)

range R

period P

θ   V

angular spatial frequency = R/P
when θ = 90°
induced contrast frequency = V/P
induced angular velocity = V/R

155

Figure 7.4 Measurement of angular velocity. Four motion-detector neurons, A, B, C and D, have different retinal spacing of their inputs, so they are tuned to different spatial frequencies. They have an increased response (A1, A2, A3) to increasing temporal frequency. a) The high-level neurons that detect appropriate coincidences (A1, B2, C3) are detectors tuned to different angular velocities. b) The higher-level neurons that detect appropriate coincidences (C1, B2, A3) have an increased response to increasing velocity irrespective of spatial frequency.

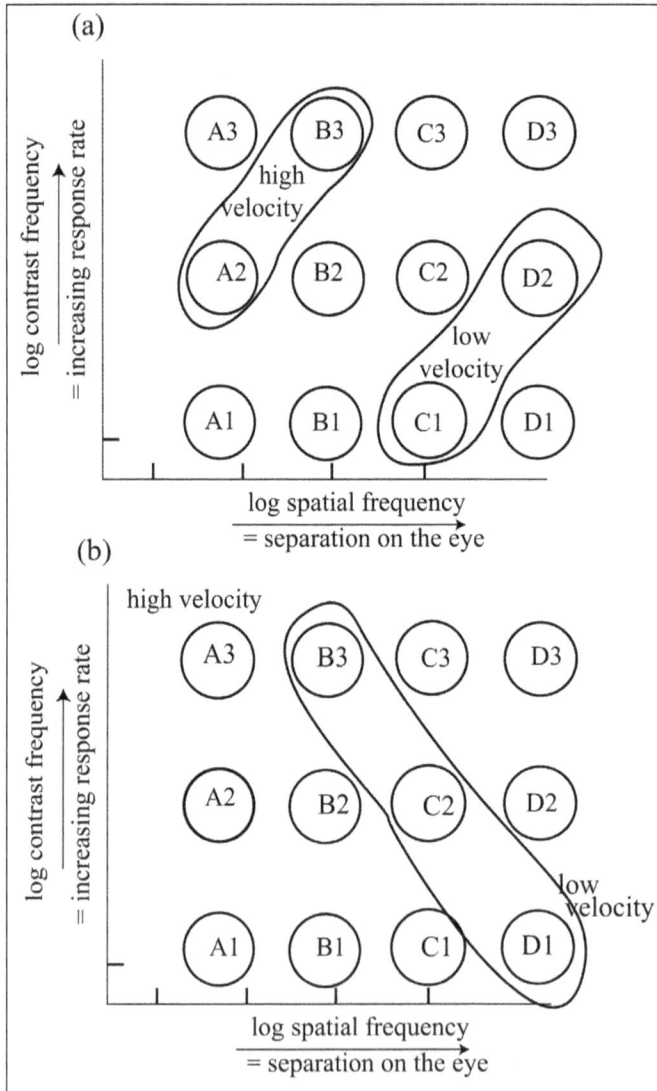

More likely there are several motion-detector channels in parallel, with different spacing on the retina (Figure 3.4). Consequently, they have response curves tuned to different spatial frequency peaks (cells A, B, C and D in Figure 7.4). Each of these types occurs with different contrast frequency peaks (cells A1, A2 and A3 in Figure 7.4). We can imagine various combinations of spatial frequency and temporal frequency tuning to the optic flow. If there is a higher-level neuron that detects the coincidence between a strong response of detector B3 with a weak response of detector A2, the insect responds to the high angular velocity (Figure 7.4a). If there is another higher-level neuron that detects the coincidence between a strong response of detector D2 with a weak response of detector C1, the insect responds to the low angular velocity. Even two different channels with different spatio-temporal constants would allow some estimation of the perceived angular velocity.

We need a mechanism, however, that yields *an increasing response to increasing velocity irrespective of the spatial frequency* of the pattern. This could be achieved by a higher-level neuron that detects the coincidences between responses of detectors D1, C2 and B3 (Figure 7.4b).

Significantly, when the flying bee measures the perceived angular velocity to get the range or the distance flown, she appears to be unaware of the direction of the perceived motion, but at the same time, her body position is stabilised around three axes by the quite separate optomotor response that is sensitive to the direction but not the perceived angular velocity. The automatic stabilisation of yaw, pitch and roll in the flight posture is separate from the choice of direction. At present, some models of the control of flight in the fly appear to put the whole signal through a single channel. Further channels—for example, from the halteres and antennae—also play a part.

The channels that detect looming, avoidance of collision and active control of direction are all faster than the optomotor response. As in all insect motor systems, in the bee, there are slower tonic and faster phasic (twitch) motor neurons in the bee's control of neck movements, but much remains to be analysed in these responses.

These ideas omit the fact that the compound eye is an ideal device to measure the angular spatial frequency in a local region of a scene. For example, at every saccade of the head, each contrast in the image stimulates a modulation detector. We know that these responses are summed in local eye regions in the bees, which can learn local modulation quantitatively and remember the different modulation in separate regions of the eye when landmarks are distinguished. The sum of the modulation is a measure of the average spatial frequency in each local region. We do not know whether insects make use of saccades in this way.

In the fly, the fields and directional sensitivities of the different neurons that measure direction of motion and its temporal frequency are well known from recordings in the lobula plate, but the eye regions and neurons that measure range and apparent angular velocity have not been distinguished.

Direction and speed over the ground

The tendency of insects to fly upwind has been known for more than a century and it was not a mystery until work by the Reichardt group, from 1956 onwards, turned the emphasis to the optomotor response. Kennedy (1940) made the basic discovery that the control was entirely visual, showing that the speed and direction of freely flying mosquitoes in a wind tunnel could be controlled by letting them see a moving pattern. In slowly moving air, the insects overtook the apparent movement of the ground, which ensured that they flew upwind at a preferred ground speed. If the wind was strong, the insects turned around and flew with it or settled. The results were later repeated on aphids, moths and several flies including *Drosophila*. In the early work, the flight height was not controlled, unless there was an odour plume. A wide range of behaviour—from the migration of butterflies or locusts to the dispersal of aphids or the behaviour of flies in tunnels—was governed by the visual response to the perceived motion of the ground. Some insects that migrate, and swarming locusts, turn to fly downwind and are assisted by it. In a locust swarm, the individuals at the top fly faster, overtake their fellows and then descend at the front of the swarm. In light wind, the swarm tends to drift downwind and compensates for any change in speed over the ground.

Within the range of wind speeds that they can exceed, flies or bees flying upwind in a horizontal wind tunnel have a preferred upwind speed relative to surrounding objects, irrespective of the pattern that they see. When there is a sudden step in the period of the pattern on the tunnel wall or floor, they maintain the same perceived angular ground speed as before.

Bees fly faster when higher and slow down as they approach a landing. They fly slower in narrower tunnels, faster in wide ones or in the absence of strong visual feedback. They are slowed down by the angular velocity component of the visual feedback irrespective of the spatial frequency of the pattern. The preferred angular velocity is in the range 250–750 degrees per second and is controlled from moment to moment by the range of nearby contrasts. Two important points follow from all this: first, the optomotor system does not control the flight velocity or direction; and second, because the dance cannot code the height at which they fly or the effect of the wind, the bees' dances convey the perceived distance to the goal over the ground, which makes the instruction in the dance independent of the wind, but the recruits must fly the same route at the same height as the dancers.

In each situation along their route, flying bees have a preferred forward flight speed and a preferred flight height, and these are both related to the perceived angular velocity. There are other factors—for example, they learn how to negotiate a convenient route and when to slow to anticipate a turn; and all flying insects have airspeed detectors in the antennae and head hairs. They might also follow an odour plume at a certain height.

Following an odour plume

Although the primary stimulus is the odour of water, a flower, scent mark, dung pile, carcass, male's pheromone or animal sweat, many insects including bees fly upwind towards an odour source by visual control from the feedback from the motion of the ground. In general, the odour receptors initiate the search but they are not directional. When the insect comes to the edge of the odour plume and loses the scent, it casts about or scans from side to side or up and down under visual control until it finds the odour again and turns upwind. There is no question of holding a station by the optomotor response or of using an odour gradient. Aquatic animals behave similarly.

Other ways to measure range

It has been known since Wallace (1959) that grasshoppers sway from side to side to measure the range before they jump. In 1977, I proposed that to measure the range of nearby objects, flying insects also made use of the apparent motion caused by similar relative movement in flight because they lacked other means. There are two mechanisms involved—one based on induced lateral motion, which is hard for humans to do because they fixate with a moveable eye, and the other on observing the parallax as an object appears to move against a background, which is the way that a one-eyed person usually estimates range. Without parallax, many kinds of insects can catch a mate or prey as seen against the sky and mantids and locusts can also estimate range by peering from side to side at an object against a featureless background.

In 1986, we began a series of experiments that demonstrated range discrimination in a number of ways. The first experiment was to train flying bees to discriminate paper 'flowers' raised on stalks (Figure 7.5a). The bees were able to distinguish the correct length of stalk and land over the one with a reward of sugar solution; other flowers had only a drop of water. The flowers were randomised in size and position but the rewarded flower always had the same length of stalk. Later, we put paper discs on flat shelves of perspex (Figure 7.5b) and also showed that bees could discriminate different horizontal ranges of black discs of randomised size when they were on a vertical surface.

Figure 7.5 The discrimination of range irrespective of the position, absolute size or angular size of the disc. a) The first experiments were done with discs of different sizes on stalks of different heights in a box over which the bees could fly. b) Later, the discs were arranged on thin, clean sheets of perspex. The bees were trained to land over a disc of random size at the specified vertical range, while the heights, positions and sizes of the other discs were repeatedly randomised during the training. The reward of a drop of sugar solution is always over the target disc and drops of plain water are offered in random places to ensure that the bees do not search for the reward.

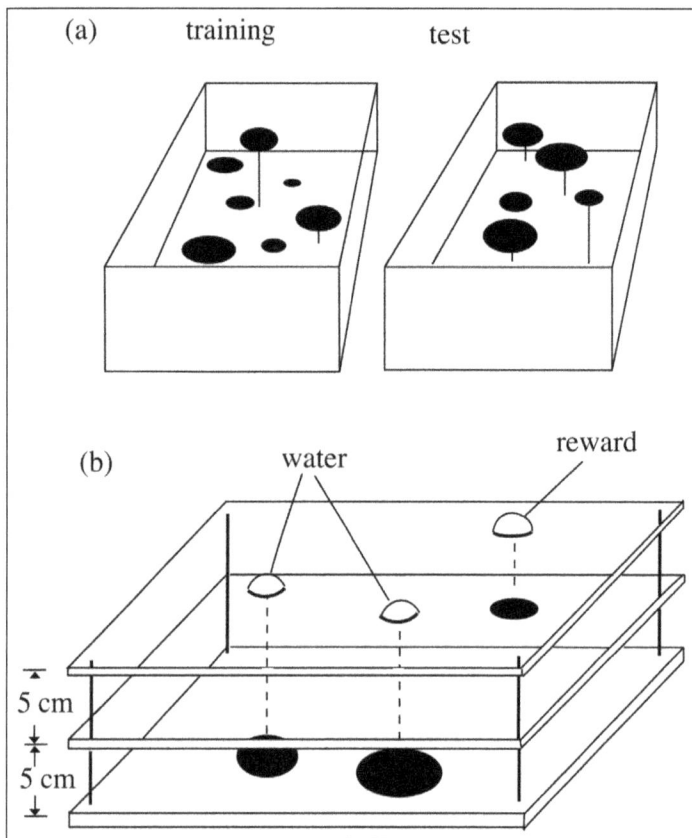

Source: Horridge et al. (1992).

With a similar arrangement, bees discriminate the absolute size of a black disc, presumably by combining the range with the angle subtended at the eye. When the angle subtended at the eye and the range of the target are randomised, bees learn to discriminate the correct absolute size and, when size and range are both randomised, they can learn to select the correct angular size. Bees can also do these tasks looking horizontally at vertical targets. It appears that absolute size is important in the three-dimensional world of bees and wasps, especially for control of flight height, locating self relative to landmarks and recognition of familiar places.

Control of straight flight

A human learning to steer a car or a boat with a wheel or a tiller must learn which direction of steering causes which direction of turning, and further, what happens when going backwards, which many insects, including bees, can do. Learning by trial and error is involved (or not, as the case may be). Insect flight in a straight line is dependent primarily on visual feedback from the surroundings and fixation on a target, although factors such as odour plumes can initiate it. In free flight, the fly *Drosophila* makes saccades with angular velocities between 1500 and 4000°/s randomly in either direction. The fly, like the bee, can integrate successive turns in free flight, steer towards local contrasts and return later to its original track.

Figure 7.6 The essentials of the flight simulator with the fly mounted on a torque meter in a white drum, and a motor that moves the black bar as instructed by the computer. The only measurements are of yaw torque, drum velocity and drum position; therefore, any theory must be expressed in these terms. The fly can learn to fly on a straight course in this apparatus and also to select one of several different patterns.

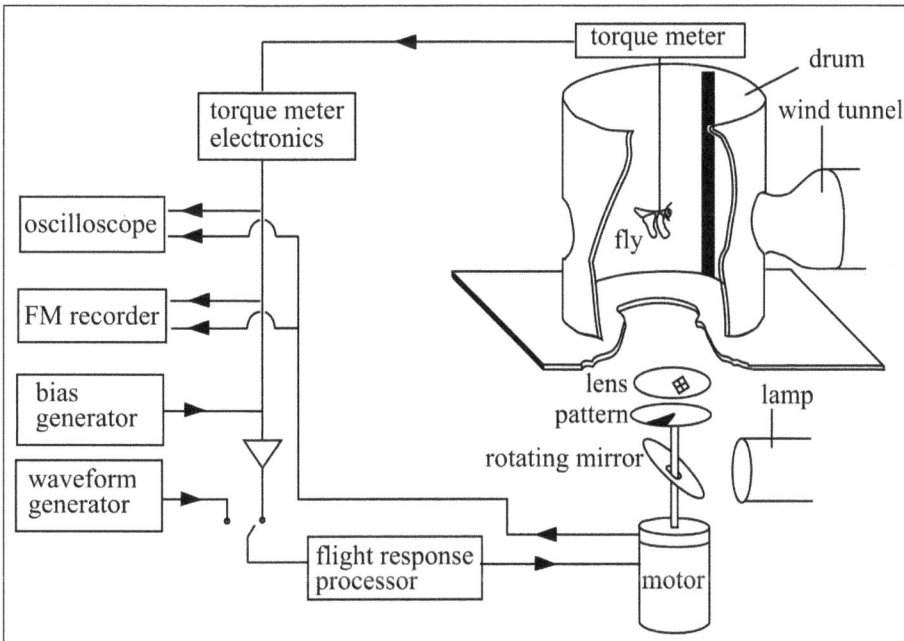

Source: After Wolf and Heisenberg (1991).

To analyse the mechanism of steering, Wolf and Heisenberg (1990) fixed the *Drosophila* within a drum on a device that measured the turning torque generated by the wings in the horizontal plane and converted this torque into the expected rotation of the visual field that the torque would cause (Figure 7.6). Neither the head nor the body moved, but the fly's efforts to turn now appeared to it to have been effective. The coupling ratio was not critical. There was continual jitter and fluctuation of the output. The fly's effort to make a saccade appeared on the recording as a brief pulse of torque, which was converted to a drum rotation up to 60° in the opposite direction (Figure 7.7), but there was no optomotor response to the movement of the drum that was initiated by the saccade.

Figure 7.7 The behaviour of *Drosophila* in the flight simulator, with a single black strip visible to it (Figure 5.9). a) Traces of yaw torque (left scale) and pattern position (right scale) showing torque spikes and small adjustments of torque between them. b) A reconstruction of the trajectory that the fly would have made if free, assuming constant forward flight velocity and adding the turns.

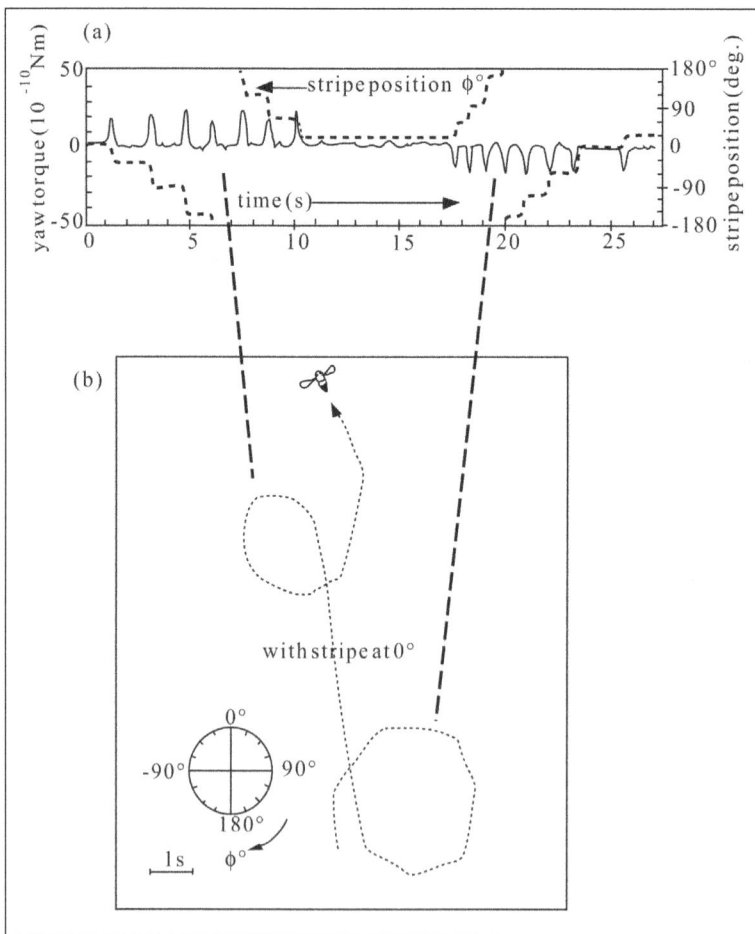

Source: After Heisenberg and Wolf (1992).

On the other hand, an unexpected imposed rotation of the whole visual field by as little as 0.1° causes a torque response with 40–50ms latency and high gain. If one wing is damaged, causing continual turning in flight, the fly readjusts again to the new relation between the eyes and the torque. The visual input generated by a saccade informs the fly within 50ms whether there has been a change in position of contrasts in the visual field apart from that which was expected, and it adjusts the output. Any low contrast in the visual field is sufficient for visual stabilisation on a straight course. There is no fixed cancellation of the optomotor response to allow for voluntary turning, for that would fail to compensate for damaged wings.

The fixed fly can also learn to control the position of a vertical bar, which it prefers to bring to the front and then fly towards it. If the bar is displaced sideways, the fly exerts torque to bring it back to the front (Figure 7.7). The angular motion input resulting from the saccades is ignored, probably because it is too fast. Careful analysis of the wild type, supported by work on the double mutant *rol sol*, suggested that the single bar was controlled by its position, but the background was controlled to reduce its perceived motion, with a slip speed, so displacements were not fully corrected. If the coupling between the fly's effort to turn (the torque) is reversed, with a bar moving in front of the background of the drum (the figure/ground stimulus), a normal *Drosophila* can learn to control the position of the bar but not the position of the background. The double mutant *rol sol*, however, lacks the optomotor system but still retains the faster non-directional system that normally detects the position of a bar and, with this, it can learn to stabilise either the bar or a textured background despite the reversed coupling.

If the apparatus is rearranged so that the movement of the bar is controlled by the forward thrust in flight instead of by the turning torque, the fly learns the new controls in a few seconds of trials. The fly can also learn to control its visual field by pushing with its legs on a table below or in response to warmth as reward or cold air as a punishment. Long ago, I showed that the leg posture of locusts and cockroaches was learned by operant conditioning in a similar way. We can conclude that when there are several motor outputs to legs and wings, it is necessary to have a learning mechanism so that they function effectively in an unpredictable world. Even humans relearn quickly if placed on a bicycle or stilts or if their vision is inverted by prisms.

Visual measurement of distance flown

In their dance, bees signal the direction and distance to a worthwhile food source from which they have just come. Despite the observations of his students, von Frisch believed that the bees measured the energy used to reach the food source, irrespective of the distance over the ground, although the latter was the most reliable cue in a windy world. Recent work has shown that bees use their eyes to measure the distance they have travelled over the ground, not the energy used. Esch and his colleagues made use of the fact that bees coded the outward distance from the hive to a reward by the number of abdomen wobbles in the dances on their return. They flew bees to a reward in a balloon. As the balloon was raised, the bees were obliged to fly higher and this reduced the optic flow, so their dances indicated a shorter distance. Also, flights between the tops of tall buildings were coded as too short by the dances of the returning bees. The oxygen usage was strongly influenced by temperature but was related only indirectly to the distance coded by the dances.

When trained bees flew down a long tunnel with patterned walls, along which small dishes were distributed, they learned to fly for the correct distance to the particular dish where they expected the reward of sugar to be (Figure 7.8). They flew the same total distance if an extra length was added at the entrance of the tunnel (Figure 7.8a). They learned the distance on the way in, not on the way out (Srinivasan et al. 1996). There was no effect on the distance flown by the bees when the period of the pattern was changed on the inside walls of the tunnel or when a wind was blown along it, although a wind changed the duration of the flight. The bees do not count landmarks along the tunnel nor do they fly at a constant speed irrespective of the wind, but if bees trained in one tunnel are tested on a wider tunnel, they fly too far. Conversely, if given a narrower tunnel, they do not go far enough (Figure 7.8b). These effects show that the bees measure distance by integrating the apparent velocity of the optic flow at the side of the eye over time. If the optic flow is made useless by putting stripes horizontally along the sides of the tunnel, the bees do not know how far to go and also they fly at great speed.

Recent measurements by Srinivasan, Zhang and others have shown that outdoors the bees give about 1.6ms of waggle dance per metre travelled, and 1 metre of flight in a tunnel 30cm wide is equivalent to 25m outside. In terms of angle, one millisecond of waggle encodes about 18° of image motion on the eye in any situation.

Figure 7.8 Bees learn to fly the correct distance in a 3m tunnel to the dish that contains the reward. a) Results of training followed by testing with a 1m extension of the tunnel. They fly the same distance as before but to the wrong place, showing that they do not use landmarks. b) There is no effect of changing the period of the stripes in the tunnel, so they do not count stripes.

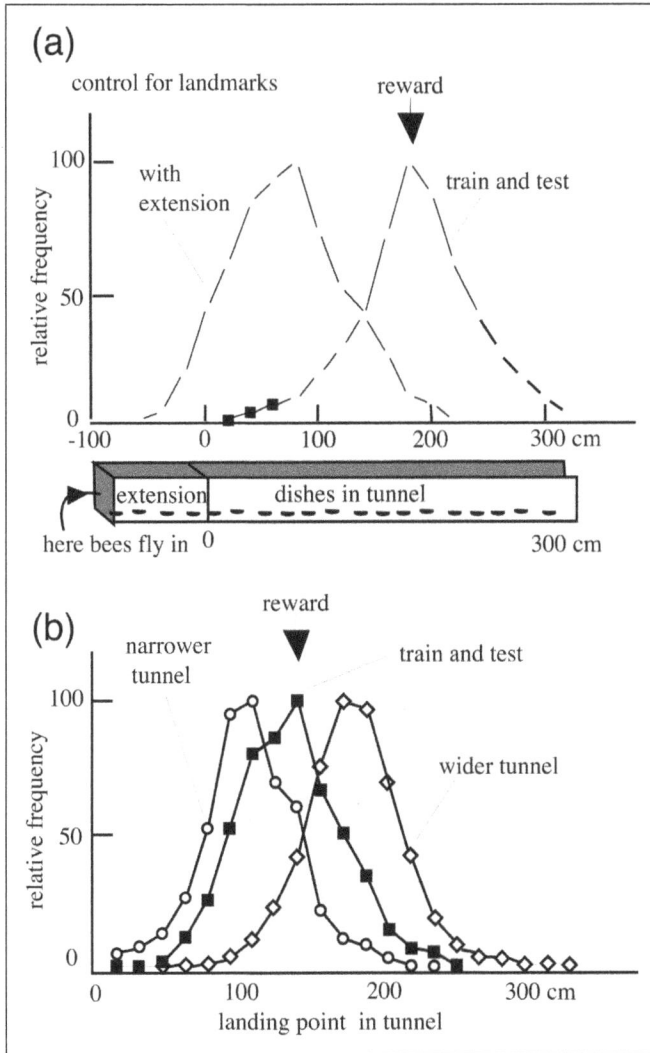

Source: After Srinivasan et al. (1996).

Figure 7.9 Bee flight paths in a large, featureless tunnel. a) and b) Positions of the reward holes on the end wall. c) Flight between two black lines. d) Flight deflected by the black line. e) In side view, they drop to the floor if it is featureless. f) They fly low to avoid a wind. g) The textured floor. h) and i) Sudden change of the spatial frequency makes them aim (h) too low or (i) too high, without change of flight speed.

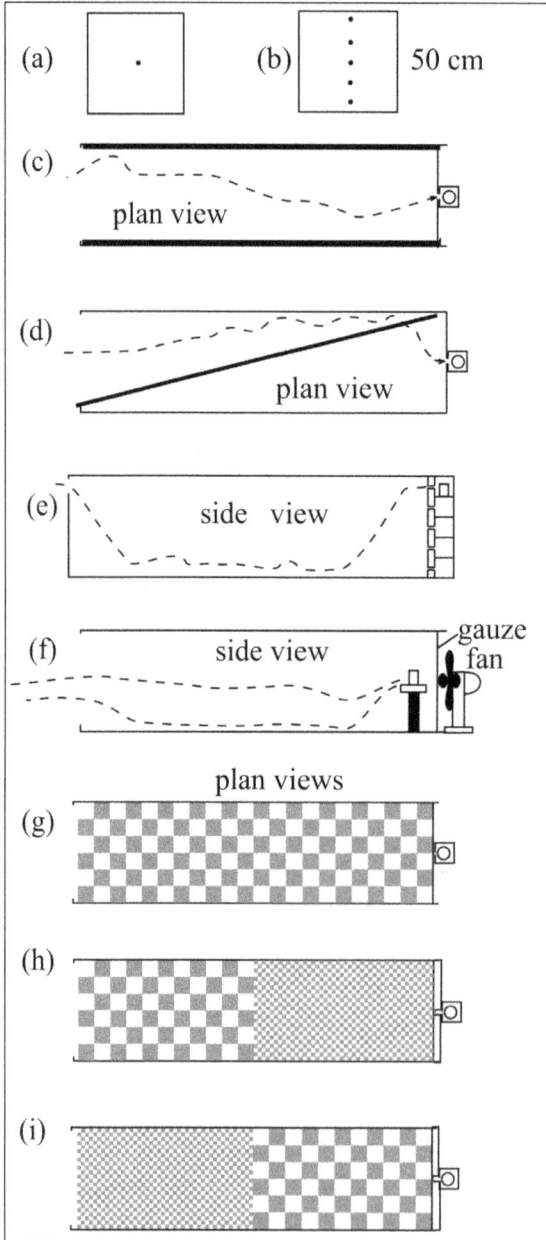

Original unpublished observations

Control of flight altitude

In their normal terrain, bees have a preferred flight height, which depends partly on what they have experienced in previous flights at the same place—for example they will fly low or in the shelter of a wall to avoid a strong wind (Figure 7.9f). When bees fly along a familiar route, they have learned the familiar average period of the pattern below them and use it to bring them back to a preferred flight height. When bees were trained to fly into a large featureless tunnel, 1m high and 1m wide, they flew along the centre near the floor if it was white, but at a height of about 50cm if there was a pattern on the floor, irrespective of the pattern. They learned to fly higher or lower if they anticipated meeting obstacles in the way and they learned to follow a black line sloping up or down (Figures 7.9c and 7.9d). When given a choice of several holes at different heights, but only the top one with a reward, they learned to fly at the height of the anticipated reward hole (Figure 7.9). When they had thoroughly learned the correct height to fly over a coarse pattern (Figure 7.9g), the floor below the reward hole was unexpectedly changed to a pattern with a smaller period (Figure 7.9h). The bees then aimed too low until they relearned the correct height of the reward. They therefore learned the perceived temporal frequency of the ground at their preferred flight speed. Bees learn to go to a reward at a certain height but they cannot recruit to a food source that is high above the ground, so the dance apparently contains no information about altitude.

Control of flight speed

In each situation, bees have a preferred speed of passing the panorama that they see. They fly faster in a head wind, fast when familiar with the terrain and slowly in an unfamiliar place. Outdoors, they fly fast when high up and slower when closer to neighbouring contrasts. They fly slower when exploring or uncertain and faster when certain of their route. As the bee flies along, it adjusts its speed and direction by the perceived optic flow, forming a feedback loop that again changes the optic flow. In the tunnel experiments, they fly slowly because the walls are close and appear to pass faster, but by the time the measurements are made, they have also learned the dimensions of the tunnel. When there is a constriction in the tunnel, the bees slow as they pass it then return to the same speed as before, but the perceived angular velocity remains approximately constant. Because continuous learning is involved and the internal state of the bee is unknown, it is difficult to discover how the preferred speed is decided.

The front and the sides of the eyes are clearly instrumental in measuring the optic flow as the bee flies along, but the optic flow has a distribution around the eye—slow at the front but faster at the sides and below. The directional effects also differ with the region of the eye, as lateral motion at the front causes turning, at the side it affects only the speed, and ventral motion makes the bee turn and fly upwind. In experimental unmanned planes and helicopters, it is sufficient to control landing by the optic flow in a solid angle looking forward and downward, and to prevent collisions via an eye with 360° vision looking towards the horizon.

As well as visual control via the optic flow, bees and flies have head hairs and a special mechano-receptor system at the base of the antennae, both of which are sensitive to air speed. How the three systems interact has not been studied.

Hovering

In several groups of insects, some adept fliers can hover in flight while they examine an object visually, feed from a flower, lie in wait or guard a nest entrance. It is no more of an achievement than flying; exactly the same parallel mechanisms are in action, including learning each familiar situation. In various hovering insects, a suitably tuned optomotor response for stability against unexpected perturbations, fixation on a target to stabilise the direction of looking and a shift sideways or a measure of target size for range estimation have all been described as contributory mechanisms. Expansion of the image on the retina is a sign of an approach towards something. Locomotion in any direction is reduced to zero by detecting and moving away from each centre of expansion, combined with keeping landmarks at fixed positions on the eye.

The interesting questions about hovering to fixate on a contrast are how much it improves vision and to what extent stationary images at the front of the eye can be better processed and remembered. Some dragonflies have more than one fovea. In some situations when insects hover, the image is fixated with deliberation by the fovea and this behaviour is somehow related to the improved discrimination of a mate, a flower or prey. Almost always these insects 'turn and look' at a specific object; however, it has not been shown that they partition their visual world into separate objects.

Male hoverflies and dragonflies, hovering in wait for a passing female, fly in exactly the appropriate direction to intercept her. The response is simpler than it appears at first sight, because the line of interception is selected on the assumption (perhaps learned) that the target is a female of standard size flying at a predictable speed. Bees commonly use landmarks at the side or the lateral parts of a target to localise a reward that is in front of their eyes, but they turn to centre their vision on a spot of blue colour, a radial hub, a source of parallax or an expected cue. All of these reactions are strongly influenced by learning during the course of the experiment.

Landing

As a flying or swimming insect approaches an object, it slows long before the legs are extended for landing. When a fly sees something in its path, it unfolds its forelegs, brings the other legs down and then extends the forelegs to break the shock on contact. Any strong addition into the flow field at the front of the eyes is an adequate input. Although the necessary trigger for the landing response has been studied in detail in the fly, we are still not clear how it controls its own dedicated motor pattern. There appear to be fixed or learned motor sequences for the initial stages of landing, triggered by looming and net darkening, which are preset for the task. Flying flies held by the thorax go through these motions repeatedly but tethered bees soon learn that they are fixed.

The motion perception for the landing response is tuned to a higher contrast frequency than that for the optomotor response. There are several vision mutants of *Drosophila* in which either the landing or the optomotor response is ineffective—again suggesting they are separate systems. Selective habituation of neurons suggests that the optomotor and landing pathways have separate motion-detector pathways. The landing is best stimulated by fast motion of a single edge or by a spatial period greater than 20°, whereas the optomotor response is most sensitive to slower motion of intermediate periods. Although it is an attractive idea, it has not been demonstrated that any insect computes the time to contact when landing and slows accordingly.

A honeybee coming in to land on a flat, patterned surface detects the increased speed of the ground beneath it and turns to land at right angles to an edge. As it loses height, it slows to its preferred angular velocity over the ground. Analysis of numerous landings by high-speed photography shows that the average bee keeps the average angular velocity of the surface constant as it approaches, until the flight speed is zero at touchdown. There is one input variable, the perceived angular velocity, and one controlled variable, the flight speed over the ground— both ending at zero. For most of the way, the angular velocity is maintained at 400–600°/s, but there is great variability in the response and a lot of unknown factors. Obviously, the relations between the instantaneous perceived angular velocity, flight height and instantaneous airspeed are not as consistent as the averages. How this translates into other situations, like landing on a thin twig, has still to be worked out.

Scanning and peering

When dropped on the ground, many insects make exploratory head movements. Before they reach out to step across a gap, young praying mantis make a few lateral scans of the head—sometimes called peering—direct their gaze at the other side and prevent head rotation while they move the head sideways to measure the range.

When the mantis sees a potential prey, it behaves quite differently: it freezes and uses triangulation by the two eyes to manoeuvre very cautiously into the predetermined range for a strike with a foreleg. A mantis can track the movements of a small prey on a featureless background by smoothly tracking it with movements of its neck and legs, but if the prey is on a textured background, the mantis tracks it with short jerks of the head, like saccades, that overcome the visual stabilisation by the background.

Before they jump, grasshoppers, locusts and young mantids make a similar lateral movement of the head as they measure the range ahead (Figure 7.10a). The velocity at take-off and the distance jumped depend on the range measured visually. Peering or scanning by standing insects appears to be the same mechanism as range measurement and segmentation of the three-dimensional cluttered world by flying bees: it involves non-directional relative motion irrespective of pattern and it is also sensitive to the parallax when an object moves across the contrasts of the background.

A fovea at the front of the eye is an advantage for peering because it increases the number of receptors involved. Locusts have a small, forward-looking fovea.

Two types of experiments show that the measurement of range from the relative motion on the eye is done by non-directional (that is, scalar) motion detectors. When a locust peers from side to side to estimate the range for a jump, the target can be moved as the locust moves its head (Figures 7.10b–d). The added motion reduces the apparent motion on the eye and the locust jumps short, according to the motion it sees. When the added motion is so arranged that the apparent motion is reversed (Figure 7.10d), the locust will still jump according to the amplitude of the net motion. The visual system measures the range irrespective of the direction of the induced motion on the eyes.

Flying insects also scan in flight and, like peering in mantids and locusts, the scanning detects the range, but we have no evidence that they see shapes. They identify biologically important objects by detecting cues. A good example of peering in flight is the flight of a fast fly, weaving from side to side as it goes, or the lateral movement of a hoverfly or hovering dragonfly when an intruder comes into view.

Another kind of scanning is seen when bees fly over striped patterns in search of a reward. They tend to follow the edges, as a way of reducing the relative motion and flicker on the eye, whether trained to the pattern or seeing it for the first time. We know nothing about neural mechanisms of saccades, peering or scanning, or their use out of the horizontal plane, and few species have been studied.

Figure 7.10 Range measurement by the locust before it jumps. The locust peers by moving its head sideways by a distance (h), keeping the midline of the head pointing directly forward. In (a) the target is stationary and the angle, α, indicates the range. In (b) and (c) the target is moved when the locust peers so that the angle, β, is too large and γ is too small. In (d) the target is moved to the other side so that the angle, δ, is in the opposite direction to α. The locust still estimates the range as R, so it uses the other eye. e) The jump velocity is related consistently to the apparent target distance.

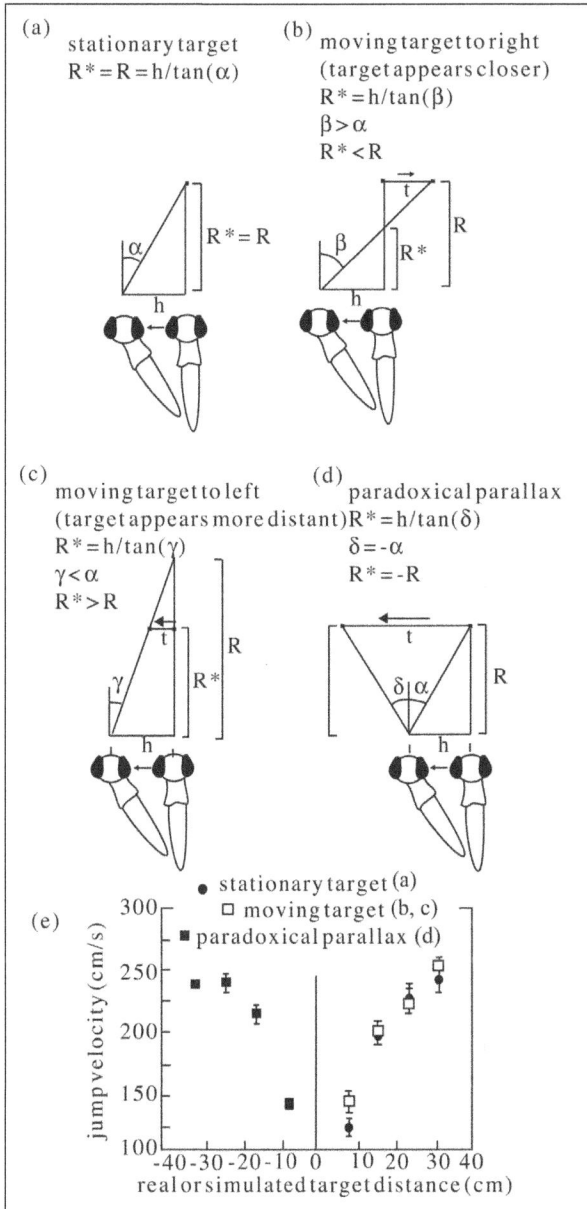

Source: Partly after Sobel (1990).

Figure 7.11 a) Sideways motion of a hovering insect induces opening or closing parallax as well as relative motion. b) When an insect is approaching an object or coming in to land, the nearer edge presents an increased velocity coming into view and also the closing parallax as contrasts in the background are obscured. Either stimulus is accepted as a measure of range, but the insect never lands on the far edge, which presents opening parallax.

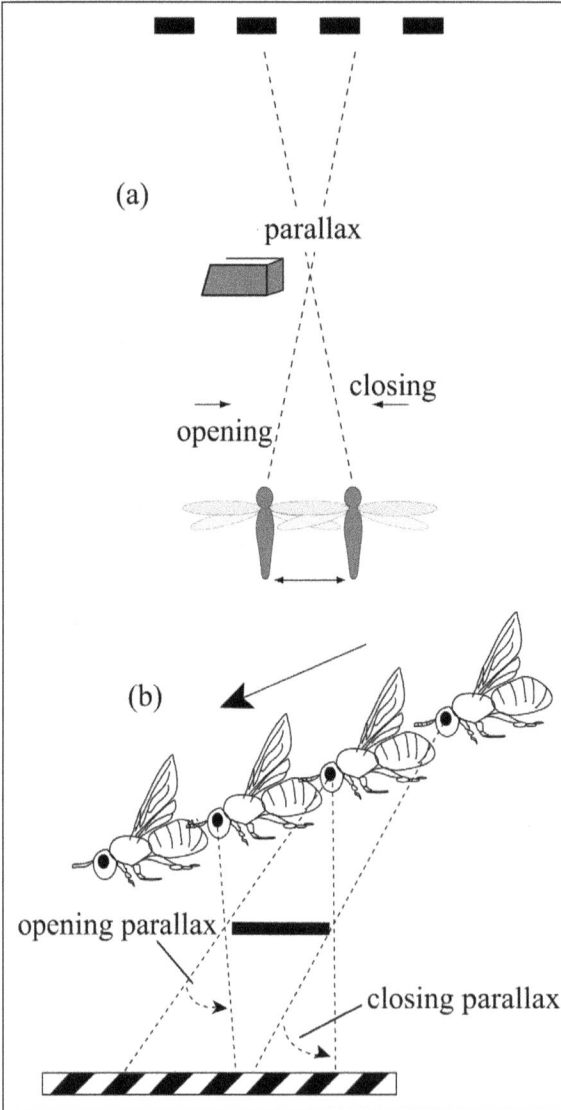

Self-motion generates parallax

When an eye moves, nearer objects appear to move faster against the slower motion of the more distant background. Relative motion between two objects caused by sideways motion of the observer is called parallax. For centuries, this was the term employed by astronomers to describe the apparent motion of near stars against the positions of more distant stars caused by the motion of the Earth around the Sun. Two components of parallax are important in ordinary vision: a) the covering or revealing of the more distant contrasts by the nearer object (Figure 7.11c), so that there are local changes in intensity, contrast or colour; and b) the step in relative velocity at the step in the range (Figure 7.11b). Insects tend to fixate on a contrasting edge and head towards it, but it is less certain whether they perceive the edge as part of an object.

When locusts prepare to jump, they peer equally well at textured or black targets, seen against a textured background, and at the edge of a wide target (Figure 7.11a). When the difference in perceived velocity is made smaller by bringing up the background closer to the target, but keeping the target range constant, the locust jumps harder as though it interprets the smaller parallax as a measure of a greater range. It therefore assumes that the background is stationary and far away (Collett and Patterson 1991).

Bees can be trained to come for sugar to a platform bearing a pattern of stripes or random contrasts, seen against a similar more distant patterned background. They discriminate the platform by parallax (Figure 7.12). Their behaviour towards a moving contrast on a moving textured background also reveals that they detect the edge by the difference in motion, not merely by increased flicker. They land on a boundary that provides closing parallax (Figure 7.11b) and will not land on the far edge of a patterned surface beyond which they can see a more distant patterned background (opening parallax).

Bees can also be trained to come to an edge between two moving patterns or between a moving and stationary background. They land at right angles to the line of increasing velocity, whatever its direction. Once again, we see that bees recognise velocity and edges separately irrespective of direction or pattern and measure range by using a non-directional measure of angular velocity. They have several separate and parallel inputs for different features, but there is no evidence that they recognise separate objects by anything other than simple feature detectors.

Figure 7.12 Demonstration that bees detect the foreground moving over a patterned background when the bee itself moves. a) A sheet of perspex is raised above a 50/50 black-and-white background of random pixels at an adjustable height (h). A small target of random pixels is placed on the perspex sheet. The target is rewarded with a drop of sugar solution and drops of water are placed elsewhere on the perspex sheet. The target is moved at intervals and can be found by the flying bees only when raised above the background. b) As the target is lowered towards the background, bees trained to a target at a height of 5cm progressively lose it, but performance returns when the target is again raised to 5cm.

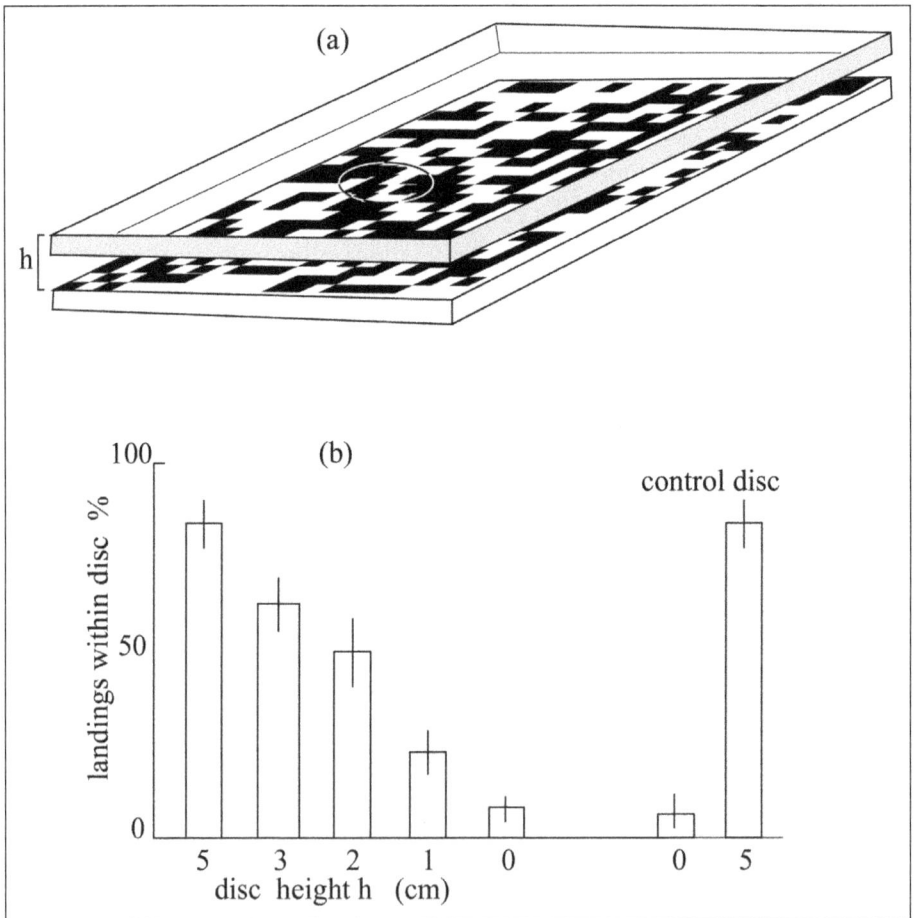

Source: After Srinivasan et al. (1989).

Mechanisms of piloting

Two things can be said about mechanisms. First, we are nearing the stage when the different behaviour patterns are sufficiently well distinguished for us to restrict experiments about mechanisms to one or another of them, so that we have some chance of separating the parallel inputs and solving them one at a time.

Second, like all the rest of the insect visual inputs, the piloting mechanisms are likely to be simple feature detectors that operate independently but in parallel. I expect that angular velocity at a place on the eye will turn out to be measured separately from the direction of motion at the same place, that the optic flow at each place on the eye, range and flight speed will be measured as scalars, not as vectors; that direction of motion is likely to be a direction, not a vector; that all the measurements will be irrespective of the pattern and they will be correlated by coincidences, not by reassembly. Time will tell.

Endnotes

1. A large part of the results in this chapter will be found in the publications of Srinivasan and also of Martin Egelhaaf. For information about the practical applications to flying vehicles, see Javaan Chahl on 'Google'.

08

THE ROUTE TO THE GOAL, AND BACK AGAIN[1]

Besides piloting, bees steer towards particular goals. They are, after all, heading somewhere. One of the lessons from human navigation is that several mechanisms are used and any cue can be useful. While circumventing obstacles, landmarks must be remembered, the general direction must be maintained and irrelevant things must be ignored. In early studies of insect navigation, with a single effect attributed to a single cause, the bees' use of several mechanisms in parallel led to confusion. As usual, positive results were explained by the first idea that came to mind and unsound theory was only slowly recognised, so controversies persisted unnecessarily.

Before the days of radar and satellite navigation, the human navigator had five main ways to navigate: by recognition of landmarks on a map; by dead reckoning, which summed the distances and the known currents noted in a log book; by compass directions; by sampling the depth, temperature and bottom of the sea; and by finding a position and direction from the sun and stars. He used all five, with a few extras such as the direction of the ocean swell or the smell of land. Insects also have several interacting mechanisms: 1) a measure of distance over the ground; 2) direction relative to the sun's position; and 3) the recognition of places or landmarks, which requires an efficient learning, discrimination and forgetting system. Less clear are 4) cues from odours, and 5) the Earth's magnetic field.

Early observations

About 1880, the famous entomologist J. H. Fabre took marked mason bees in a box for a few kilometres from their nest and noted the direction in which they disappeared. Although 20 out of about 40 set off in the right direction, less than half eventually found their way home and the fate of the rest was unknown. Fabre, obviously an optimist, inferred an inner sense of direction! Darwin suggested that they used the Earth's magnetic field, but experiments on the bees with little magnets failed.

In a beautiful experiment, George Romanes (1885), who was a physiologist with a critical mind, took a hive of bees from far away to a house that he had rented for his family near a bare stretch of sand on the Cromarty Firth in Scotland. The hive was opened and the bees explored the garden and surrounding land, and all returned to the hive in the evening. The hive was then closed and the next day it was taken at most 250 metres across a stretch of empty sand where there were no landmarks, and again opened. This time every bee that emerged became lost. It was an excellent example of the power of an experiment in which the bees failed in a test. Romanes knew nothing about the sun compass of the bees. He concluded that the bees had no special sense of direction and they relied on local landmarks that they learned quickly. Modern critics will conclude that the bees did not compensate for the wind blowing over the featureless sand and the direction of the sun is little use if they have no reference point.

In the next decade, Albrechte Bethe (father of Hans Bethe), an indefatigable experimentalist, studied how bees found the entrance of their hive. He moved a hive sideways and put another in its place. The bees went to the introduced hive but soon came out again, then flew in again and out again, over and over. A few bees, however, started a procession that landed at the entrance of the new hive then walked home. When their hive was moved back by a short distance, the bees hovered in the air at the place where the entrance had been and eventually found it.

When their hive was moved back by 2m, an enormous swarm of returning bees congregated in the air at the place where the entrance would have been. When the hive was replaced, the whole swarm precipitated themselves into it. Bethe concluded that the bees were not guided by the sense of smell or hearing and that they could not recognise the hive by sight. To show that the bees were not guided by vision, he left large pieces of coloured cloth and paper near the hive or covering it, then changed the colour, only to find that the bees could still find their way. He went so far as to cut down a large tree that was close to the hive, and which the bees had to fly over, but with little effect on their homing ability.

Bethe then carried some marked bees from the hive into a meadow on one side of his house and some others into the narrow streets of the town of Strasbourg on the other side. Most of them headed in the correct direction when released and returned to the hive from places up to 3km away. From all this, Bethe concluded that *a totally unknown force* guided the bees for distances up to 3km. This was the period when x-rays, radio waves and radioactivity were discovered.

The age-old way to find a hive of wild bees is to catch a wild bee, feed it with honey until it is full and note the compass direction of the 'beeline' it takes when released. Then, at another place, release another satiated bee. The hive will be found near the place where the two tracks cross. Bethe explained the beeline by the directing power of his new force. He was stumped, however, when he took some bees in a box and released them far away, finding that they flew high into the air and returned to the box from which they had been released. Of course, these bees had been moved to a place with no familiar landmarks and started to make exploratory flights.

At the time, with almost no data, there was an argument between the empiricists and the intuitionists as to whether bees were automata or rational (von Butel-Repen 1900). The ability to learn and be adaptable made them look rational. It was argued that if they were not rational, bees would never adjust their daily tasks to the changes of the natural world about them. Because bees that are moved to a new place do not take a beeline home, von Butel-Repen 'consigns the hypothesis of Bethe's unknown and mysterious force absolutely and irrevocably to the realm of absurdities' (Forel 1908:259). Forel (1908:229) was also very short with Bethe's hypothetical force: 'I am obliged to combat his conclusions as preconceived, one-sided, and of an absolutism quite contrary to the scientific spirit.' Forel concluded that the bees familiarised themselves quickly with the appearance and direction of landmarks, both distant and nearby. At that time, researchers were familiar with the flights of exploration and saw the inexperienced bees point towards the hive as they flew in circles above it, as if to keep it as a reference point.

In 1905, the hive that had served for Forel's experiments in 1901 was moved further away from the garden table where the family often ate breakfast on warm summer days. Forel recounts at length how in previous years the bees had never come to the table, but that year they discovered the marmalade—a recent introduction to Switzerland brought by English mountaineers. Experimentally, the table or the sweet things were moved or placed at times under covers. Forel showed that when forced to be versatile in their search, the bees quickly learned to give up their rigid return to the place of the previous reward. It would be another 50 years, however, before the mechanism of navigation came into view, and a further 50 before there was some understanding of how honeybee vision would be sometimes limited and rigid and sometimes flexible and adaptable.

In the century that followed, arguments that had started with Fabre, a creationist, Romanes, a physiologist and intuitionist, and Bethe, a physician and mechanist, would continue in France, Germany, the United States, Switzerland and eventually Australia. It is of interest that the arguments were not resolved earlier. It is amazing that so little was noticed by so many interested parties who had been doing experiments for a century or so. These men were not stupid! There were many anomalies they could have noticed, many simple experiments they might have tried, but, no, they repeated the partial explanation they had. The reason for the delay was not entirely the pig-headedness of the professors.

With hindsight, we now know that new physical principles remained to be discovered and the researchers were misled by the bees' ability to switch between different mechanisms. Nobody could imagine the answers. Lack of imagination and observation were not the only limiting factors.

Figure 8.1 Navigational options available to bees. a) When the path is shifted sideways, the sun compass retains the direction. b) If the reward is shifted, the bees search in the place indicated by the landmarks. c) They search for the reward after going the expected distance. d) They compensate after being deflected by an obstruction. e) They search for the reward at the place where the landmark has the expected angular size. f) The beeline.

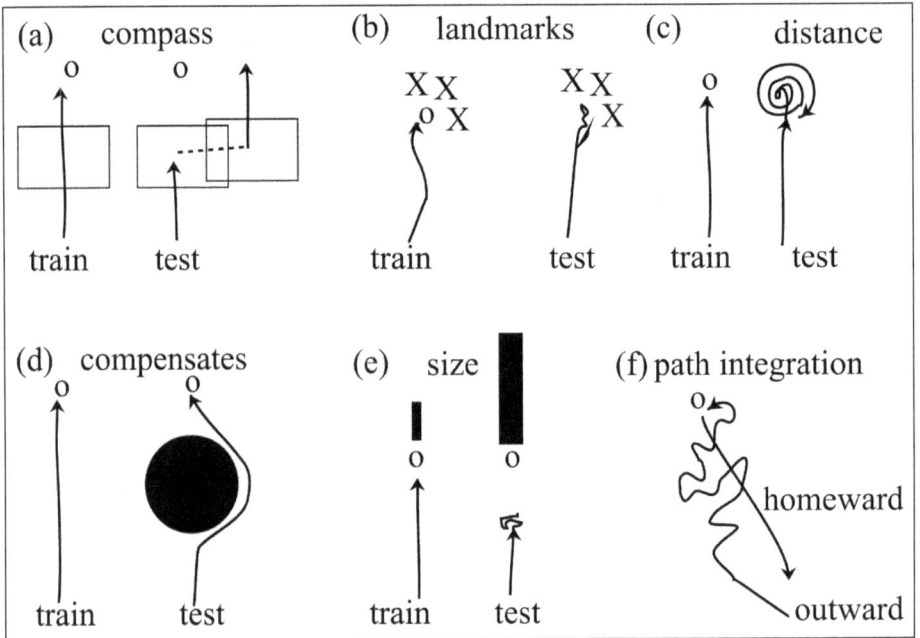

Dead reckoning

By the beginning of the twentieth century, it was known that ants used landmarks, made directional odour trails and had at least one sense of direction. In a classical experiment, Pièron (1904) allowed ants of the genus *Messor* on their way to the nest to walk on to a piece of paper. With the ants on it unaware of the move, he then moved the paper sideways. The ants continued along the former direction (Figure 8.1a). When they had walked for the distance that had previously brought them to their nest site, they searched for the entrance (Figure 8.1d). They clearly knew the remaining distance to their home by dead reckoning and ignored landmarks, but their cue for the direction was unknown. A different ability—to repeat a set of manoeuvres by going through the same sequence of turns as previously—was called the 'muscle sense' and was distinguished from dead reckoning.

Many arthropods are known to keep an internal record of their path as they move about. As a part of their repertoire, as they progress, bees and ants integrate every turn and distance in their outward path so they have a continual measure of the direction and distance of home. To some extent, cumulative errors are avoided by turning alternately left and right.

To fly directly home along a novel track—the 'beeline'—bees require landmarks, especially if there is a wind. Experienced bees and desert ants set out from the nest with an internal representation of the direction and distance to their goal (Figure 8.1f). As they go along, they check their direction by the sun and landmarks and measure the distance covered over the ground visually. As early as 1872, Lubbock (1893) was intrigued by the ability of a wasp to head directly to its nest after entering his room repeatedly via another way. Bethe (1898) used the beeline as support for his mysterious force that attracted the bee to the hive. Without considering dead reckoning, many have claimed that such action proves that bees have an internal map.

Having deviated around a new obstacle, desert ants and bees take the new direction directly towards the nest (Figure 8.1c). Experienced bees integrate the turns and motions of their own path and at all times know the direction of the hive or the goal. Their direction can be changed by an internal rotation of 180° to reverse the path either to food or to home. While returning home, ants that use vision remember the track for use on the next trip. Bees store vector memories that can be activated by the odour of the corresponding food source. They continually update the track direction from the sky compass and learn the relations between their path and the landmarks that are useful to them, particularly at the ends of the path. Under an overcast sky, desert ants (*Cataglyphis*) use dead reckoning along the ground, but experienced honeybees switch to the exclusive use of landmarks. If obvious landmarks are moved while the bees are in the hive, so that at the next flight they go in the wrong direction under an overcast sky, they follow the landmarks and return home by a direct route, but they have inferred the wrong direction of the sun from the displaced landmarks so they dance as if the sun had been moved and their recruits are misled.

von Frisch was in error in assuming that bees measured distance by the energy used on the outward journey. His bees flying uphill or carrying small burdens reported longer distances because they flew lower and registered more optic flow, not because they were working harder. The experiments of Harald Esch and Srinivasan in the 1990s showed that honeybees measured the distance travelled over the ground by integration of the optic flow, and those of Wehner and his colleagues showed that desert ants used the number of strides as well as the visual flow field.

Angular dead reckoning also plays a part under overcast skies and under forest canopy. The angular component in dead reckoning can be measured from the rotation of contrasts in the surrounding panorama, irrespective of landmarks or of the compass. The fly *Drosophila* flies in shade and is not known to learn landmarks. When Martin Heisenberg placed one in a uniformly striped drum, it learned to face one way when the light was blue and at right angles when the light was green. The motivation to learn was provided by the temperature, which was controlled by the orientation of the fly. The regular stripes allowed the fly to measure the angles through which it turned but they provided no fixed landmarks. When the colour changed, the fly turned through the appropriate angle relative to the pattern. The fly acted as though it integrated its angular velocity relative to the drum and at all times kept a memory of its direction relative to its visual surroundings.

An important part of angular dead reckoning is the memory of the retinotopic position of an outstanding contrast or landmark on the eye, so that if disturbed, the animal can turn itself until the landmarks return to the same position as before. When this performance was analysed in the crab *Carcinus*, it was found that the positions of edges and areas were detected separately, and that vertical black/white edges were not necessarily distinguished from white/black edges (Horridge 1966a). This behaviour is a simple form of the visual recognition of a place by ants and bees by comparison with a memory of the retinal positions of two or more cues (Figure 8.1e). When dropped on a beach, a crab detects the direction of the movement of the sun within 10 seconds. Burrowing crabs show that they are aware of the direction of their burrow at all times when they are out of it.

The motivational state of the bee

Researchers were often puzzled by conflicting results before it was realised that bees must be in the appropriate motivational state for study. Primarily, they must be forager bees and known to be experienced or otherwise. Not surprisingly, when bees and visual ants that are on the move are displaced and released, they continue in the same compass direction as before. Those caught at the nest entrance when setting out are motivated to take their accustomed outward route. Conversely, when caught fully satiated at the foraging place, they are motivated to head homewards. Only when they are captured as they arrive home do they have no preferred direction.

Bees are motivated to learn only when they meet an unexpected difficulty or when presented with alternatives, one of which has the expected reward. Most learning involves active participation, but bees fixed in the opening of a small tube will learn passively in a single trial that one odour but not another is associated with a reward. Bees in flight presented with a choice hover and look first one way then the other. If at first they fail to find the reward when arriving at the expected place, they will hunt about for it (Figures 8.1b and 8.1c).

Landmarks

Landmarks are the most important guides for bees and wasps to find their foraging place and return to their nest. We can understand why this is so because they work on overcast days when no sun or blue sky is visible, and dead reckoning is not much use in a strong wind.

Local knowledge is acquired visually when bees explore an unfamiliar place. When a hive of bees is taken to an unfamiliar site, the emerging bees explore in the immediate vicinity. If individual exploring bees are removed from the hive during this process, their ability to return depends on the number of hours that they have previously explored. As days pass, their familiar area expands. In a short textbook, Rabaud (1928) described the use of landmarks and of the position of the sun by visual ants, but made no mention of the dance of the honeybee—probably because von Frisch had described the round dance as a mechanism for alerting the recruits to look for nearby food sources by their odour. Rabaud quotes results from Romanes, von Butel-Repen and Yung—all of whom conclude that bees learn the visual appearance of landmarks and improve their memory in successive journeys. Many observers had noticed that shifting a single prominent landmark might have no effect and the intuitive conclusion was that the insects memorised the general layout of conspicuous objects around the goal, particularly those on the skyline, and made the best visual match that they could. In fact, they recognise much less than this.

Baerends (1941) obtained new insights from intensive studies of individually marked female digger wasps (*Ammophila*) that carried caterpillars back to their nests. A female could have four to six nests at the same time, with an egg in each, so that she was obliged to visit them in turn. The wasp was familiar with local landmarks over quite a large territory, and when holding a caterpillar and carried in a box to another place, she had no difficulty taking the direct route to whichever of the nests was the former goal. It was 60 years before a similar ability was also accepted in the honeybee to remember the routes to several foraging places, and even then only after long arguments and the introduction of new radar-responder technology that recorded the tracks of individual bees in flight.

In recent decades, researchers have discovered several ways that bees use landmarks in the field. First, the bees learn only the tracks they need, not the whole surrounding district, unless forced to do so. If satiated bees are removed from a reward to the north of a hive and carried an equal distance to the south of the hive, they take a long time to get home or they become lost. On the other hand, bees can be trained to go to a reward that is 10m from the hive in a direction that is changed every 10 minutes. Then, when released from a box, they return to the hive quickly from any direction. If the hive is moved, however, they lose it. The performance of a bee and the number of available routes available to it therefore depend strongly on what has actually been learned, making the memory of the landscape an elastic concept.

There is abundant evidence, from artificial landmarks in tunnels, mazes, tents or open featureless fields, that every familiar landmark tells the bee which way to go and how far to the next landmark. As found by von Frisch (1965), the memories of landmarks include directional vectors, as shown by the directional dances on overcast days. The track is therefore a chain of measured vector stages between landmarks, as well as a total vector and distance conveyed in the dance. In addition, the angular dead reckoning might help.

Bees are extremely sensitive to a whiff of the odour with which they mark a food source and to small differences in the distribution of light, polarisation and colour. Also, when forced, they learn to use remarkably small landmarks— down to a single small black spot. When the sky compass and all landmarks are removed, however, there is still some behaviour that can be attributed to the direction of the Earth's magnetic field (Lindauer and Martin 1968).

Signs along the route

Bees flying towards their goal don't only look ahead. Miriam Lehrer demonstrated that bees located themselves vertically relative to markers in their lateral vision when they were presented with a spatially complicated set of choices ahead of them. The bees were trained to look for a reward in one of 89 holes in a round target (Figure 8.2a). When presented with this target alone, the holes are in front of the eye, but the bees are unable to remember which one to enter and explore them at random. They have machinery to locate the centre of contrast in a local region of the eye (Figure 9.19) but not for memory of spatial layout of round holes. When given a single stripe at the side of the target as a marker, however, they locate the correct level and on this level they enter the hole at the correct range from the stripe. They use the map coordinates, not the map. They perform horizontally and vertically better with a stripe at each side. When the indicator stripe is moved in a test, the bees return to the hole that is indicated by the new stripe. In locating themselves relative to a stripe, the bees measure range from relative motion, not from apparent stripe size, and transfer the information to both eyes. Experiments with coloured stripes reveal that this use of markers in the peripheral field is colourblind and is done by the green channel of receptors, like vision of motion. These experiments are important evidence to show that the eye does not fix the spatial layout of the target into memory. Instead, lateral regions of the eye detect cues of range and direction of neighbouring landmarks. They triangulate but do not read the image.

Figure 8.2 The position and range of a landmark at the side direct the bee to the correct hole among many. a) Front view of the target; one hole leads to the reward. b) The bar at the side. c) The whole apparatus can be rotated to test the bees with a clean target and a different position of the bar.

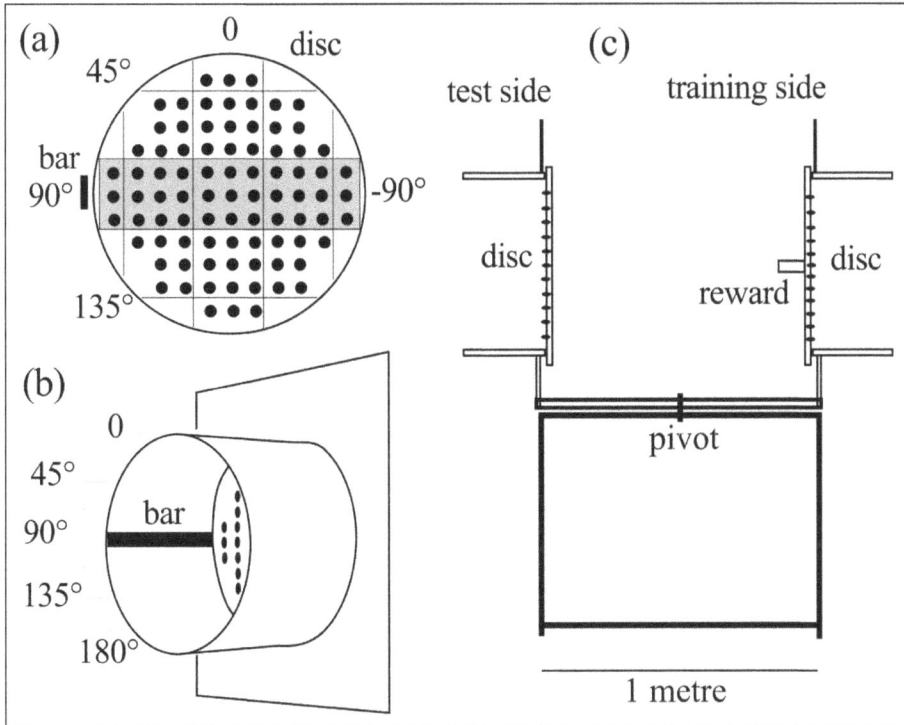

When bees are using the Y-choice apparatus, it is obvious that the narrow entrance hole excludes naive recruits, and even trained bees will not fly through a familiar hole that has been reduced in size. In 1995, I trained bees to pass between two black bars at each side before they passed the baffle (Figure 8.3a). The bars pivoted about their centre. Bees were trained with all bars vertical in one arm of the apparatus but inclined at an angle in the other arm. Left and right sides of the apparatus were interchanged every 10 minutes and there was no other cue. The bees quickly learned to fly between the bars with the orientation that was rewarded and could detect a difference of 15°.

In the next experiment, with the baffles in the normal position, bees were trained to discriminate between a vertical and a horizontal coarse grating (Figure 8.4b), and then tested with the grating replaced by a vertical and a horizontal black bar on each side of the holes in the baffles (Figure 8.4c). Although the task looks simple to us, the trained bees are unable to use the orientation cues of the bars, because they are in an unexpected place. When the bees are trained to fly between the oriented bars alone, however, they perform very well (Figure 8.4d).

In fact, they prefer to learn the bars at the sides rather than the gratings in front, as shown by training them with both in place (Figure 8.4e) and testing them with the gratings alone (Figure 8.4f).

Figure 8.3 Detection of an orientation at the side. a) The Y-choice apparatus modified with bars at the side. b) When trained to fly between bars at an angle of 30° to the vertical, the bees respond to an inclination down to 15°.

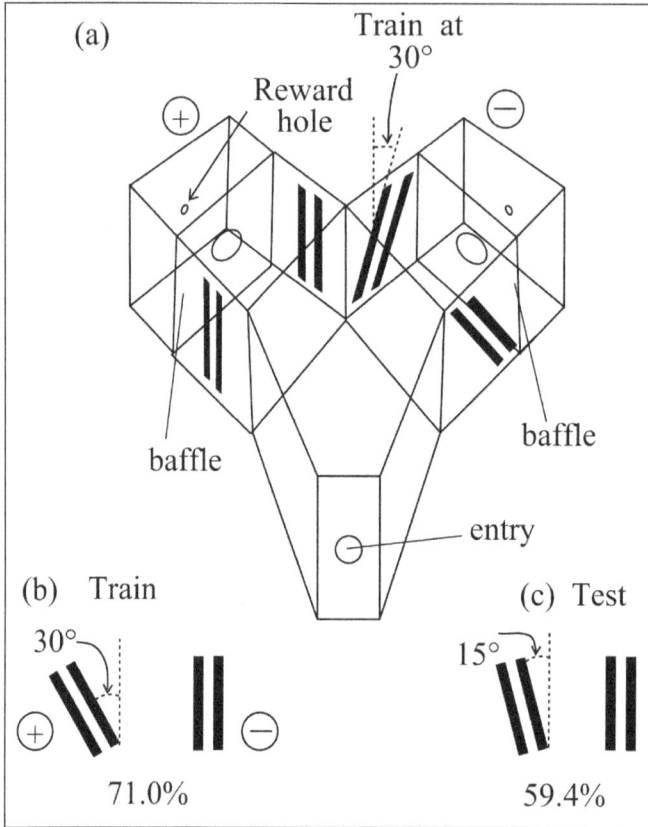

From Horridge (1996b)

Figure 8.4 Orientation cues in front and at the side. a) The Y-choice apparatus modified with bars on the baffles. b) The bees discriminate the gratings very well. c) Bees trained on the gratings fail when tested with the bars. d) They discriminate very well when trained on the bars alone or (e) with both bars and targets together. f) Bees trained on the bars plus the gratings, as in (e), do not recognise the gratings alone.

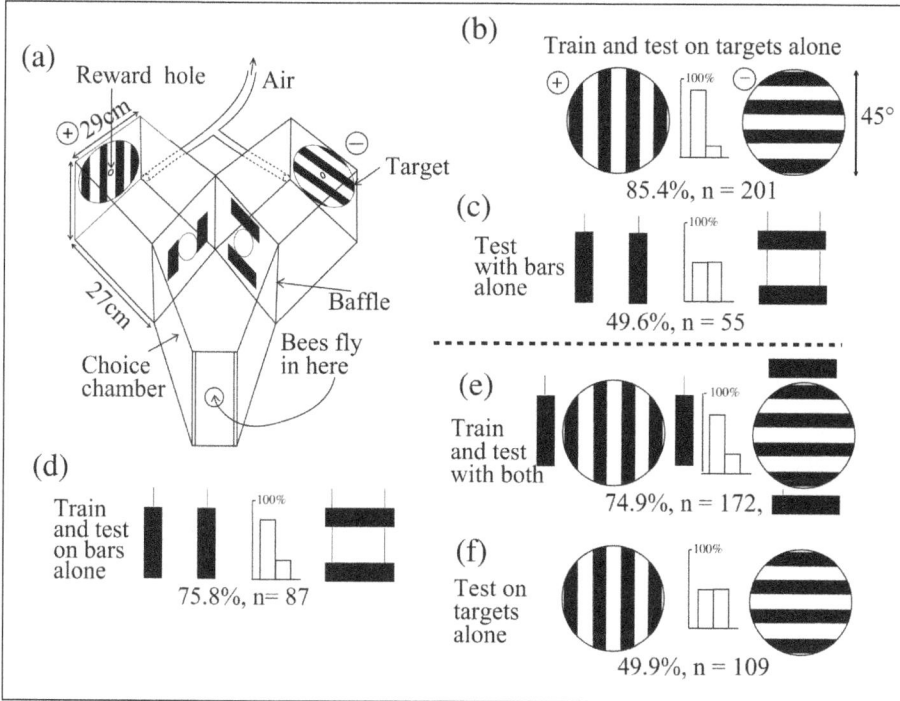

From Horridge (1996b)

Maze learning by bees

Ants find their way through a maze by laying down directional odour trails that carry a signal like an arrow on a road. Walking bees will learn the correct route to get a reward and then walk back for more. After all, bees have to negotiate through a nest in a hollow tree and sometimes fly through a dense wood. A maze was at first just an apparatus to reveal the performance and was only recently used to find out where the bees looked, to analyse what cues they recognized and how long they remembered them. Flying bees can use an odour or a visual cue to tell them which way to turn at a choice point in a maze (Weiss 1953). When the cues were identified in the 1990s, Zhang Shaowu started an investigation of maze flying by bees using boxes that could be arranged side by side on the floor, with holes communicating between neighbours. A maze is constructed by blocking some of the holes so there is only one correct way through (Figure 8.5a). The boxes can be replaced in tests to prevent the bees using odour cues. Bees learn to fly though holes marked by a cue. The cue can be an instruction to turn right or left at the next choice or alternative cues can be displayed on the alternative holes. At the end, the bees escape by a back door.

Figure 8.5 Mazes. a) A maze made from square boxes; the dashed lines are deadends. b) A zigzag maze made from cylinders. c) A turn-left maze. d) A maze for successive alternatives. e) The series of pairs of cues for the successive choices in (d).

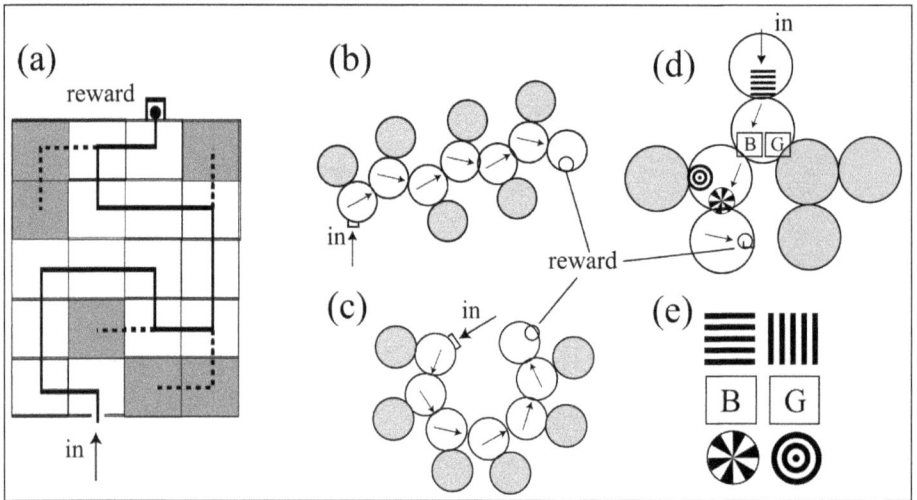

In the maze, bees can learn by trial and error to follow a trail marked by one cue at each choice point and then use the same cue through a different maze. Bees trained to use one colour through a particular maze can switch to another colour because they have learned 'any colour' not 'this colour'. They can also use an unfamiliar colour to negotiate an unfamiliar maze. If the maze has a fixed route, the bees soon learn to negotiate it correctly although the cues are removed. The cue can be indirect—for example, the bee can learn to turn right when the back wall is green and left when the back wall is blue, and bees trained to do this can thread novel mazes guided by the same colour cues. They can also learn to choose to turn left and right alternately (Figure 8.5b) or always to the right (or left, as in Figure 8.5c). Bees can also learn an unmarked maze, but they learn more slowly.

Maze learning by bees shows that the cue is an instruction to turn in a given direction at that place and then go a certain distance, and they can learn a sequence of several choices. There is no reason to suppose that this ability is restricted to small mazes. The experiments in which bees take a definite track through space show that angular orientation also follows a learned motor sequence—called the muscle sense—as demonstrated by the persistence of turns in the track when obstacles are removed.

In the next series, the bees learn to turn left or right according to whether the back wall of the box is blue or yellow. Next, one colour is placed above the entrance to a choice chamber, so that the sight of it precedes the decision, and the bees remember what they have just seen, with similar results. In other experiments, the colour cue is placed on one side of a narrow tunnel through which the bees must walk—yellow to turn one way and blue to turn the other. The trained bees are tested with the cues on the opposite wall of the tunnel and it is shown that in this situation the bees can transfer a colour cue from one side to the other (Zhang et al. 1998). In these and similar experiments, the choice can be learned according to the time of day.

The experiments became more sophisticated. A sample cue, A or B, was exhibited outside the entrance to the choice chamber. The bees must look at the cue and then inside the choice chamber they must choose the hole with the same cue, although the sample outside is changed randomly between A and B. (In other experiments, other bees must not choose the same cue.) Having learned this, the trained bees are given a sample of quite a different cue outside the choice chamber, such as a horizontal or a vertical grating, and inside they must choose the same cue (or not the same cue, if they have learned to avoid the cue).

Finally, one of two cues—for example, horizontal or vertical bars—is displayed over the first hole as an instruction for how to make the next choice—for example, between blue and green (Figure 8.5d). The correct choice—in this case, blue—then instructs the bees how to make the next choice, between circles and sectors. The correct choices are in the left column in Figure 8.5e. The trained bees are able to use this series of cues starting at any point in the sequence.

In all these experiments, the mazes were fixed and the bees were successful, so it was difficult to say what the bees in fact detected and remembered because they were not tested. It was all performance, no analysis. The bees detect a cue and act on it, and a delay is essential to correlate the cue with the reward. The bee learns the minimal cue. For example, it learns that there is a colour, not *the* colour, just as, in experiments with patterns, it learns less than the whole pattern. The bee learns whether to follow or avoid the sign on the door, sometimes irrespective of what the sign is.

There is no case for inferring that the bees learned a 'concept' of sameness or difference because they were not tested to see what they had in fact learned. In other examples where the performance looks remarkable at first sight, analysis reveals that the bees have learned a simple cue that is just adequate for the next choice—not a general solution that suggests some kind of insight. The descriptions of the performance in mazes scarcely began to analyse the feature detectors and cues or what the bees really detected (Chapter 9).

The celestial compass

At the beginning of the twentieth century, workers on ants described the use of the sun, as well as landmarks, to provide a reference direction. Ants with large eyes can be expected to use landmarks, the sun compass, dead reckoning and odour trails, but the majority have not been investigated. Felix Santschi (1872–1940), an Italian physician, described several examples between 1900 and 1920, when he changed the perceived position of the sun with a mirror and showed that the tracks were deflected accordingly. The best example, however, was the desert ant *Cataglyphis*. Santschi showed that when placed at the bottom of a featureless drum, this ant could use a small patch of the blue sky to take the correct direct route homewards. A ground-glass screen spoiled the performance, but Santschi did not know why; it was the polarisation of the sky. Extensive analysis of this splendid animal in recent times by Wehner and his group at Zürich has shown that its navigation mechanism is very similar to that of the honeybee.

Aristotle, and presumably many beekeepers, thought that the dancing forager bee led the recruits back to the food. In 1923, von Frisch already knew, from a famous little book by M. Maeterlinck, that this was not true. He described the round dance (Figure 8.6a) as a signal for nectar and showed that returning foragers excited recruits to go out and search nearby for food with the same odour. He missed the sun compass, thought that the figure-of-eight dance signified pollen and turned to other topics. After 20 years, while living in seclusion from the Nazis in his country house beside the Wolfgangsee in Austria, he turned again to the study of the dance. For some years, he believed that the recruited workers were guided by the odour that the returned foragers carried from their foraging. After numerous experiments, he discovered the relation between the figure-of-eight dance and the direction and distance of the food source relative to the direction of the sun (Figure 8.6). When the direction of the sun was used as a compass, to find the direction on the Earth's surface, the dancing bees made an adjustment according to the time of day. They recognise the sun as a bright light free from UV polarisation.

After returning for more food at least once, the forager dances on the vertical comb in the hive or on a flat surface outside the hive, in either a circle or a figure-of-eight double loop with the straight piece in the centre giving the direction. On a horizontal surface, the direction of the central run in the figure of eight points in the direction of the food source. On a vertical surface, the angle between the central run and the vertical is the angle between the direction of

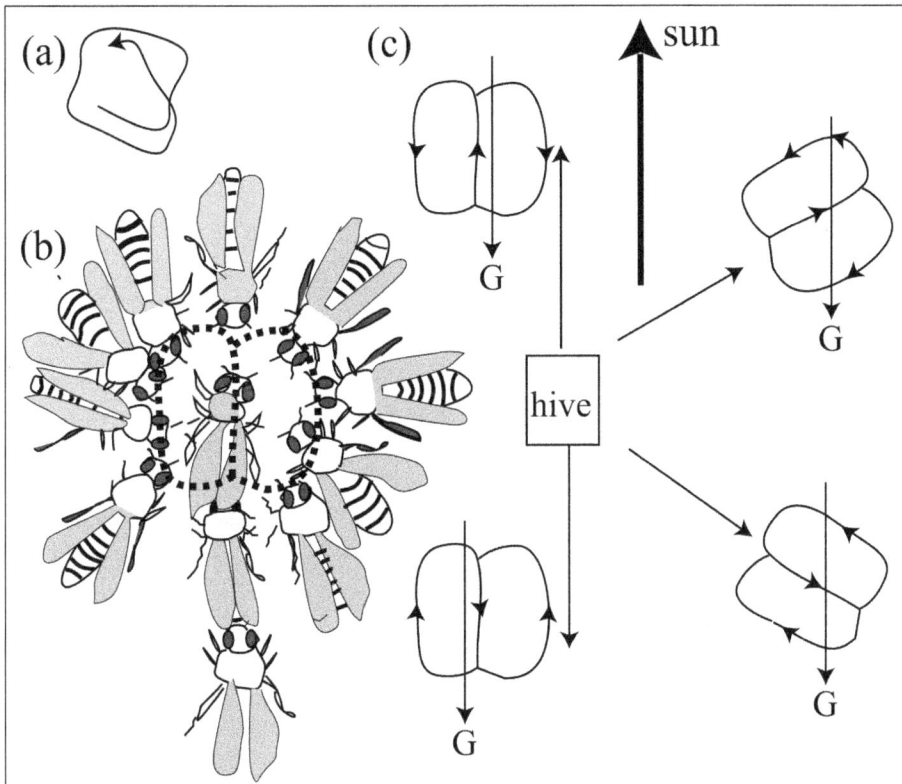

the food and that of the sun (Figure 8.6). The bees detect the direction of gravity by hair sensillae in the neck joint. They detect the vibration of the dancer by Johnston's organs at the base of the antennae.

Figure 8.6 The dance of the returning foragers. a) The round dance. b) Potential recruits stand close behind the dancer, while others stand around within the area of vibrations of the comb. c) The direction of the central bar in the figure of eight relative to gravity on the vertical surface indicates the direction of the food source relative to the direction of the sun.

The dancer indicates the direction to the final goal as learned on its homeward flight, together with the total length of the outward flight, and these two components can be influenced separately. For example, a segment of the outward flight along a narrow tunnel, which increases the optic flow for a given length of flight, increases the apparent distance to the goal, but has no effect on the direction of the homeward flight.

The distance measured on the outward track is the sum of the optic flow over the flight irrespective of the real track. The bee sees its surroundings passing by, integrates the angular velocity over the whole length of the outward flight and in the dance reports only the final total. This is exactly what the other bees need in order to follow for the right distance, as long as they take the same direction and have similar preferences for flight height and treatment of obstacles. Odours are not essential for success. Of course, if new to the track, recruits require the sun or blue sky to follow the vector in the dance and then might need an odour to discover the real place. Experienced bees can recall the whole track from the odour of the food on the dancer. If new to the track, newly recruited bees quickly link the landmark vectors and separation distances into their memory. Experienced bees can switch between the memory of the total vector and the expected landmarks, which explains why omission of a landmark might have no effect and why they can ignore landmarks and take the beeline home.

In 1996, David Sandeman showed that the figure-of-eight dance on the surface of the comb was not a run in which other bees followed, as described in most accounts (Tautz et al. 1996). The dancer faces in the appropriate direction but in fact takes only one step while waggling the abdomen from side to side. The further away the food source, the longer the waggles last. With a source 1200m away, this step takes 1.2–1.8s to cover 8mm. The duration of the waggles and the direction of the food are conveyed by vibrations through the comb and the legs of the follower bees, not by following the dancer. In other situations, however, longer directional runs of a few centimetres are made over hard surfaces or the backs of other bees.

The calculation of the direction outside allows for the movement of the sun, which is in opposite directions in the northern and southern hemispheres, as migrants to Australia discover for themselves. When transported across the Equator by air, the experienced bees do not learn to compensate for the reversed direction and become lost for ever. The direction of compensation was learned once in these bees' lifetime, but they were not given a second chance. The hive survives because young bees learn the task afresh.

There are two components to navigation by the sun compass—one innate and one learned. Lindauer found that bees that had seen the sun only in the afternoons could immediately use the position of the sun if released in the morning, so they had an innate expectation of where to find it in relation to the time of day. Bees are genetically programmed to expect the sun to be at a constant position in the eastern sky in the morning and at another position in the western sky in the afternoon, although the azimuth position of the sun in fact rotates from east to west at an average of 15° per hour. Inexperienced bees, shut in a box during one of these periods, make predictable errors when released. About noon, the bees tend to come to training sites less frequently. Inexperienced foragers learn the motion of the sun and get a better measure of compass direction with the aid of landmarks, as indicated by the improved alignment of the axis of their dance. To exclude the sky compass, experiments must be done indoors or under a heavily overcast sky—as was done by accident for a century or so—forcing the bees to use landmarks.

The recruits are influenced by the odour of the food source on the dancer. From earlier Russian work, von Frisch knew that, even without a dancer, bees could be induced to leave the hive and search in the correct places when the odour of familiar flowers was blown into the hive. While much of Europe was in ruins, von Frisch published quite a lot about this—and the Nazis probably considered his work useful for directing bees towards pollinating useful crops.

The bee wars

Soon after von Frisch's publication of the figure-of-eight dance, bitter controversy broke out when Adrian Wenner (1967) reported that bees were able to locate the food source when exposed only to its odour in the hive. As said, this was already in the non-English literature. Wenner found that the bees that followed the dance were unable to locate the food source without assistance from its odour. They were presumably inexperienced bees, while those working for von Frisch were better trained. Wenner found it impossible to publish further work until after the Nobel Prize was awarded to von Frisch, Lorenz and Tinbergen in December 1973. Wells and Wenner (1973) then correctly pointed out that von Frisch's original claim—that the dance alone was sufficient to convey the direction and distance—was not controlled against memories induced by odour cues.

This little 'bee war' aroused unexpected indignation, but not on account of the merit or otherwise of the experiments, which all had typical weakness of design, confusion arising from multiple causes, ignorance of the previous experience of bees and intuitive inferences with no confirmatory tests. Wenner was rightly indignant that his publications were blocked (how or where has not been revealed), but he should have read the German literature. No doubt, von Frisch was indignant that his authority was questioned and he saw nothing new in Wenner's claims. After all, von Frisch had made a new observation; Wenner had not.

More of a colonial skirmish than a war, this difference of opinion was at the time not seen as another incident in the long defeat of Kantian intuition by the requirement for empirical proof. John Stuart Mill was not mentioned. To observe the performance and infer that the mechanism was inadequate; experiments were required. James L. Gould took the matter up. He arranged for the dancers to display the direction in the dance according to the gravity stimulus, while the followers interpreted the direction according to the celestial stimulus and were therefore fooled into searching in the wrong direction. This showed that the dance alone could be an effective directional signal as well as a message to leave the hive. After that, Gould lost interest in the odour stimulus, but the controversy was kept going and served a useful function as a warning that proper controls were advisable.

In his later polemics, Wenner produced no new data to show that the dance was ineffective, only arguments that it had not been proved to be effective, except as an urge to make the bees go and search. When work on the measurement of range, optic flow, bee navigation and piloting began in Canberra about 1988 and again in 1995, I was astonished that the data that had been argued about were so incomplete. One of my former students, Professor David Sandeman, who worked mainly on Crustacea, by chance studied the bee dance and found that it was not a walk or a dance, but just one step with a waggle (Tautz et al. 1996). After 1995, the correlation of the number of waggles in the dance with the bee's odometer—as tested in tunnels, over water or in the open—produced a wealth of data that might have been gathered earlier with less talk and more business. One millisecond of waggle encoded about 18° of image motion on the eye—leaving Wenner without a leg to stand on. As further tests showed, the bees could use dance, odour, landmarks or the sun compass, depending on the situation (Vladusich et al. 2005, 2006).

The signposts in the blue sky

In August 1869, at the top of the Aletschhorn in Switzerland, the English scientist J. Tyndall scanned the sky with his Nicol prism and found that the direction of maximum polarisation of light from the blue sky was always perpendicular to the direction of the sun (Figure 8.7d). This observation was explained by elongated and horizontally floating dust particles that scattered the light preferentially, and was eventually published in *The Forum* for February 1888—such was the speed of scientific advance in those days. Scattering is inversely proportional to the fourth power of wavelength. At large angles to the sun, the sky provides sufficient ultraviolet to be useful for detecting small dark objects against the sky with the improved lens resolution of the bee ommatidium that the shorter wavelength allows.

Unaware of the earlier work by Santschi on ants, von Frisch discovered that a small patch of blue sky light was sufficient to direct the orientation of the figure-of-eight dance. A physicist colleague in the faculty at the University of Graz, Han Benndorf, advised von Frisch to consider the polarisation pattern of the sky and, in 1945, the observation was published. Knowing the time of day and the direction of polarisation, the bees had another compass.

The mechanism within the eye took more trouble to unravel. In each ommatidium of cockroaches, butterflies, dragonflies and the honeybee, there are one or two retinula cells with sensitivity peak in the ultraviolet. These cells are probably responsible for UV-specific behaviour, such as the escape response towards the open sky. In the drone bee, which pursues the virgin queen against the background of the sky, and dragonflies that catch flying prey from below, the dorsal part of the eye is predominantly UV sensitive.

Figure 8.7 The solar compass and its detector. a) The position of the polarisation-sensitive (POL) area. b) The pattern of the planes of the orthogonal pairs of UV receptors in the dorsal rims of the bees' eyes. c) The angular sensitivity curve of the single receptor has a wide skirt that integrates over an extended area of sky. d) The pattern of the e-vector of polarisation of the blue sky. e) The orientations of the detectors that were inferred from the behavioural experiments.

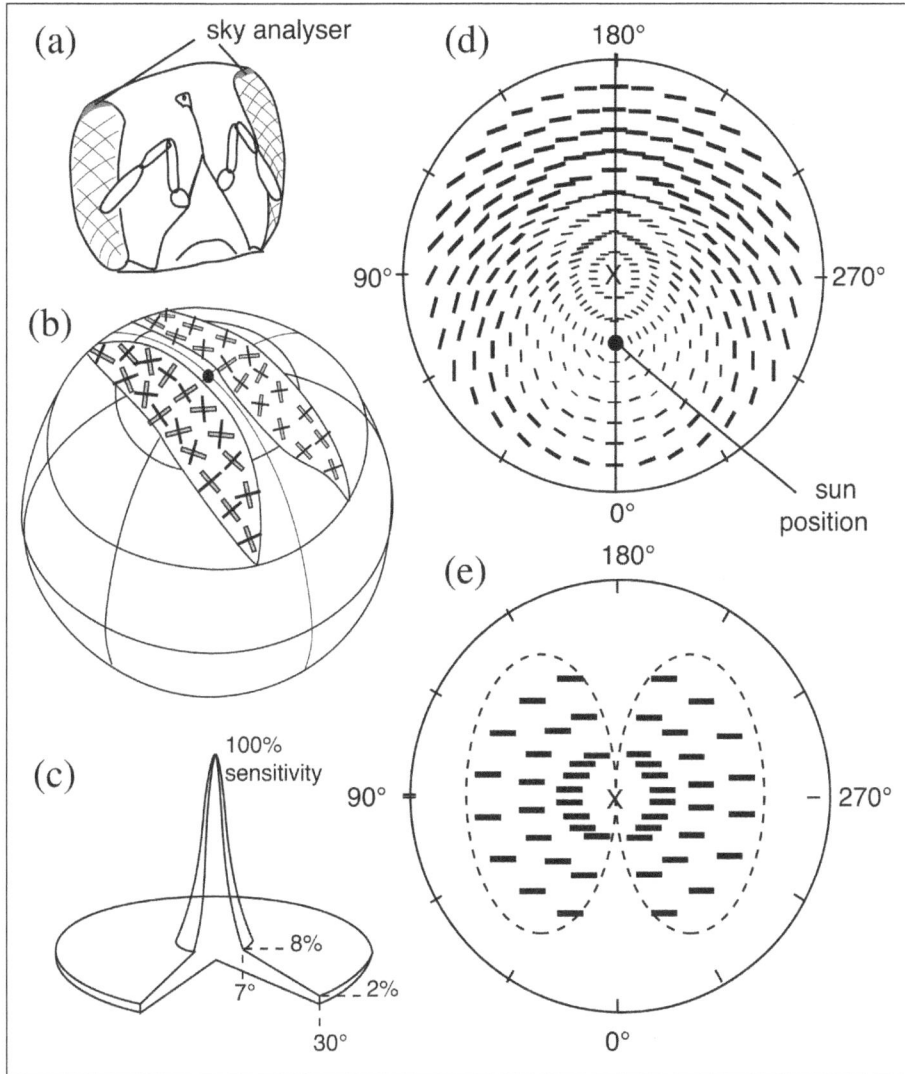

Scientific theories often appear in imperfect form and are refined. From Santschi (1911) onwards, knowledge of the mechanism of the bee eye for detecting the polarisation plane was a real dog's breakfast. A selective scattering or reflection to convert the polarisation pattern into an intensity pattern *outside* the eye was suggested in the late 1950s, but none could be found. Other mechanisms *inside* the eye—for example, a Nicol prism in the cornea or cone—were suggested in the early 1960s. Later, one such was in fact found in some fossil trilobites. These theories were forgotten when the intrinsic properties of the visual pigment rhodopsin were discovered. By illumination of vertebrate rods and cones from the side in tissue slices, it became clear that molecules of rhodopsin absorbed more light polarised in one direction than in the perpendicular direction—that is, the molecules were dichroic and were lined up (as in Figure 5.3). In the 1960s, electron microscopy revealed the structure of insect rhabdomeres as organelles made of parallel microvilli. It was generally accepted that the rhodopsin molecules were oriented in the plane of their lipid membranes and it was assumed that the whole compound eye detected the plane of polarisation by having different retinula cells with microvilli (and therefore rhodopsin molecules) oriented in different directions and repeated in other ommatidia. To mimic this, von Frisch had models made with a hexagonal mosaic of polaroid so that when held up to the sky they showed different intensities in each piece of the mosaic. They were even proposed for military use, to direct missiles.

There was much discussion about how the responses of the retinula cells could be integrated. There was further discussion about how the rhodopsin molecules in the microvilli could be oriented to have more than double the sensitivity in the best direction than in the worst direction, because the value of only 2 is the maximum to be expected from rhodopsin molecules that lie randomly in the plane of the membranes. Later, the basal (ninth) retinula cell of the bee ommatidium was proposed as the sensor because it was sensitive to ultraviolet and had a high sensitivity to the plane of polarisation (explained by filtering by cells above it). There was a lot of discussion about twist in the bee rhabdom, which, if true, could abolish the sensitivity to the plane of polarisation. There were experiments with discriminations of patterns composed of polarisation directions. Direct recordings showed that all bee retinula cells had some polarisation sensitivity, but the ninth (basal) cells remained mysterious. There was great excitement in the 1970s as these discoveries were worked into a mechanism for navigation.

The efforts were not entirely rubbish. The data were valid, but the conclusions drawn at the time were led astray by the historical context. The explanations were based on known components, but ignored the possibility of a completely different explanation, which arrived too late to be in the excellent book by James L. Gould (1982).

The dorsal rim receptors

Many insect groups have a specialised region along the dorsal edge of the compound eye (Figure 8.7a) where the microvilli in the rhabdoms are aligned in cells with poor optics, large fields, blue or UV sensitivity and high polarisation sensitivity—first properly described in the cricket. During the 1980s, it became apparent that in the bee the ommatidia along the dorsal rim of the eye have poor spatial resolution but are very sensitive to the plane of polarisation in the ultraviolet. The critical experiment showed that bees could not dance or navigate correctly when only this dorsal band of ommatidia was covered—a splendid example of evidence of absence. These special receptors sensitive to the polarisation plane are used to determine the position of the sun when it is not visible.

The dorsal rim cells have angular sensitivity fields with a wide skirt around a central peak (Figure 8.7c). Single cells therefore integrate the polarisation over fields at least 45° wide. The ratio of absorption, and therefore sensitivity, in the plane of the e-vector to that at right angles (see Figure 5.3a) can be as great as 15. This shows that the rhodopsin molecules are lined up in the parallel microvilli. The receptors in the rest of the eye have a ratio less than 2. The 140 or so dorsal rim ommatidia of the honeybee (Figure 8.7b) look upwards, and each contains nine long, straight retinula cells.

Tom Labhart (1980) found that the axis of sensitivity to the polarisation plane had a special pattern in the dorsal band (Figure 8.7b). So, as Wehner, Rossel and colleagues in Zürich concluded during the 1980s, the dorsal rim cells acted as fixed-feature mini-detectors for an expected visual task. Two types of cell at each place had orthogonal axes. In tests with the polarisation pattern of the sky, the dorsal band acts as a functional unit and the detector axes appear to be parallel (Figure 8.7e). All the bee has to do is rotate itself until this sensitivity pattern of the dorsal rims of its eyes makes the best fit with the pattern in the sky, then it is facing the sun. The best fit would depend on the coincidences of a number of independent feature detectors, any of which can be omitted. Either eye would serve. In the locust, the signals from the dorsal rim cells reach the central body in the protocerebrum, where they are laid out as a close-knit topographic representation of the e-vector panorama in the sky (Heinze and Homberg 2007). Presumably the bee is similar.

Observation of bees suggests that they read the compass direction from the sky without turning themselves. To do this, there would have to be an internal analyser of neurons in the brain. Three such macro-analyser neurons have been found in the central body of the locust brain; they are binocular, with large fields. In all three, light polarised in one plane is inhibitory to light polarised in another, with peaks of maximum sensitivity 60° apart. The function is clearly not to see the polarisation pattern of the sky. The task is to show the direction to go, for which two other sets of data are required—the land coordinates from the distant landmarks and the expected position of the sun at the time of day— both of which are learned by each individual bee. It is likely, therefore, that individual bees also learn the polarisation patterns that are useful to them, from the coincidences in a distributed array of high-level neurons.

The visual estimation of distance flown

By an accident of the terrain, von Frisch (1965) placed a food source at the back of a projecting ridge of a mountain and found that the dances of the returning foragers indicated the direct direction to the food source, and the total distance flown around the detour on the way home. The bees following the dance fly off in the correct direction, meet the detour, which they circumvent, and find the food at the expected distance flown. Therefore, the information they received was just what was needed. These results were confirmed for other detours.

von Frisch noted that the bees reported a shorter distance when flying downhill than when flying uphill and for many years he believed that they measured the distance by the energy expended on the homeward flight—an intuitive inference that became fixed in textbooks. Harald Esch, who worked with von Frisch, never believed this proposal, but he had no opportunity to prove otherwise until long after he was obliged to move to the United States. While Srinivasan was first experimenting with bees flying along tunnels in 1995, Esch and Burns (1995) published the result of a now-famous experiment. They trained bees to fly to a food source on the ground. Then they raised it with a balloon. Later, they flew bees between the tops of tall buildings. The homeward distance reported in the dance was less as the bees flew higher above the ground, showing that the perception of distance was dependent on the scene. The bees measured the distance they perceived visually over the duration of the flight. They had a visual odometer, but clearly they did not indicate a fixed unit of distance for each waggle in the dance.

The next year, Srinivasan published the account of how bees, flying along a tunnel towards a food source, measured the distance to the position of the food source by integrating the visually perceived angular velocity of the walls and floor (see Chapter 7). The distance to the place where the bees searched was independent of the pattern on the walls and tests with a wind along the tunnel, or moving walls, showed that the bees summed the apparent angular velocity over the flight, not the total number of edges or time passed.

The next logical step was to calibrate the dance in terms of the perceived motion. In 1995, there was new enthusiasm among grant-giving bodies for collaboration between distant laboratories, so it was easy to bring experts on the waggle dance to Australia and send experts on bee tunnels to Europe, resulting in a flurry of detail published by Srinivasan, Collett, Esch, Zhang, Tautz, Vladusich and Lehrer and their colleagues at both antipodes. Outdoors, the bees gave about 1.6ms of waggle dance per metre travelled, and 1m of flight in a tunnel 30cm wide was equivalent to 25m outside. In both situations, one millisecond of waggle encoded about 18° of image motion on the eye.

Later, a tunnel provided a convenient way to add a large deviation at right angles to the path to a distant goal. In the dance, the bees that were deviated by a tunnel measured the total optic flow on the outward journey and the compass direction of the hive in a straight line from the food source. Bees would not fly through the tunnel on the return to the hive or switch between tunnels at right angles, showing the influence of their dead reckoning.

When bees fly slowly against the wind or faster with the wind, they still measure the impression of the distance travelled over the ground. As the contrast is reduced—for example, at dusk—the odometer continues to function normally down to contrasts of about 20 per cent of that in sunlight. When flying over water, the odometer registers less than the usual distance (Tautz et al. 2004).

When desert ants travel up and down hill, they remember only the horizontal component of the distance walked. Bees flying in tunnels measure and remember the total distance travelled, even in the vertical direction in a vertical tunnel. This is reminiscent of the earlier finding that flying bees measure the angular velocity of passing contrasts irrespective of their direction of motion. Some species of stingless bees direct the recruits to the correct height of a food source by scent marks.

The bees appear to reset their odometer at the transition at the entrance to a tunnel. A landmark placed in the tunnel before training improves the accuracy of measurement of distance in the tunnel, and when the landmark is moved, the search place moves with it. They reset their odometer at each landmark and the flight is divided into sections, each with identified beginning and length. The bees would not search beyond a landmark that was placed in the tunnel after training, just as they would turn away from a landmark that displayed an unfamiliar cue. Conversely, bees overshot the goal when a familiar landmark was removed. In a tunnel, the landmark positions overrule the visual odometer (Vladusich et al. 2005), which is what happens in their natural terrain, but paper tunnels are far more convenient for manipulation of the visual scene and for experiments with positions or numbers of landmarks or a controlled wind.

Practical route finding by foraging bees

Last century, Lubbock described how displaced wasps flew higher and higher until they recognised a distant large landmark then headed towards home. The pattern of the horizon is certainly important. The largest landmarks are preferred—even distant mountains. When there are two or three similar landmarks, bees learn how many to pass, but normally they judge distance from the perceived ground speed.

Bees that are recruited at the dance fly out from the hive using the sky compass, learning landmarks as they go. They can do this under an overcast sky by inferring the sun's position from known landmarks. At the distance indicated, they search for the food scent picked up from the dancing bee. They also look out for other bees to follow and they will land beside bees that have settled. The results from maze learning show that bees learn routes from very small cues, but Romanes' (1885) experiment shows that the sky compass and dead reckoning are not sufficient by themselves in a windy place.

Experienced bees in a busy visual environment use the sun compass if it is available and the shifting of a landmark under a blue sky does not disturb them. They can use one set of local landmarks at one location and another set somewhere else, and learn to visit each foraging area at the appropriate time of day. They can select an odour cue according to time of day or location, but different colour cues only according to location. When some of the local landmarks that mark a goal are displaced, the bees search in a spot that shows that they are placing the remaining landmarks in the expected directions relative to their own eye. There is strong evidence of assistance from the horizontal vector of the Earth's magnetic field. Similarly, they perform faster if they regularly fly along a particular path.

By definition, a landmark cannot be recognised by its position; it must be recognised uniquely by its shape, size or colour and then it indicates a position. Bees can fly towards and past an obvious isolated landmark that they have learned, then to another one and so on to the goal, then directly back home. Commonly, if they fail to find the reward, they return to the previous local landmark and make their approach again. Some landmarks are like beacons on which they rely. With these, if the landmark or the feeder is moved, the bees show that they have not learned the compass direction and, not finding the goal, they cast around until they find it. They then calculate a new compass direction towards home and remember the additional information needed to correct the error they made.

Each bee repeatedly flies along the same track and the landmarks fall on the same regions of the eye each time. On the other hand, the sun and its polarisation pattern move across the sky, so the bees must adjust for the time of day. A landmark is a local cue indicating which direction to fly relative to the sky compass at that place. Bees that are displaced to a new location under a clear sky continue in the direction they were going but quickly begin to search for familiar landmarks. Bees displaced under an overcast sky can recognise tall landmarks from unfamiliar directions, which is the only way they can head towards a food source or home. In these experiments, it is essential to distinguish between bees returning to the hive or food source and those setting out.

From comparisons of the turn-back-and-look behaviour with route finding, bees and wasps apparently take snapshots with the side of the eye and bounce from side to side of an imaginary corridor that they have committed to memory. Experienced bees also use the apparent size of the landmark, having first learned where it is located. As Collett showed, if landmarks were made larger, the insect searched further away from them; if local landmarks were moved, the target was sought at a place from which all landmarks had the best approximation to their expected size, appearance and directions (Figure 8.1b). Movement of local landmarks relative to distant ones confuses the insects. All of the evidence put together shows that the actively moving insect quickly discovers which way to go to bring individual landmarks into the desired arrangement as seen from the goal. Many insects apparently can do this while flying in circles. Perhaps a single saccade can put a snapshot of all surrounding landmarks into the 360° visual system. The landmarks change as the insect moves along the route home and natural changes in landmarks are tolerated. The mechanism is flexible because there are alternative cues (Figure 9.21).

Detailed study of what cues bees really use in order to recognise a landmark shows that they detect landmarks in the same way that they discriminate patterns in the Y-choice maze. They detect only a few simple features that are summed into cues and they recognise them in the places where they were in the training. The enormous number of observations can probably all be explained by the same few feature detectors and cues listed in Chapters 9 and 10—perhaps with a few other undescribed cues.

The orientation flight

The first flight of a young bee from the hive is an orientation flight in larger and larger loops up and away, returning within a few minutes. Older bees that emerge to an unfamiliar scene from a displaced hive make a new orientation flight. If carried away from the hive, most bees that have taken only one orientation flight return eventually, but bees that had no orientation flight always become lost. After the orientation flights, bees can distinguish some landmarks by colour, height of the centre, angular size, range, orientation and by their angular directions relative to each other, but not by compass direction from the point of choice. This implies that near the hive the landmarks do not carry an attached homeward direction towards the hive.

When a hive is moved, all the bees in it must make new orientation flights. Experienced bees learn the direction of the hive from any point in their area and, when the sky is clear, they progressively shift their reliance from landmarks to the integrated vector path as guided by the sun compass and distance over the ground. On overcast days, landmarks take precedence and if the landmarks are moved the bees are either fooled or they make new orientation flights. The orientation flight is a performance, not a mechanism. How their several navigation systems are coordinated has not been analysed.

Turn-back-and-look flights

Turner (1908) described the curious way that bees turn back and look (TBL) to acquire or update their memory of the location on leaving a nest or food site:

> When a bee had discovered one of my honey producing artefacts and collected therefrom, it would make a flight of orientation and then fly home…After the association had been well established, the bees usually departed for home without making a careful flight of orientation. If however I had made a marked change *in the position* of the artefact since the last visit of the bee, then a careful flight of orientation was always made.

It was not clear until the 1990s what the insects were doing. The details are more obvious in wasps than in bees. Wasps leaving a new food source back away and swing from side to side in flight, along a series of successive short arcs centred near the goal, and learn the location by making inspections from successively greater distances along a line from the target in just one or two directions (Figure 8.8). The detail suggests that the wasps store a series of landmark sizes, directions and ranges, which they can use in reverse order to return to the food—a kind of inversion of the desired flow field. The ground speed increases as the insect backs away in successive arcs, so the turning speed is visually controlled at 100–200°/s irrespective of the radius of the arc. The body angle is fairly constant relative to the landmark. The null in the visual flow field falls on the landmark and on a particular eye region as the insect swings around in flight, while the foreground and background move in opposite directions. They do not zoom and loom. Shadows, which being flat have no motion parallax, have no effect. Parallax can be a crucial cue, if available. The relative locations and ranges of nearby landmarks are learned during one manoeuvre, but there is insufficient time to learn the pattern around the goal. For a bee, if the TBL manoeuvre is prevented, the size of the target becomes the cue.

When the approaching bee centres its vision on a symmetrical target or on the colour of the goal, it can learn its size, minimum cues and colour on the approach and then depart without making a TBL manoeuvre, but the bee can also learn the cue or colour of the goal in a TBL if the cue is put in place just before the bee leaves. If shown one colour, cue or size of target on arrival, and another on leaving, they retain the first cue only, so the TBL is not primarily for learning colour, pattern or size, but for range, landmark size and depth. In the Y-choice maze (Figure 9.1), bees do not make TBL manoeuvres as they look first at one target and then at the other, because they already know where they are.

Some of the angular manoeuvres are repeated when the insect returns, so that it places itself in the same postures as before, looking in the same directions, and the flow field swirls around the same null points, landmarks fall on the same eye regions, while the goal is brought to the null point at the front. There is no evidence that the area around the goal is compared with a retinotopic memory.

Figure 8.8 The turn-back-and-look behaviour. a) When departing for the first time, the bee flies backwards in a convoluted path while facing the target. b) The track without arrows. c) At the second departure, the bee departs sooner. d) When familiar, the bee leaves without looking back.

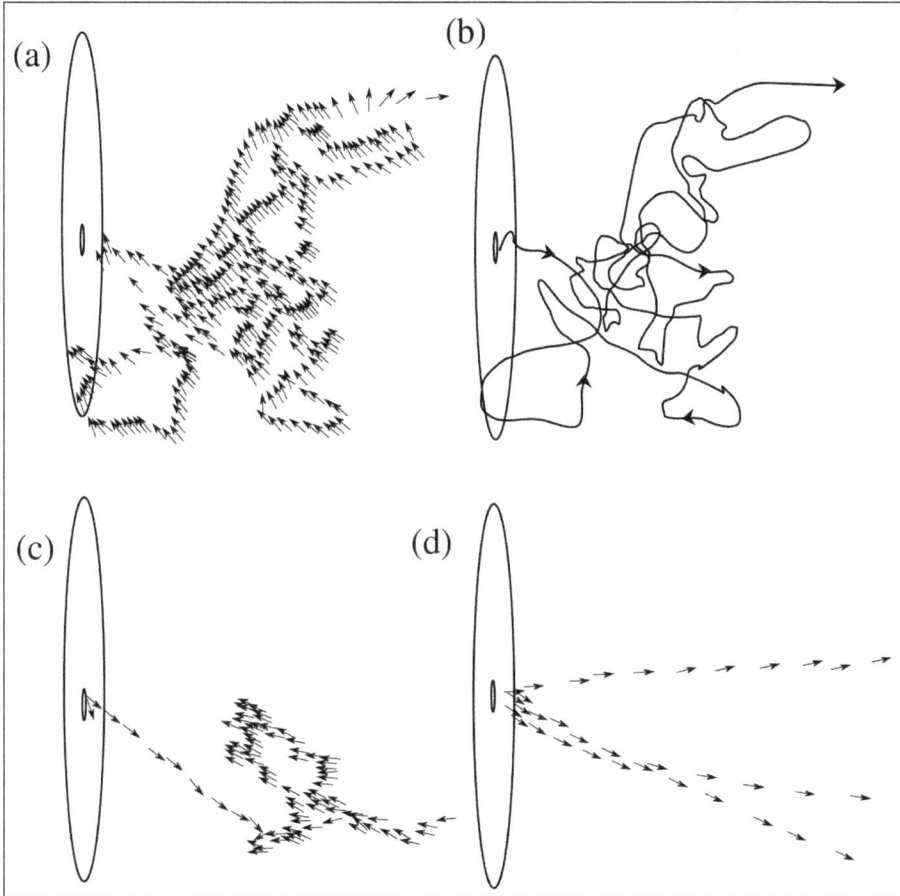

In later visits, these actions progressively disappear and experienced insects arrive and leave directly. The nature of the manoeuvre and its effect on the visual flow field suggest that a sequence of ranges, sizes and locations on the lateral part of the eye is remembered. If the goal is moved, the bee returns but cannot find it, but if the pattern on the goal is modified the bee might not notice, because the place is learned, not the pattern. It is just one of many acts of looking that suggests that patterns are equivalent to landmarks.

Navigating with one eye

Many of the visual tasks of ants and bees can be carried out with one eye. This is not surprising because to a large extent each eye appears to act independently of the other, and even each local region of each eye detects separate cues or landmarks. Naive bees trained to measure a distance along a tunnel with one eye can transfer the information to the other eye. They can also be trained to measure one distance along a tunnel with one eye and a different distance with the other eye. There remains some uncertainty about how the eyes interact in the natural situation, because bees in flight see a different scene with the two eyes, but they turn around for the return flight (Srinivasan et al. 1998). When one wall of the tunnel is blank and the other displays a pattern, the bees estimate the distance of the reward along the tunnel almost as well as when both walls are patterned.

There are strange discrepancies between investigations of this topic and it seems that insufficient notice has been taken of the overlap of the visual fields of the two eyes. Early work showed that the ant *Cataglyphis* could navigate by the polarisation of the sky or by landmarks with only one eye and could then find its way home with that eye. When trained to home with one eye and then tested on the other eye, they can navigate by the polarisation of the sky but not by using landmarks (Wehner and Müller 1985). Similarly, bees can learn a colour or an orientation cue when presented to one eye but cannot transfer when tested on the other eye (Giger and Srinivasan 1997), and they can use the position of a laterally placed bar but they do not recognise it with the other eye. They can, however, transfer relative motion of the cue and the eye between the two eyes (Lehrer 1994) and also a colour when it is a cue for the direction of the next turn (Zhang et al. 1998). The experiments are descriptions of the performance, not an analysis of mechanisms, there have been too few tests to reveal the cues and each group of trained bees was given only one kind of test, which was successful, so there is the probability that they learned during the test procedure.

Cognitive maps

If a cognitive map is defined as any internal representation of an extended world, however rudimentary, most active animals have it built into the their visual responses. If, however, a map is defined as an internal representation of the geometric layout of objects in the surrounding environment, such that an animal can place itself in the right place by recognising landmarks or other cues and then take a novel shortcut to its goal, it is simply a matter of the scale of the map. In a test, an animal with an internal map must be able to take the shortest distance along a novel track to any goal in its territory.

On a small scale in the Y-choice maze, bees quickly learn the layout of the whole apparatus and when they receive no reward at one target they quickly fly to the other target without going back to the last landmark, as though they know very well the relative locations. On a small scale, they learn a map of sorts when the location of the goal is randomised, which suggests that to make the bees learn a large-scale map, nothing more is needed than training with the reward in many possible places. In a maze, the bees that take the wrong turning soon learn to take the alternative turning, as though they have an appreciation of the spatial layout of the maze. Again, they learn a sort of map when necessary.

Bees can learn to go to one feeding place in the morning and another in the afternoon. If bees that are departing from the hive are taken to the wrong feeding site and are allowed to feed, they set out in the correct direction when they return to the hive, showing that they associate the landmarks they see with the compass direction to the hive. If they are taken to the wrong place and not fed, they usually do not fly off in the direction of the other place, but they usually return to the hive and then set off again or they go scouting around for another food source, then fly their beeline home.

As seen in Bethe's old idea of a homing force, there have been contending opinions for more than a century. So far, there is no suggestion that ants have maps of their territory. For wasps, the detailed studies of Baerends and many others before him showed that wasps carrying food could head directly to one of a dozen goals when displaced to anywhere in their territory, probably by use of landmarks learnt in numerous exploratory flights. In 1986, Gould produced evidence that displaced bees did not merely continue along their previous compass course, but were able to 'make use of novel and efficient routes on the basis of map-like cognitive representations' of local landmarks. The next year, Gould went too far in claiming that the dance of a returned forager was interpreted by the recruits in terms of their own internal map based on landmarks.

Most unusual and out of character, one of Gould's former students questioned the conclusion and suggested that distant landmarks indicated the direction of the hive. Also, the idea of an internal map was so objectionable that two professors, not known for previous amicability, collaborated to throw it out. In 1990, they jointly reported new experiments in Europe and the United States in which marked satiated forager bees continued along their compass directions when displaced. These bees eventually searched around or flew up high and circled before returning home, so were apparently using distant landmarks as a backup. It was concluded that bees used only the local landmarks close to their day-to-day tracks, but when lost, they revealed reserve memories of distant landmarks.

In 1998, Randolf Menzel found that when the foragers were fed at one site in the morning and a different site in the afternoon, they were able to take a novel shortcut back to the hive when displaced from the hive, but not when displaced from the feeding sites.

Notwithstanding this earlier conclusion, in 2000, Menzel and others found evidence of memory of a wider area within which bees could return to the hive from any point, and in 2006, they measured the time it took for bees to return home after being displaced. Bees that had been trained to a feeder that was regularly shifted in any direction at a distance of 10m from the hive all returned quickly, but bees familiar with only one flight track took longer. Finally, they brought into use a method for continually tracking the position of a bee in flight by attaching a radar transponder to it—invented by Osborne et al. (1996). After being displaced under an overcast sky, bees can use familiar landmarks to take a novel shortcut to where they ought to be. The results reveal 'a rich, map-like organization of spatial memory in the navigating honey bees' (Gould 1986). The displaced bees can choose between at least two goals. They can take straight and rapid flights directed either to the hive or first to the feeding station and then to the hive. In the featureless landscape used, moveable tents act as landmarks, but apparently the varied textures of the ground provide sufficient information for navigation when the landmarks are shifted.

To my mind, these results show that the bees learn as much as they need in order to know the direction of home at all times. When the destinations are shuffled, they learn the positions of all possible places to look, exactly as in the Y-choice apparatus. They are able to extend the scale of the exploratory flights and build up a memory of the vector directions to two or more goals from a larger number of landmarks, as inferred by Baerends for the way that the experienced wasp can take a caterpillar to one of several nests from any point in her territory, and also demonstrated by the way that bees fly through complicated mazes.

The present opinion seems to be that when bees make orientation flights, they are learning to associate the directions of landmarks with the sun-compass direction of home. The beeline home therefore does not prove the existence of a two-dimensional map in the brain of the bee. Departing bees, going either way, have a strong internal signal for distance, which is all used up in arriving bees. Then, after finding the food site—sometimes by scent or by seeing other bees feeding—the new bees learn the local landmarks at the food site and also associate them with the direction of home. We have no idea how or where this sequential memory is coded.

The reader might note that researchers on navigation were interested in the mechanisms, as well as the performance (that is, what the bees could do).

Endnotes

1. For further information, see the recent works of Menzel, Srinivasan or Wehner.

09 FEATURE DETECTORS AND CUES[1]

This chapter traces the effort from 1990 to 2008 to identify and characterise the parallel pathways of feature detectors and cues at the heart of the mechanism of visual processing in the bee. Bees have a few different kinds of feature detectors in large arrays that respond to parts of parameters in the pattern. The features in the parameters are edges or areas of black or colour; that is all. The analysis has been done with patterns subtending 30–45° at the eye, so the responses are limited to a small part of each eye. The responses of each kind of array of feature detectors are summed into a cue that lies within the bee and the cue can be learned together with its position on the eye. The bee tends to look at the reward hole at the centre of the target and the summation is done separately in each eye, so each eye picks up one set of cues from its own side of the pattern.

The common cues are area, modulation (total edge length), position of the centre of an area, radial edges, average edge orientation, tangential edges and the absence of a cue. The coincidence of the different cues in each local region of the eye can be remembered as the label on a landmark, whether or not an isolated landmark lies in that direction. Visual recognition of the place of the reward is nothing more than the coincidences between a few landmark labels at large angles to each other and the corresponding positions of simple parameters in the panorama.

In the most recent period, since 1990, the cues have been identified and characterised in the bee's visual system. Each cue has its own story. Perception of the configurational layout in patterns of this size (30–45°) was ruled out on logical grounds because the responses of edge detectors were summed on each side of the target so that the orientations of separate edges were lost, and it was ruled out experimentally because there were many quite different patterns for which bees showed equal preference, despite being trained to go to one of them.

Figure 9.1 A summary of the visual processing system for some of the cues in a local region of the eye. Most of the input is from green-sensitive receptors. The lamina detects the rate of change of intensity for the feature detectors for contrasts. The three types of orientation detectors are summed together. Cues relating to bilateral symmetry and the position of the centre are not illustrated here.

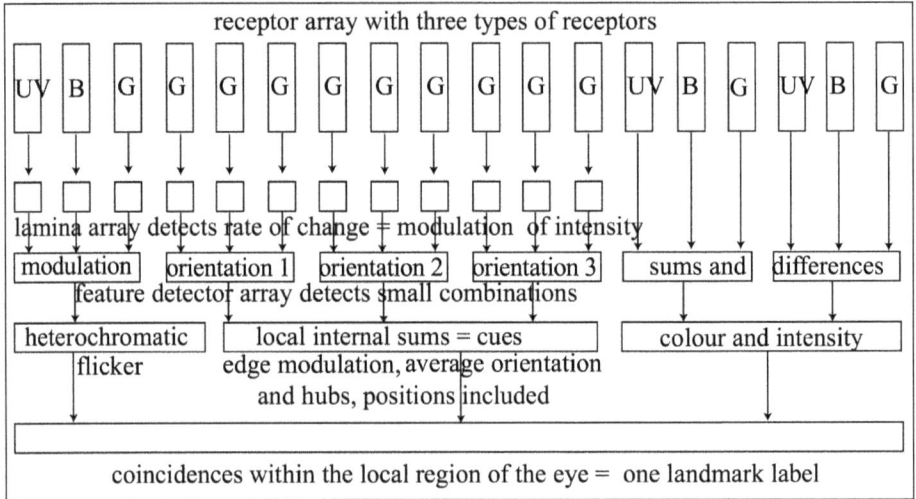

receptor array with three types of receptors

| UV | B | G | G | G | G | G | G | G | G | G | G | UV | B | G | UV | B | G |

lamina array detects rate of change = modulation of intensity

| modulation | orientation 1 | orientation 2 | orientation 3 | sums and | differences |

feature detector array detects small combinations

| heterochromatic | local internal sums = cues | colour and intensity |
| flicker | edge modulation, average orientation and hubs, positions included | |

coincidences within the local region of the eye = one landmark label

Box 9.1 Glossary of terms

The parameters are outside the eye as part of the pattern or panorama. The image is the distribution of excitation on the retina. The feature detectors behind the eye respond to the parameters in the image. A cue is the sum or count of the responses of one kind of feature detector in a local region of the eye, and is therefore a quantity inside the bee. The cue is derived from a part of the image in the local region, but the process of summation destroys the local layout. If rewarded, the bees learn the cues in their retinotopic positions. A landmark is recognised as the coincidence of several different cues in a local region of the eye.

The feature detectors are the units of perception of modulation, edge orientation, black, white or colour. They are small, about 3 ommatidia across on the retina, and all respond independently in parallel. The responses of the feature detectors are summed to form cues and the bee remembers the totals and their positions, not the individual detector responses.

The field of a filter or neuron is the region in space and time within which a signal is detected.

A fixed pattern, as opposed to a shuffled one, has the pattern fixed as seen from the choice point of the bee.

A generalised parameter is one that is recognised in a context other than in the training pattern. Originally, it was merely in a different position on the target, but later it was in a different pattern.

The image is the pattern of excitation in the array of receptors in the retina.

The label is the coincidence of cues in a local region of the eye, by which the bee recognises a landmark and its position.

The modulation of a receptor is the change in the light intensity in the receptor and the consequent electrical signal. The motion of the eye over contrasts generates the modulation of the receptors. The modulation of a pattern is roughly equal to the total length of edges in it.

Orientation of an edge is usually the angle to the vertical in a vertical plane.

Bees can be trained to remember the retinotopic position of some cues within the local region of the eye.

A parameter is a scalar or vector measurement of some aspect of the pattern outside the eye—that is, the area, total length of edge or averaged edge orientation.

The patterns are displayed on the targets during training and tests.

Place for bees is a geocentric term, like the place on a map; position and direction are usually retinotopic terms for the direction relative to the axes of the head. Location or position refers also to the position of a parameter on the target, a shift in position of a pattern or a shuffle of the locations of boxes, targets or bars during training and tests.

Point of choice is the place where the bee detects a cue and makes a choice by moving away from or towards the reward or the next target.

A sign stimulus is an older and more general term that is not restricted to vision—for example, it also applies to the call of a bird. It is the human idea of the essential stimulus outside the animal, not the parameter that is eventually identified, and certainly not the cue formed by the feature detector responses within the animal.

A template is a hypothetical mechanism that detects a fairly complicated pattern that has been identified by the human observer; it can be innate or learned. In vision, a spatial copy is usually implied. Templates are useful in pre-programmed robot vision.

Figure 9.2 Representations of the cues that are similar to common parameters, in order of preference during the learning process. The cues in fact consist of excitation in groups of neurons, but this illustration might assist the reader to understand the text. The bees do not see the parameters; they detect edges and areas with feature detectors and the cues are the various separate sums of the responses.

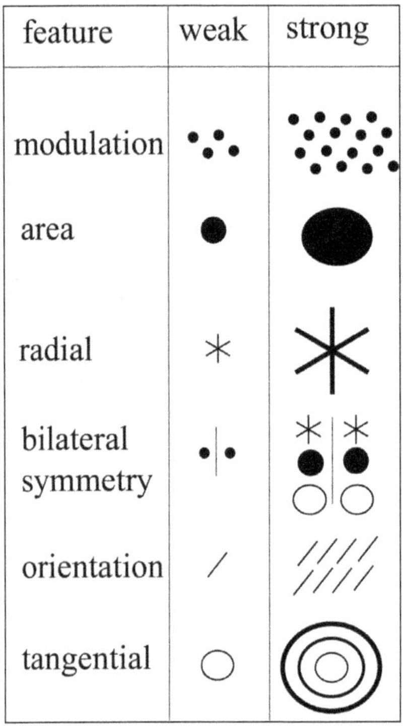

A new apparatus for measurement of resolution

To measure the resolution of bee vision required a new apparatus in which flying bees chose between two targets at a known distance (Srinivasan and Lehrer 1988). In earlier experiments, the bees made their choice with the external panorama around them and, except in Wehner's experiments (Figures 1.1c and 4.4), they learned the cues at an unknown range while they prepared to land on the reward hole. The Y-choice apparatus (Figure 9.5) can be used for a great variety of experiments in carefully controlled conditions. The bees detect the patterns on the targets at a fixed range and fixed angular size, so calculations of resolution are possible. By chance, in the apparatus constructed in Canberra, the angular sub-tense of the target from the point of choice was 35–50°, and later this turned out to correspond with one local region at the front of each eye. The bees therefore learned two landmark labels that were usually identical if the pattern was the same on each side of the reward hole.

Figure 9.3 The Y-choice apparatus. The bees enter through the hole 5cm in diameter at the front and pass through one of the transparent baffles. The targets and their patterns with the reward change sides every five minutes, to prevent the bees from learning which side to go. The air pipe extracts odours. As in all the figures, (+) and (–) indicate the rewarded and unrewarded patterns.

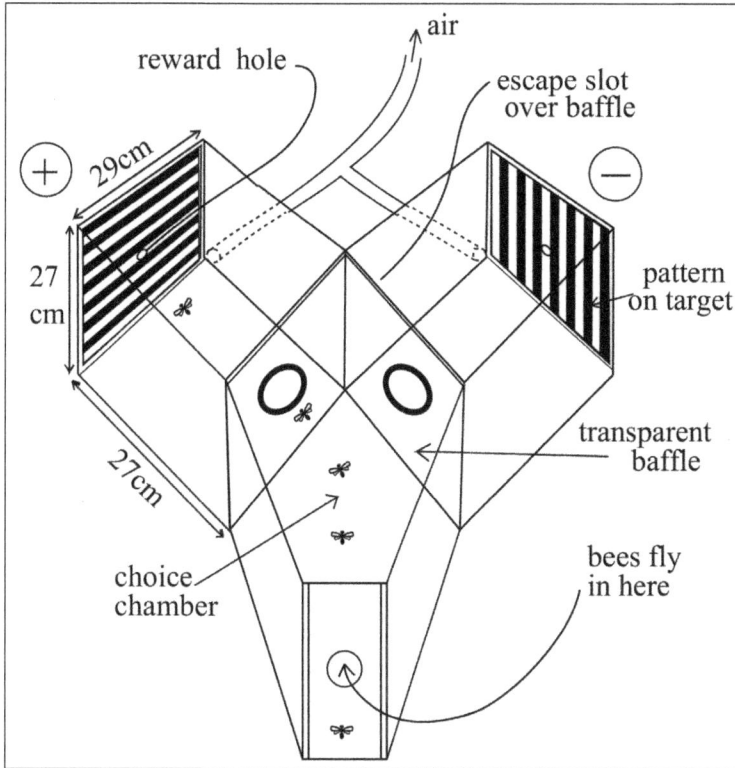

At first it was not realised that this new apparatus restricted the visual angle of the bees like blinkers, so that they were restricted to one or two local regions of the eye. The bees could no longer learn several landmarks and, unlike bees trained on large targets, they became tolerant of test patterns that differed from the training pattern. This change in their behaviour made possible the identification and characterisation of the cues.

The apparatus was placed under a shelter so the bees could not use the sun compass when inside and was lined with clean white paper, which was originally intended to help the bees find the patterns, but left plenty of vertical and horizontal edges that the bees could use to stabilise themselves in flight and orient themselves with reference to the patterns. The bees must look for the expected positions of the parameters by use of a frame of reference within the box—usually the internal edges and the reward hole at the centre of the target. I did not realise the significance of these changes in the task until the position of the hub in radial patterns was recognised as a parameter—about 1998.

After 1995, a transparent baffle was placed across the entrance to each arm to slow the bees and define the range at which they made their choice. No adverse effects of the baffle were noticed, but it gave the bees more time to look and favoured the use of the front of the eye. The bees came more easily into the apparatus if the entrance had the bees' attractant odour. After 1995, a stream of air was drawn out at each side so the bees made their choice in clean air without interference from new odours. During training, the two sides were interchanged every five minutes (10 minutes before 1995) so the bees could not learn which side to go and must look at the targets, which could both be seen from the choice chamber. The bees learn the geometry of the inside of the apparatus while in flight. Usually two hours of training is sufficient and it is important to consider why training takes so long.

To investigate what the bees had learned, they were given a variety of different tests with unfamiliar patterns on the targets. Several different tests were intercalated so that they saw a given test only once or twice a day. When they arrive, they look at one test pattern and then at the other if the first is not recognised. By watching them in the choice chamber, one can see whether they decide quickly or whether they spend a long time looking. In each experiment, the aim is to see whether the bees can do the task or not after a reasonable period of training. The test patterns changed sides after five minutes, which allowed only one visit on each side, then training resumed for 20 minutes before a different test was displayed. It is preferable to use many tests in a sequence so that the bees cannot learn any one of them. Usually a small group of individually marked bees was trained on a Monday morning and each experiment lasted all week.

Figure 9.4 Orientation detection at two steps in visual processing. a) A circuit that detects the coincidences between receptor responses. b–d) Feature detectors for edge orientation. They are symmetrical about one axis, with a field size of 3°, and are therefore 3 ommatidia long. e) A modulation detector of the same size (compare with Figures 4.1b and 4.1c). f) The circuit that detects a cue by the coincidences of feature detectors. g) Responses of a detector neuron in the lobula of the bee to moving edges at different angles. The stimulus was either motion at right angles to the edge or alternating phase of bars, as shown in the insets. The field size was about 20° across.

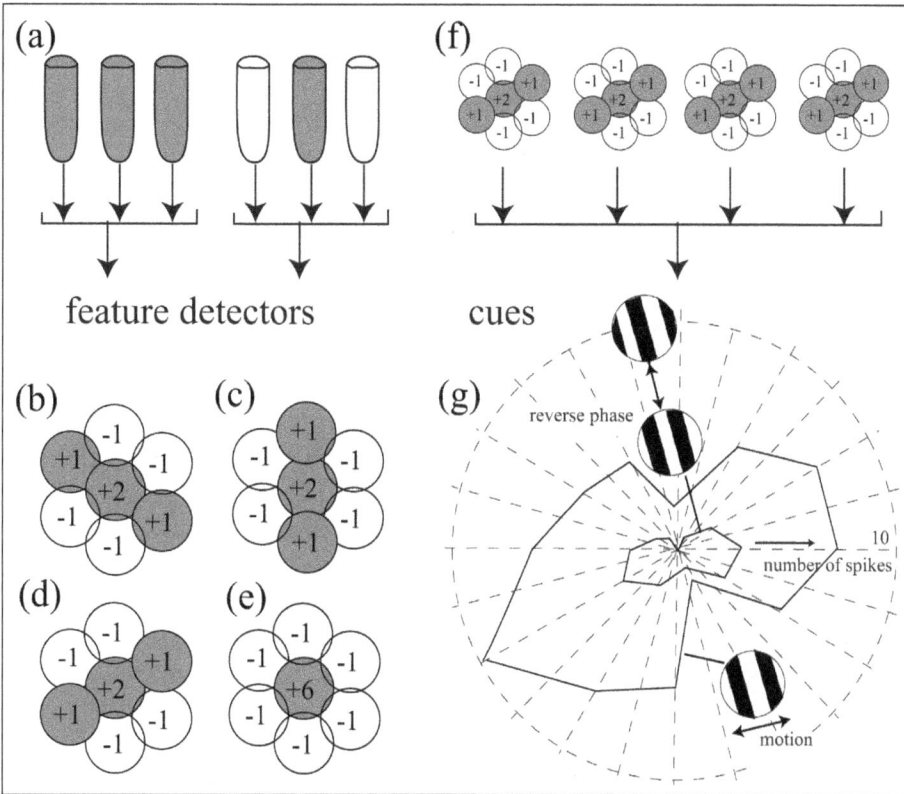

The cues

The modulation cue

Long ago, Hertz showed that bees distinguished between many patterns by something related to the total length of edge (Figures 1.2 and 1.3), a finding later confirmed by all who examined this point. The bees must have made a quantitative measure of something, but no-one asked over what area of the eye the measurements were made. It seemed to be assumed that the bee detected an object or shape, then measured its edge length. Until about 2003, when the feature detectors were measured, it was not clear what the bees really detected.

Resolution tests going back to Hecht and Wolf (1929), and also repeated several times, gave an absolute limit of the resolution of the modulation caused by a regular grating near 2°, irrespective of the orientation. The most likely modulation detector is therefore a single ommatidium with a symmetrical inhibitory surround (Figure 9.4e), which is smaller than that proposed by Jander (Figures 4.1c and 4.1d). The modulation cue would then be the total number of simultaneously excited modulation detectors in a local region of the eye, maybe in the field of a neuron. The difference in modulation between two patterns is a preferred and common cue that is measured quantitatively. In a test, the trained bees make the same measure irrespective of a change in the pattern (Hertz 1933; Horridge 1997a).

There is experimental evidence that modulation is detected in colour, but it is not clear whether there are heterochromatic detectors or separate blue and green detectors—or both. Bees discriminate between a horizontal and a similar vertical grating in colour with no green contrast; therefore, modulation is discriminated via the blue as well as via the green receptors. With a grating at 45° versus one at 135°, with no green contrast, bees fail to discriminate even if the period is large, because there is no modulation difference and the orientation detectors have inputs only via the green receptors.

Discrimination of oriented edges

Before 1988, bees were trained either with very large patterns, with a single pattern versus a blank, or with landing on the patterns as the criterion of success. Therefore, the significance of edge orientation was not discovered because it was not a preferred cue when modulation or an area of black was available. In 1990, with the Y-choice apparatus (Figure 9.5) without the baffles, bees learned to discriminate between two gratings—one with horizontal bars versus a similar one with vertical bars when the stripes were continually shuffled in position and width during the training and tests to eliminate the position of black as a cue (van Hateren et al. 1990). Memory of eidetic images was therefore not tested, let alone refuted. The authors assumed that the orientation of edges was learned, although they already knew that in this situation the bees learned the difference in the modulation caused by scanning the eye across horizontal versus vertical edges (Srinivasan and Lehrer 1988). The part played by the orientation of the central axes of the bars, as suggested long before by Wehner (1967), was not considered. For 10 years, no-one worried about the missing controls because the randomisation technique suggested many new experiments. Later, the difference in modulation was eliminated by the use of oblique gratings at 45° versus 135° to the vertical, so orientation detection was substantiated.

At this point, let us look at the unregenerate beliefs of the time (Srinivasan 1994):

> The 'vocabulary' of the visual system continues to be a mystery, and we are still largely in the dark as to what 'tokens' the visual system uses to represent, analyse and recognize patterns…While there is little doubt that bees use some kind of neural 'snapshot' to remember and recognize patterns and landmarks, it is hard to imagine that this is all there is to pattern recognition.

The concept of a snapshot came from the work of Collett, with several artificial landmarks at different positions relative to a reward out in the open. It was given a new meaning by the demonstration that orientation was recognised in brief 2ms flashes of illumination, so the relative motion of the eye and pattern was unnecessary (Srinivasan et al. 1993b).

Srinivasan then pointed to a number of publications from 1987–93 suggesting that the visual processing of the bee resembled that of the mammalian cortex. Unfortunately, none of the examples he quoted now substantiate this claim. Illusory contours have not been confirmed, despite efforts. Simultaneous recognition of two or more different orientations is explained by the large size of the targets that spread over adjacent local regions of the eye. Different cues are processed in different pathways that learn separately.

Srinivasan concluded that

> recognition is probably mediated by at least two kinds of processes. One kind…involves the participation of long-range mechanisms which evaluate general properties of the object, such as colour, size and orientation. These mechanisms…rapidly exclude objects which do not possess the right attributes. Another kind of process involves mechanisms which operate at short ranges and require fixation. These mechanisms are more precise and work on the basis of a memorized template. (Srinivasan 1994)

Srinivasan had a concept of mechanisms in parallel, but both of his processes were guesses and both required a global perception of the whole image.

Edge orientations are not the basis of pattern vision

In the work on discrimination of the rotation of a square cross in the 1960s, the patterns were huge, subtending 130° at the eye (Figure 4.2a). The response to rotation of a cross was not a sine-squared function of the angle. The smallest detectable rotation of the cross was only about 4° and the edges could be cut into steps with no effect, so that edge orientation was not involved. Instead, the bees learned the positions of separate areas of black in neighbouring regions of the eye (Wehner 1967, 1969).

In 1994, however, Srinivasan et al. discovered that bees were not able to learn to discriminate between a smaller square cross and the same cross rotated by 45° (Figures 4.2c and 4.2d). They suggested that the detectors of the orientation of edges had very large fields and an angular orientation sensitivity curve that was 90° wide at the 50 per cent level of sensitivity. The response of the detector to the rotation of an edge or thin bar would then be a sine-squared function that rises from zero to a saturation of 100 per cent as the angle changes from 0° to 90°. The response to the rotation of the orthogonal edge or bar would be a cos-squared function. With two equal bars in the form of a square cross, the total response would be constant so that rotation of the cross has no effect. Later, many other examples were found where equal lengths of edges at right angles on each side of the target mutually cancelled the orientation cue.

Irrespective of the theory, the data showed that different orientations in the same local region could not be detected simultaneously and therefore patterns could not be recognised by the combinations of orientations of their edges. The mutual cancellation of orientations in fact destroys pattern, including texture, but not the modulation detected by the orientation detectors. Srinivasan's mechanism was not, however, so certain. The large fields were not demonstrated and alternatives were possible. Later, we found that the fields of the orientation detectors were restricted to the eye on their own side and neurons with corresponding properties were found in the deep optic lobe of the bee (Figure 9.4g). We all guessed—wrongly as it turned out—that the responses of the edge detectors were strung together to make continuous lengths, as observed by humans. When the orientation detectors were measured (see below), they turned out to be small (3° long), independent and not strung together to span gaps. The large fields were therefore summations of many small parallel orientation detectors.

It will be noticed that neither large nor small patterns support the idea that patterns with several edges at angles to each other are discriminated by the orientations of the edges, although this is the almost universally popular belief.

Orientation of fixed bars

In 1998, after a long delay, Wehner's 1966–72 method of training on one pattern versus a blank target was repeated. It turned out to be the beginning of a new theme. The bees were trained in the Y-choice apparatus (Figure 9.5) with a single oblique black bar versus a plain white target. The bar was offset on the target to allow for a subsequent shift to a new place. Two results were startling.

Figure 9.5 Training on a single fixed vertical bar versus a blank white target; the bees learn only that there is something black in the expected place. a) The training targets. b) Reduced response with the bar moved down. c) Reduced response with the bar rotated by 90°. d) No discrimination with the training bar versus a similar bar with the edge cut in steps, so they did not look for edge orientation. e) No discrimination with the training bar versus a similar bar rotated by 90°. f) Good discrimination of the expected position of the bar.

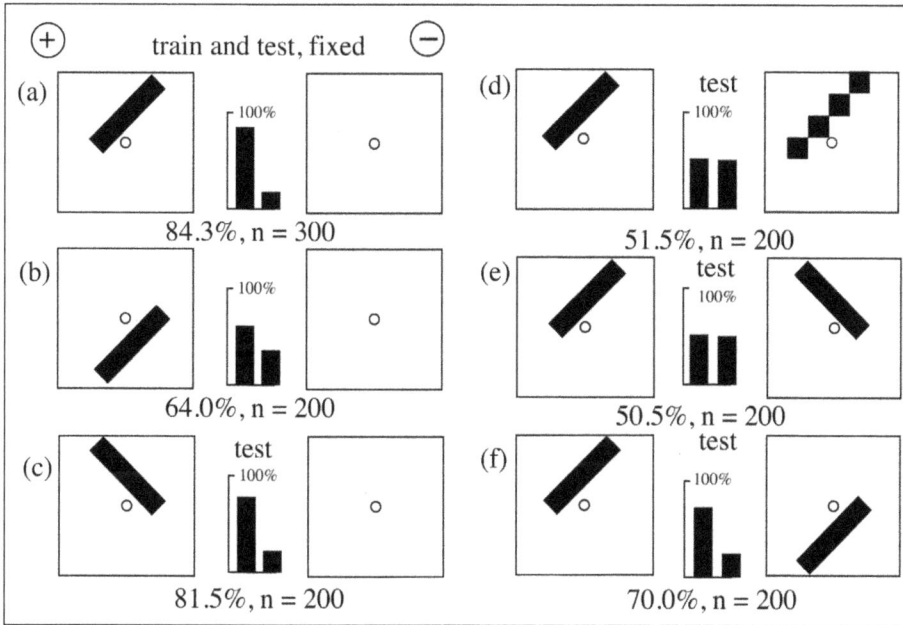

Source: Horridge (2003a).

First, when the training bar was simply moved to a new place in a test, the trained bees no longer recognised it. Second, when the trained bees were tested with the training bar versus the same bar rotated through 90°, they showed equal preference (Figure 9.5). In the training, they had learned the position of something black of a certain area and edge length, but the edge orientation and shape were not preferred parameters. The strongest cue within the bees was the position of the centre of black on the target, especially in the vertical direction. Although they could not remember the bar they had been trained on, they knew its position. This was powerful stuff that suggested new experiments that eventually showed that each cue was a measured quantity of a certain quality with its position on the eye. Of course, these are the properties of a neuron.

On the other hand, when they were trained to discriminate between a fixed black bar on one target and a similar bar at right angles centred on the same place on the other target, they responded only to orientation and its expected position (Figure 9.6). In tests, they were less able to discriminate the more the bars were displaced. When trained on a broad black bar in one position on one target versus another bar in a different position on the other, the bees learned the difference between the vertical positions of the centres and ignored all other differences. Orientation was a cue only if there was no other available.

Figure 9.6 Learning the orientation cue when nothing else is available. a) Training with an oblique bar versus an orthogonal bar, centred on a place that is neither radial nor tangential. b) They discriminate two thin oblique bars versus two thin orthogonal bars. c) They fail in tests with bars with stepped edges to remove the edge orientations. d) They fail to distinguish the training bar from two thin oblique bars with the same orientation. e) They fail in tests with both bars moved down. f) They fail when a black spot is added to the training targets.

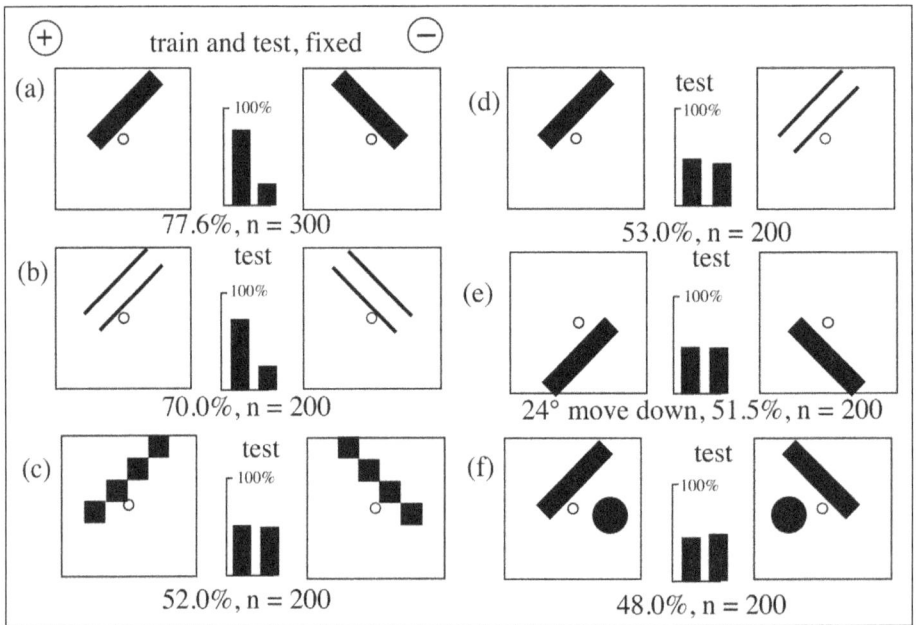

Source: Horridge (2003a).

The bees used the scene from the choice point in the apparatus to fix a frame of reference for the expected direction of the cue. In a test, they lost a cue that was not in its expected place. They therefore did not fixate on the black bars or the cues, which had no salience for them. They were not interested in the pattern, only in the cues in their expected positions as a way to identify the place of the reward. Therefore, when cues were shuffled in position during training, they were learned in the range of places where they occurred during the training.

Making a fixed edge fuzzy, even extremely fuzzy, has no effect on the orientation cue. A gradient from black to white in 20° is still detected as an edge with an orientation.

When trained on a single black bar versus another, both shuffled in position, the bees learn the orientation cue only if the bars on the two targets are in corresponding positions at all times (Figure 9.7). Each bar is then detected in the same expected positions on the eye. The local region of the eye in which the memory is formed and recovered is about 15–20° across.

Figure 9.7 The bars on the two targets must be in corresponding places for the orientation to be recognised. Every five minutes, the bar on both targets was moved to a new position, 1, 2, 3, 4, so that the bees learn not to use the location of the bar or the radial or tangential cues. a) With the bars in corresponding positions on the two sides, they learn. b) With the bars in non-corresponding positions, they fail to learn.

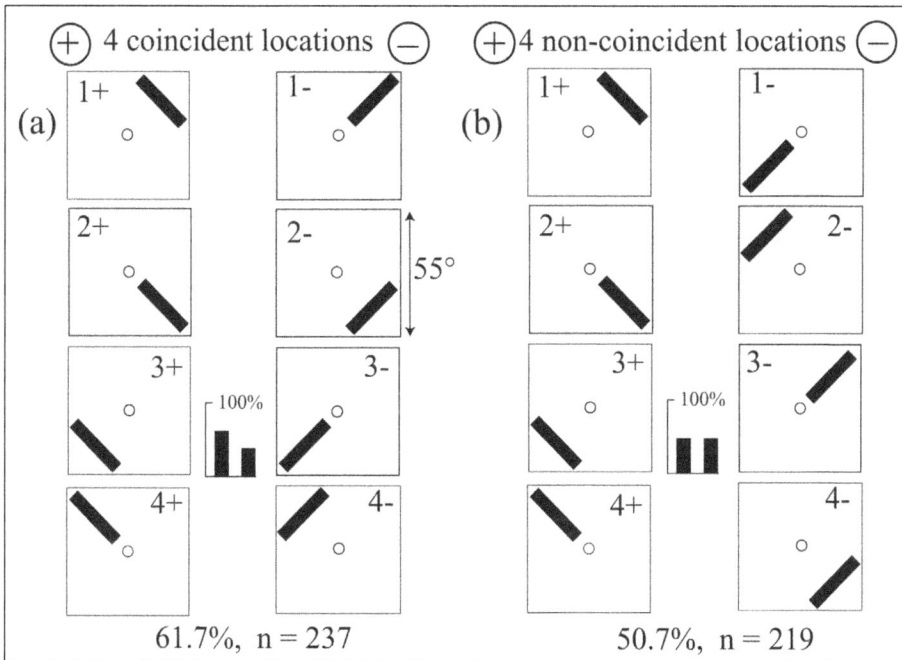

Source: Horridge (1998a).

Orientation of gratings

To a bee, a grating is a place where cues are found, not a collection of bars. When the task is to discriminate between two black and white gratings at right angles, the bees learn rapidly to a high standard above 80 per cent correct, and they are about 65 per cent correct when tested with single bars placed anywhere on the targets. The bees learned the average position of black, the difference in modulation caused by scanning and the difference in edge orientation. There was no evidence that they remembered or even detected the positions of the individual bars. When the criterion of success was landing on the reward hole, the main cue was black or white immediately below the reward hole (Baumgärtner 1928; Giger and Srinivasan 1995). On the other hand, when bees were trained on single bars and tested on gratings, the result depended on what cues they had learned. They failed when they had learned only the position of the training bar, but were successful if they had learned an orientation difference.

When the gratings have a period of less than about 10°, it is quite unnecessary to shuffle the positions of the bars. Randomising the width of the bars during the training, however, confuses the modulation cues generated by the horizontal movements of the eye. This is an important consideration because modulation is the preferred cue and the bees learn it quantitatively.

Figure 9.8 Measurement of the maximum length of the feature detectors for orientation. a) Training on orthogonal bars that are shuffled in position. b and c) Testing with rows of squares with gaps of controlled width between the squares; the orientation is discriminated when the feature detectors can bridge the gaps. d) Training with shuffled black bars versus squares of the same total area. e) Testing with rows of squares with gaps of controlled width; in each case, the limit was near 3.5°.

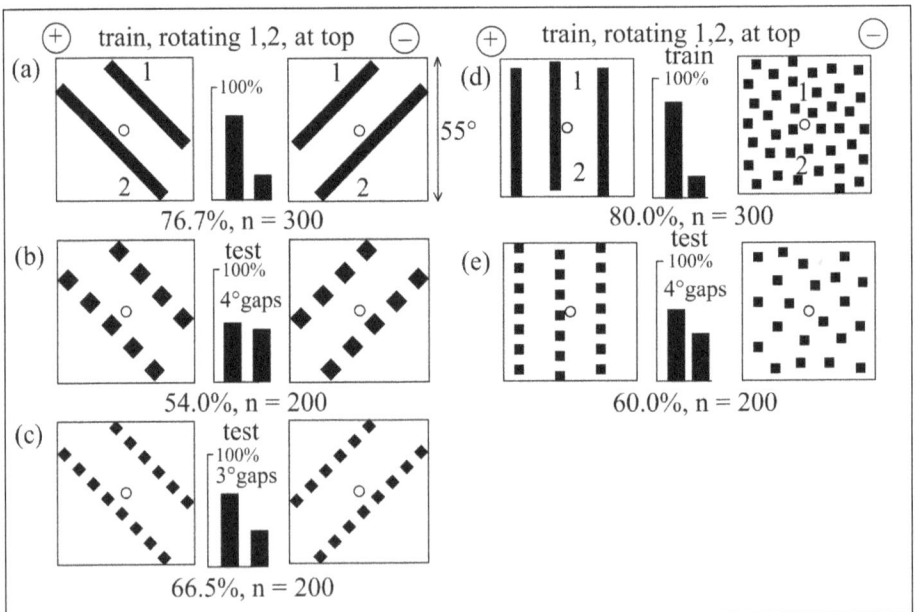

Source: Horridge (2003b).

Since 1967, it had been uncertain how fixed gratings were discriminated (Chapter 4). Bees trained with a horizontal or a vertical black-and-white fixed grating versus a grey target of 50 per cent black discriminated down to periods of about 2° irrespective of the direction of the edges, from which it was inferred that the bees learned the modulation difference by scanning, not the spacing of the bars (Srinivasan and Lehrer 1988). Although at the time the colours were used to remove motion signals, they also found that bees discriminated the horizontal/vertical orientation in the absence of contrast to the green receptors, because the cue was the difference in modulation. Later, when the bees were trained with randomly shuffled bars or gratings, it was found that the orientation cue required green contrast (Giger and Srinivasan 1996). Bees cannot discriminate equal but orthogonal oblique gratings (at 45° and 135°) with no green contrast because there is no modulation difference and no orientation cue. Black oblique orthogonal gratings at 45° and 135° present no difference in modulation induced by active vision and the resolution of them is now that of the orientation cue—about 3°. So, the preferred parameter changed as the gratings were rotated. All these data led me to devise ways to measure the size of the edge orientation detectors.

The size of the feature detectors for edge orientation

When small squares in a straight row are so close that they are not resolved separately, they are detected as a bar with an orientation, but when the squares are resolved, the orientation is cancelled by the equal lengths of edges at right angles. Therefore, when bees are trained to a given orientation, they can be tested with rows of squares at different separations to find the limit of resolution of the orientation cue. The result is the maximum size of the orientation feature detectors that can span the gap between the squares—about 3°. Cutting long straight edges into square steps that are resolved also destroys the orientation cue with a similar result (Figures 4.2b and 9.6c). A staircase that is resolved has no net orientation as it has in human vision.

In a different method, bees trained to discriminate between vertical bars and a pattern of squares (Figure 9.8d) were tested with rows of squares of controlled separation (Figure 9.8e). When the squares were resolved separately, the orientation of the row was not detected, showing again that the maximum length of the feature detector that spanned the gap between squares was about 3°. It was surprisingly small, and the edge detectors acted independently; they would not join up to span gaps.

Bees can be trained to discriminate between two equal arrays of oblique orthogonal bars with no modulation difference (Figure 9.9a) and then tested on arrays of shorter bars of similar total length. The minimum length for orientation detection at the threshold is about 3° (Figure 9.9b). In another method, bees were trained to discriminate orientation with shuffled orthogonal long oblique bars (Figure 9.9c) and then tested with the long bars versus a pattern of short bars of the same total length and parallel to them. The bees have learned only the orientation cue and when they detect it on both targets, they cannot discriminate (Figure 9.9d). The threshold is not reached until the short bars are the same length as the feature detectors—about 3°. Bees trained on Figure 9.9d failed to discriminate between orthogonal oblique bars (Figure 9.9e), showing that they had not learned the pattern or the orientation that was the same on both targets. They did, however, learn the modulation difference in Figure 9.9d, as shown by a test (Figure 9.9f).

The discrimination of the orientation cue was little affected when black was exchanged for white (Figures 4.6c and 4.6g), showing that the detectors of edge orientation were bilaterally symmetrical. From these results, and assuming that the detectors of edge orientation depend on simultaneous modulation of a few adjacent receptors, we can infer that the feature detectors for orientation are three ommatidia long (Figures 9.4b–e). This result implies that there are only three types with axes at 120° to each other (Figures 9.4b–e).

Most significantly, the feature detectors were about an order of magnitude smaller than the cues and each cue was the sum of the feature detector responses in a local region of the eye, with their average position. They remind me of the small-field and medium-field neurons of the insect optic medulla.

Figure 9.9 Measurement of the minimum length of the feature detectors for orientation. a) Train on orthogonal bars. b) Test on shorter bars. c) Train on large shuffled orthogonal bars. d) Test on the same large bars versus smaller bars with the same orientation; discrimination fails when the orientation of the small bars is detected. e) The bees trained on the patterns in (d) do not recognise the orientation of orthogonal bars in a test because it was not a cue in the training. f) Bees trained on the patterns in (d) discriminated the modulation difference.

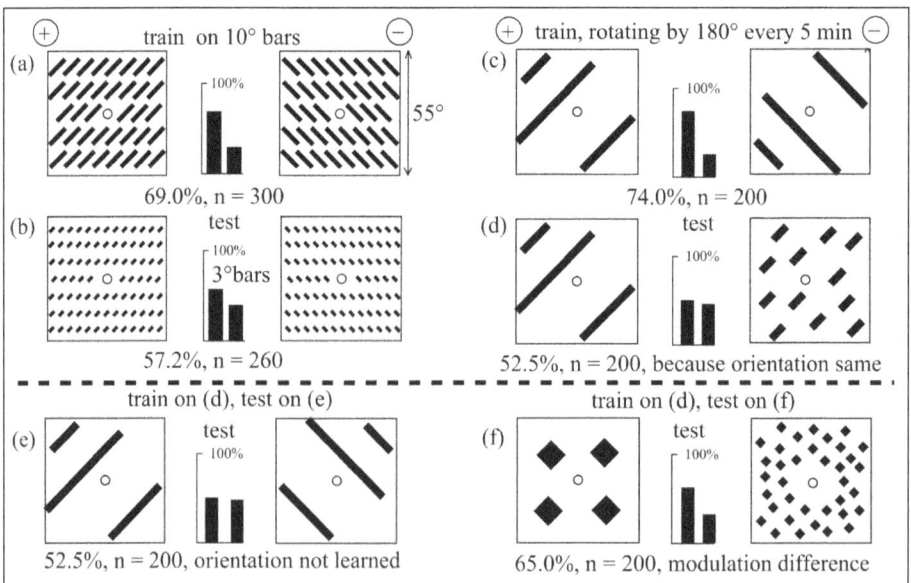

Source: Horridge (2003f).

Distinction from motion detectors

Bees discriminate between two orthogonal moving gratings that move at speeds up to 50°/s as seen from the point of choice in a Y-choice apparatus and they can do this task when illuminated by a slow stroboscope with flashes 2ms in duration (Giger 1996). The response is independent of the direction of motion of the moving grating. The edge orientation detectors are therefore quite different from the directional motion detectors described for many insects. The relation, if any, between the edge orientation detectors and the non-directional system that measures the angular velocity of the flow field has yet to be investigated.

Misunderstandings with orientation detectors

After the experiments with gratings with randomised bar positions in 1990, it was assumed that bees learned the orientation cue. This conclusion was valid, however, only when no other cue was available, because orientation was the least preferred cue. The problems illustrated in Figures 4.2–7 were long forgotten, probably because in the second half of the twentieth century one simply did not refer to previous authors if they had published contrary conclusions.

Figure 9.10 Conclusions that now appear unjustified. a) Bees were trained to discriminate between two black-textured bars raised 6cm above black-textured backgrounds, as shown in side view in (b). c) The trained bees then discriminated the bars with no green contrast. d) Bees were trained with no green contrast, but the bars differed in modulation. e) The bees trained in (d) detected a difference in the textured bars raised over textured backgrounds.

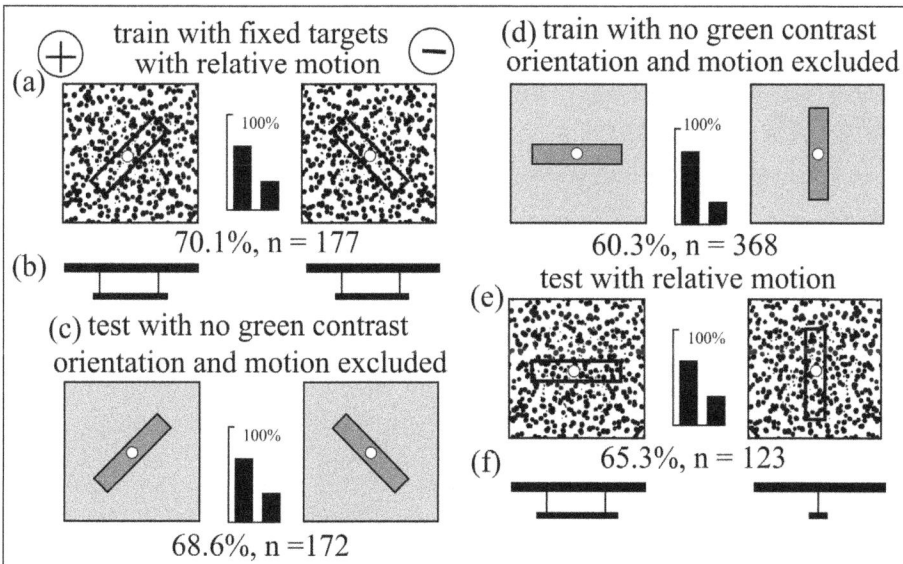

Source: After Zhang et al. (1995).

In Canberra after 1990, we looked at the detection of bar orientation at various distances from the target to compare global and local discrimination (Zhang et al. 1992), the size of the regions in which orientation could be discriminated (Zhang and Horridge 1992) and the detection of orientation in illusory edges (Horridge et al. 1992). The last cannot be repeated and inferences of global vision based on all three of these studies were premature (see Chapter 12).

The spill-over of ideas from motion perception also misled us into thinking that edge orientation could be discriminated by the relative motion of a patterned edge against a patterned background. After a preliminary training on other patterns, Zhang et al. (1995) trained bees to discriminate between a fixed oblique textured black bar at 45° raised 6cm above a textured background versus a similar oblique bar at 135° (Figure 9.10a). The trained bees could immediately discriminate between two orthogonal oblique bars in plain colour with no green contrast to eliminate cues from motion detection (Figure 9.10b). Without further tests, it was proposed that the cue was orientation, but that could not be so. At the time, it was known that a texture of square pixels, if resolved, would destroy the orientation cue. Even worse, it was later shown that lack of green contrast also destroyed the orientation cue. Later, this and similar experiments could not be repeated despite considerable efforts, but the bees were sensitive to shadows under the raised bar (Horridge 2003a), so they had probably learned the difference in position of shadows.

Zhang et al. (1995) also trained without green contrast to eliminate motion signals, but at the time they were unaware that the orientation cue was also excluded, while the modulation difference remained. They intuitively inferred that the discrimination of horizontal versus vertical fixed bars that were textured or without green contrast (Figures 9.10d and 9.10e) was due to the orientation difference, which was unlikely, but the results were easily explained by modulation differences.

As well as the mistaken use of no green contrast to eliminate motion cues and because the real cues were not identified, most of these 1992–95 experiments required re-examination. The patterns were fixed and the bees could have learned the cues of position, modulation, area and edge length, and maybe orientation. The textures used for camouflage, with pixels 4mm square, were probably not resolved at the 27cm range. Unfortunately also, the scores were too high because the test patterns were presented for 10 minutes on each side, during which some bees made two visits and could learn which side to go. This mattered only for marginal scores. Also, it was not realised that the orientation on the left side of the target was discriminated separately from that on the right side, so some of the test patterns were inappropriate. All these errors of the day were uncritically accepted at the time and they still confuse the literature because they are quoted as support for various ideas about cognition in bees (see Chapter 12).

When colour is added to the tasks of the bees, we are still not clear whether the colours of areas are detected only by tonic blue, green and UV detectors of photon flux or whether they are detected by phasic modulation detectors of ultraviolet, blue, green or chromatic contrast at the edges—or by both. Different classes of neurons adapted to either type of input occur in the optic lobe (Chapter 6). The resolution of small areas of colour is related to intensity times the area, but some authors relate the detection of colour to the contrast at the edges. The experiments require accurate calibrations and a variety of tests in colour.

Symmetry cues

Possible adaptation of flowers to bees

Flowering plants evolved long after the insect visual system, so the evolution of flower colours and shapes was presumably influenced by insect vision. The colours of flowers and the colour vision of many pollinating insects are adapted to each other, but plant communities are rarely stable for long enough for an equilibrium to be reached. Free (1970) found that bees preferred symmetrical radial patterns, then bilateral symmetry and then irregular patterns, and also that bees landed at the edge of a plain target, but on a spot at the centre of a circle. Bumblebees prefer to land on flowers that are symmetrical. Hertz, Anderson and Free all showed that bees more easily learned the radially symmetrical patterns that they spontaneously preferred.

In his earliest experiments, von Frisch found that flower-like radial or concentric patterns of the same size were easily distinguished, but that triangles, squares, discs and ellipses were not (Figures 1.2 and 1.4), and a chequered pattern of squares was not distinguished from one of triangles. Later, Hertz found three classes of patterns—stars, circles and irregular blobs (Figure 1.4)—that were discriminated from each other irrespective of the length of edge, location or orientation. For 100 years, the outstanding problem was how the lopsided visual abilities of bees were adapted to their foraging needs, and mechanisms took a back seat.

Preferences for symmetrical patterns

All the early workers made use of symmetrical shapes with radial edges for training bees. On a horizontal surface, they could be approached from any direction. They showed salience—that is, the bees found them easily on a flat, featureless white table and would learn them readily because they could fixate on them. What the bees in fact learned was a different matter. It is now clear that when the training patterns are all equally symmetrical, the bees will not learn to discriminate the symmetry, because the patterns all show it.

Hertz found that bees avoided circles (Figure 1.5). It was also noticed that when flying bees landed on bilaterally symmetrical flowers, they lined up with the direction of the axis (Jones and Buchmann 1974). Much later, it was shown that they measured the flowers for degrees of symmetry. For example, Møller (1995) found that bumblebees preferred to forage on more rather than less symmetrical flowers, and the former yielded more pollen and nectar.

In 1994, Miriam Lehrer decided to revise an old demonstration by Gertrud Zerrahn (1933) to show that bees had a preference for symmetry. We built an apparatus with 12 compartments (Figure 9.11a) and trained marked bees to come to neutral patterns. The bees entered the central arena, from which they could see into all 12 compartments. At intervals, the apparatus was rotated, so the bees could not learn the locations and had to look at the patterns. When the bees were familiar with the place, four new and different patterns (each reproduced three times) were spread around the 12 compartments with no reward, and the free choices of the bees were observed. The criterion of a choice was when a bee crossed the threshold of a compartment. With a plain black disc as the initial attractant pattern, we were able to show that bees preferred radial patterns to other patterns.

When we randomised the modulation by using one of six regularly changed 50:50 black-and-white checkerboards as the attractant pattern (Figure 9.11b), we were able to show that the bees preferred patterns of lower spatial frequency and radial flower-like patterns to random or regular textures, circular patterns and even over the checkerboards to which they were attracted in the first place. They preferred radial patterns of any sort and a vertical axis of bilateral symmetry, but avoided concentric circles. No preferences were found with other patterns, although many were tried. These results launched us into a search for mechanisms of discrimination of radial and tangential edges and symmetry.

Figure 9.11 Bees' preferences for unfamiliar patterns. a) The bees were trained to come to the apparatus with 12 partitions, which displayed various shuffled checkerboard patterns, (b) some of which were rewarded. They were then presented with several hundred choices between four patterns (three of each). Their choices were counted and reduced to percentages. c) Radial or tangential was preferred to random. d) Radial preferred to tangential. e) Bilaterally symmetrical preferred to asymmetrical.

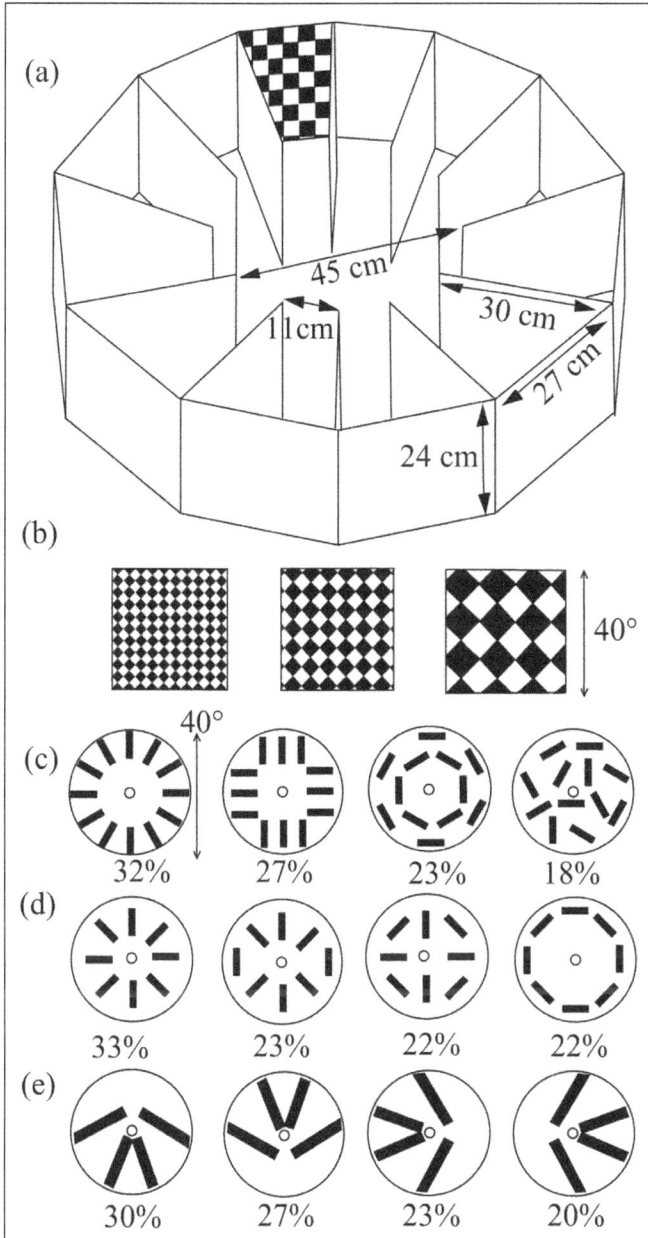

Source: Lehrer et al. (1995).

Figure 9.12 Radial and tangential cues are recognised in unfamiliar patterns.
a) In the square, training on rotated and shuffled radial patterns versus tangential
ones. b–d) Tests of the trained bees with circles, sectors or bars.

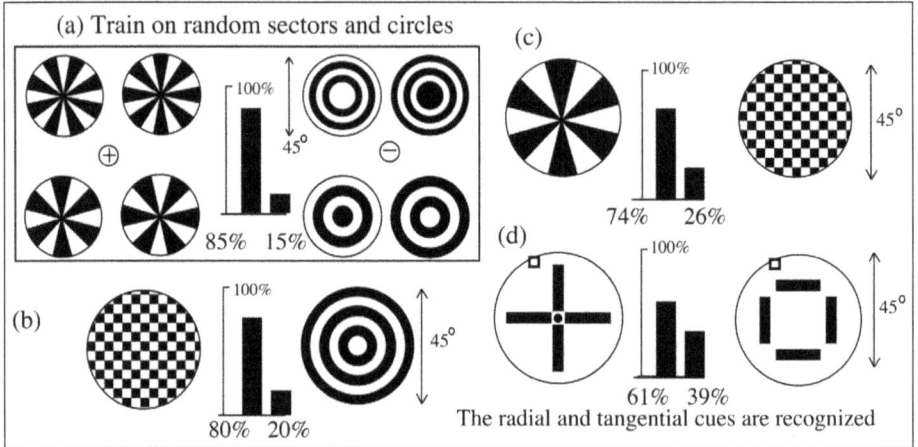

Source: Horridge and Zhang (1995).

Discrimination of sectors and circles

Following this demonstration of preferences for different parameters, bees
were trained with a pattern of black and white sectors or spokes, versus one
with concentric circles or a spiral, with no average orientation in either target.
To control against differences in areas of black, length of edge or location of
black areas, the sectors in one target and the circles in the other were randomised
by substituting a different target every 10 minutes, so that nothing remained
constant except the kind of pattern and the position of its centre (Figure 9.12a).
Bees so trained then discriminated pairs of unfamiliar patterns with radial
versus tangential contours, such as a cross and a hollow square and also parts of
circles or patterns of spokes. Rather than learning to recognise a circle, the bees
preferred to learn to avoid the unrewarded target even if it was blank.

These results, and those of Hertz with patterns presented on a flat surface, led
naturally to the proposal that bees had global filters for radial features and other
global filters for concentric circles or tangents, and that these filters detected
any part of their own pattern that coincided with their field of view. How easy
it was to imagine global filters, but how wrong!

Directing recognition with a coloured spot

The location of an added coloured spot can influence the bees to treat a bar as
a radius or as a tangent, depending on the bar location and orientation relative
to the spot. To demonstrate this, the positive target had a blue spot at the side
of the bar; the negative target had a similar blue spot at the end of the bar, with

the spots in the same position on each target (Figure 9.13a). Both targets were rotated by 90° in the same direction every 10 minutes between positions 1, 2, 3 and 4, so that modulation and the locations and orientations of the bar and spot were useless as cues. The trained bees are then able to discriminate a pattern of tangents versus a pattern of radials (Figure 9.13b).

Figure 9.13 a) A bar is recognised as radial or tangential with reference to a blue spot although the whole pattern is shuffled in location during the training. b) The trained bees discriminate tangential from radial in quite different patterns.

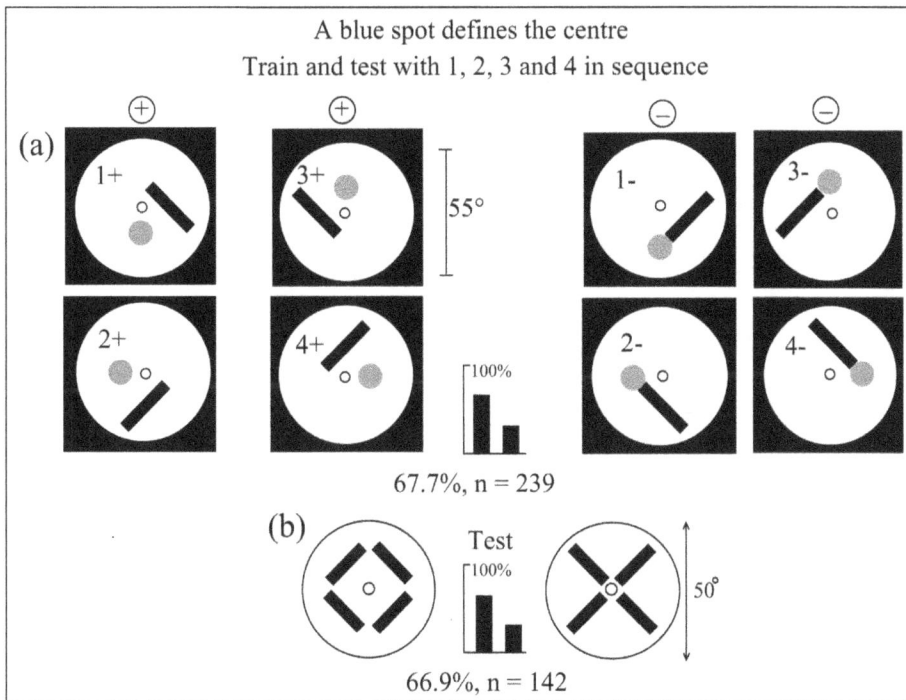

Source: Horridge (1997c).

Whether the single black bar was accepted as a radius or as a tangent depended on the bees accepting the spot as the centre. The bar could be rotated about the centre and was still discriminated as a radius or a tangent. In other experiments, strong symmetry, a strong outline of the target, occurrence of several radial or tangential features or a coloured spot and the geometry of the situation all influenced whether the bee detected a single bar as a radius, tangent or orientation.

Figure 9.14 Examples of discriminations between patterns that displayed two pairs of orthogonal bars, in which the orientation cues usually cancelled out. a–d) Patterns that differed in radial/tangential cues. e) A difference in average orientation between the two sides. f–j) Patterns that look different to humans but display no differences in cues for the bees. Possible cues are indicated as follows: H = horizontal; O = oblique + and O-- = orthogonal oblique orientations; R = radial; T = tangential; V = vertical. To the bees, the patterns in (e) differ, but (j) when rotated they are similar.

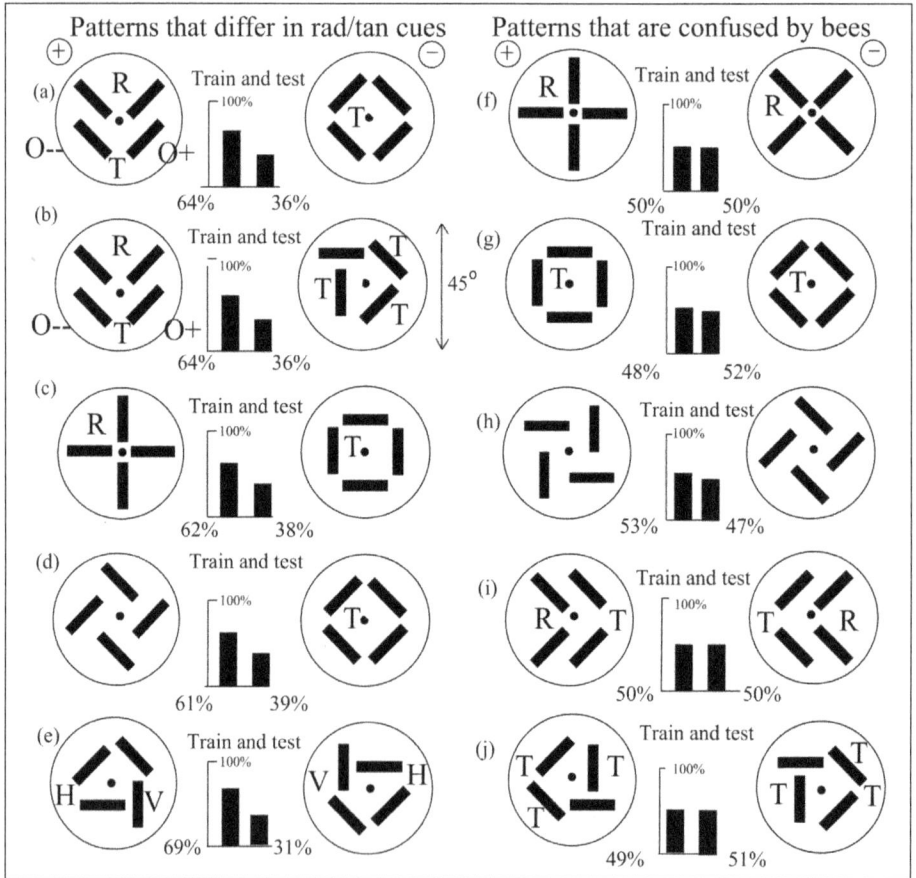

Sources: (e) from Stach et al. (2004); others from Horridge (1996a).

Strategies for listing the cues

Successive efforts progressively defined the limited number of cues in the repertoire of the bees. First, in 1995, the radial, circular or spiral patterns were rotated at intervals during training to remove cues derived from orientations, leaving the radial and tangential cues intact. In 1998, these were found to be colourblind. In 1999, radial and tangential patterns with radial symmetry based on three or six spokes were easily discriminated, but those based on four, five or seven spokes or sectors were not. This result is explained by the existence of three orientations of edge detectors.

Figure 9.15 Separate training experiments with various pairs of chevron patterns. a) and b) They detect the orientation cue on each side of the targets. c) The rewarded target is still discriminated when the other is rotated by 90°. d) A difference of 45° in the axis is not discriminated. e) and f) Discrimination requires one of the axes of symmetry to be vertical.

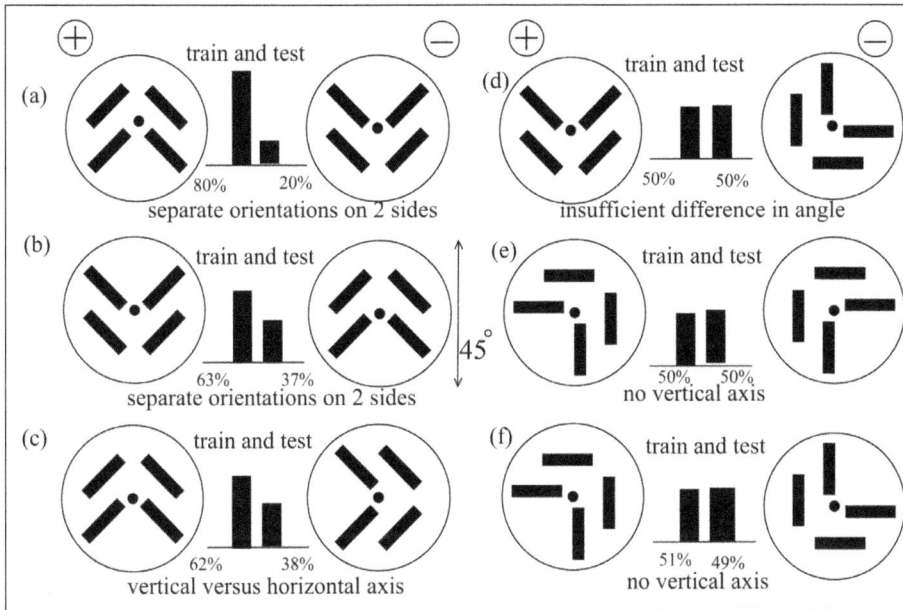

Source: Horridge (1996a).

In the second strategy, the arms of a square cross were rearranged to make many fixed patterns of two pairs of orthogonal bars, all the same size, area of black, length of edge and average position of black on the target (Figure 9.14). These patterns of four bars could not be discriminated from the same pattern rotated by 45° unless one of them had a vertical axis of bilateral symmetry (Figure 9.15). In that case, the bees could discriminate a rotation of the axis of bilateral symmetry by 90°, even if the test patterns were different from the training patterns.

The patterns with two pairs of orthogonal bars could be roughly divided into groups. The first group differed greatly in their content of radial, tangential or bilateral symmetry cues (Figures 9.14a–d) and the bees were easily trained to distinguish them from each other. Those in the second group were quite different from one another but were not distinguished (Figures 9.14f–j). It was inferred that they displayed similar cues. In tests, the trained bees accepted any of these patterns displaying the expected cues but no unexpected cue.

The bees measured the cues quantitatively. The real pattern was irrelevant. These patterns and experiments illustrate the value of Mill's rules of logic in the search for causation (Chapter 2).

Many pairs of patterns that differed from each other were not discriminated. For example, when the bees had learned the orientation cue, they failed in tests to distinguish the rewarded training pattern from other patterns with the same total length of edge and average orientation (Figure 9.9d). The bees looked for the orientation cue and found it equally in both patterns, irrespective of differences in layout. It was curious that the authors who still supported the eidetic image in the 1990s were blind to similar examples where patterns were different but not distinguished.

The third strategy turned to the recognition of position. A fixed pattern composed of two different colours was discriminated from the same with the colours reversed in position (von Frisch 1914; Gould 1986). Bees easily detected a vertical shift of the centre of an isolated area of black or colour relative to the reward hole as a cue of high priority. They also discriminated the exchange of two different colours in the left/right direction if there was green contrast or a radial pattern to stabilise the eye in the horizontal plane.

Finally, a chevron pattern with its axis of symmetry vertical was discriminated from itself rotated by 180° or by 90° (Figures 9.15 and 9.14i). The chevron pattern is curious in having no net orientation and its two radial bars cancel the effect of its two tangential bars. I concluded: 'The result with the chevron suggests that bees have a filter beyond those for circles or radial patterns, or for average orientation, and that it is related to bilateral symmetry, which is already known to have a broad biological significance for bumblebees' (Horridge 1996a). Of course, a single global filter was a bad idea, but at the time I could not model a bee filter that would detect the axis of bilateral symmetry irrespective of pattern. The widespread occurrence of symmetry in animals and plants, and the fast response to it, implies that there are many innate visual mechanisms for detecting symmetry.

More experiments with two bars at right angles

Bees detected the orientation cue separately on the two sides of the target (Figure 9.14e) and the radial or tangential edges on either side, but failed to respond to the global pattern (Figures 9.14j and 9.16). First, bees were trained with two bars on each target, alternating between radial and tangential, in corresponding positions on the two targets. The bees could not learn to discriminate because the orientation, radial and tangential cues were cancelled. The bees could not detect the consistent global pattern of an arrowhead pointing to the left on one target and to the right on the other (Figure 9.16a).

Figure 9.16 Examples of training with two orthogonal bars on each target.
a) Shuffling between radial and tangential bars in corresponding positions on
the targets; there was no remaining cue at all. b) Training with fixed patterns;
the preferred cue is the rad/tan difference. The shift of the bars causes a reversal
of choice because radial and tangential bars are interchanged.

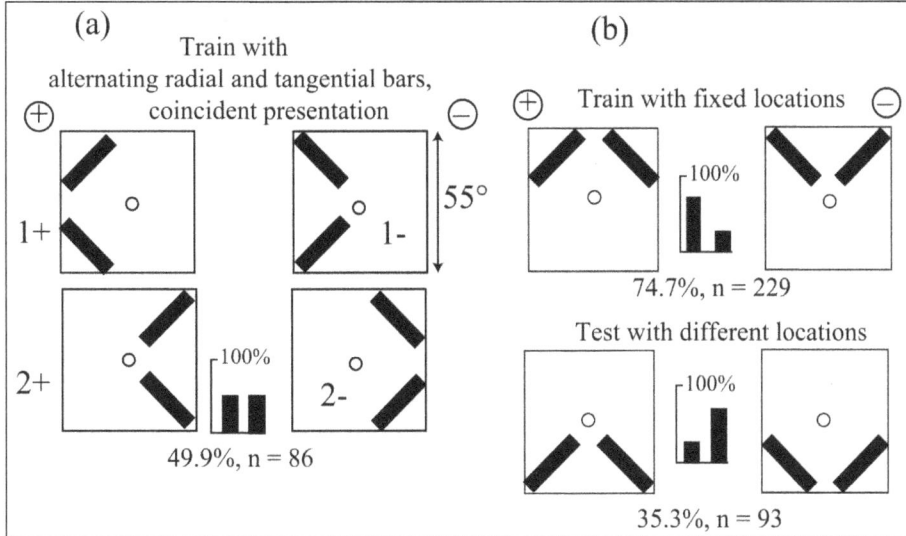

Source: Horridge (1997b).

Next, fixed bars in corresponding positions formed an arrowhead pointing
upwards on the rewarded target and downward on the other (Figure 9.16b).
The bar orientation was the least preferred cue. When the patterns were moved
down, the radial and tangential cues persisted but the orientation cues were
lost. The bees reversed their choices because the rad/tan cues were reversed
although the arrowheads were unchanged. From this, it was clear that bees did
not detect a global pattern of even two bars.

Analysis of cues in radial patterns

Bees easily spot the difference between radial and tangential edges, almost as
though the visual system operates in radial coordinates. The real pattern was of
no importance and, in tests, the trained bees accepted other patterns displaying
the rad/tan cues on either side of the target (Horridge 1996a).

Similarly, bees also discriminate between a fixed ring of spots or sectors, with
up to six spots in the ring and the same ring of spots rotated by half the angle
between the spots (Horridge 2000c:Fig. 6). The cue was the position of a key black
area, and unlike the situation with radial edges, the number of axes of symmetry
was less important for discrimination than the size of the individual areas.

When targets are rotated at random during the learning process so that positions
and edge orientations are shuffled, the cues must be presented as radial or

tangential edges, not as spots or areas of black. Radially symmetrical patterns of spokes have salience for bees, but they lose it when green contrast is removed, which shows the reliance on edges. In conclusion, there were two ways for a bee to detect the rotation of a radial pattern: by the orientations of edges of spokes and by the positions of areas with spots or sectors.

Feature detectors for radial and tangential edges

When analysed by the methods used for the orientation cue—by cutting the edges into short lengths or into square steps that were separately resolved—the feature detectors for radial and tangential edges were the same as those proposed for the orientation cue, 3° long, and therefore spanning three ommatidia in a row (Figure 9.5).

Figure 9.17 With radial spokes, bees detect the cues of 'black', 'radial' and 'position of the hub'. a) Training pattern. b) With black on both sides, the score is reduced, so black contributes. c) With the bars rearranged, the bees detect little difference. d) They detect radial on both targets, but not much difference. e) Square crosses, or angles at 90°, make a difference. f) The trained bees notice a difference in the position of the hub down to 5°.

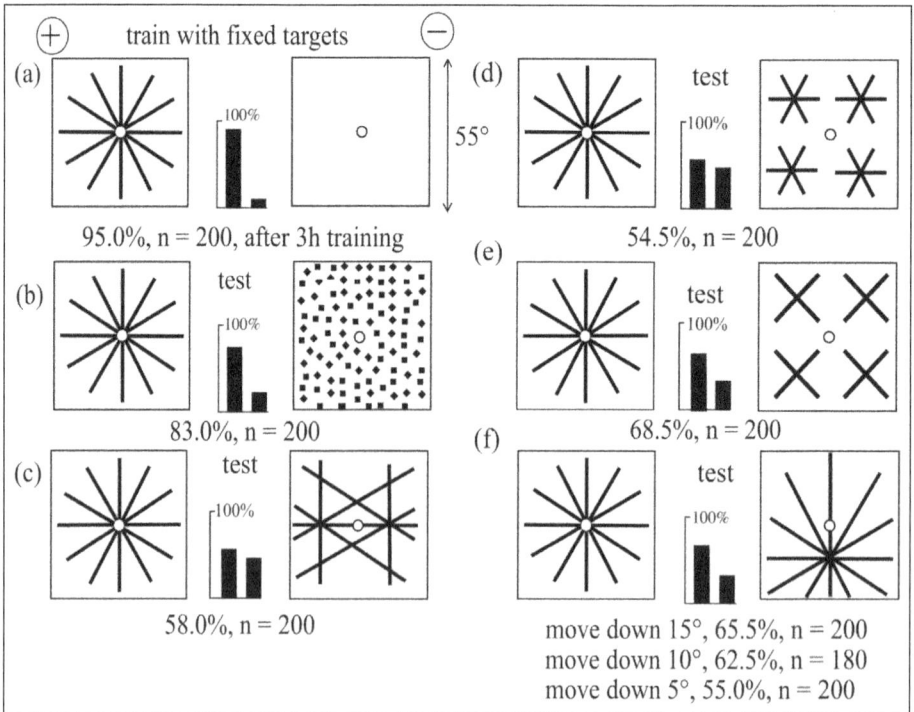

Source: Horridge (2006a).

Figure 9.18 Bees learn the position of the hub with concentric circles. a) Training pattern. b) The trained bees notice a difference in the position of the hub down to 5°. c) and d) With quite different patterns, they prefer the hub at the centre.

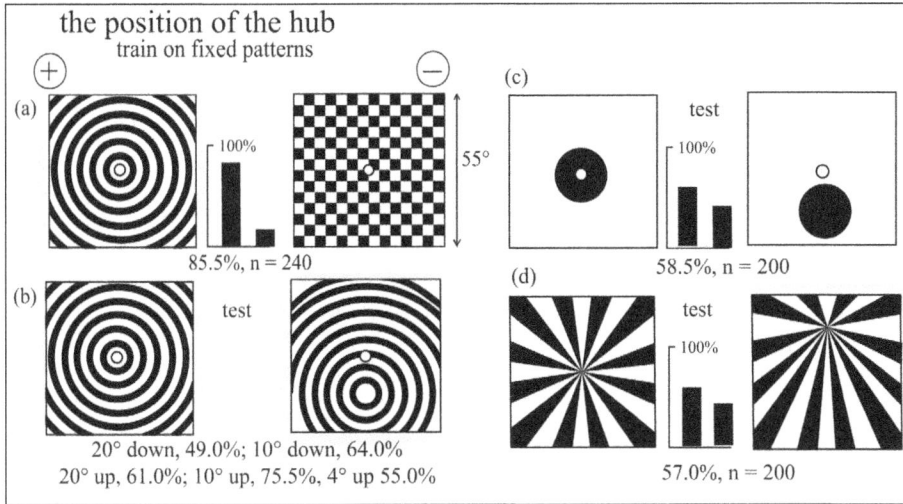

Source: Horridge (2006a).

A new cue: the position of the centre of symmetry

Bees could also learn the position of the centre of a ring or of concentric circles when trained versus a blank or a neutral pattern (Figure 9.18). They even detected the position of the centre of concentric curved lines, which implied that they detected the convex or concave curvature of the edges. There was no evidence for the idea that the bees detected the layout of whole rings or long curved edges, and much evidence against it.

When trained with a symmetrical pattern versus a blank target and then tested with two patterns at different heights (Figure 9.17a), the bees discriminated the expected position of the hub by as little as 5° (Figure 9.17f), in some cases with unfamiliar test patterns (Figure 9.18). A pattern of spokes or rings also stabilised the vision of the bees in the horizontal plane so that the position of a plain black area could then be learned (Figures 1.6b and 1.6c).

Bees discriminated half of a pattern of radial spokes or concentric circles from the other half, cut either vertically or horizontally, and irrespective of scale. This was the observation showing that radial and circular patterns were not detected by pre-formed combinations of orientation detectors or global filters like templates, because with a single output such filters could not distinguish the separate halves of the pattern. Instead, the bees detected edges as radial or circular by the coincidence of numerous local edge detectors converging to a hub from anywhere in the array, irrespective of the real pattern (Figure 9.19).

The binding that defined the cue as radial or tangential was therefore not hard-wired but depended on the coincidences of responses of similar edge detectors anywhere in the local region of the eye. This is a diffuse mechanism with no template and no memory of the layout of the pattern. It also explains the mutual cancellation of radial and tangential edges or orthogonal orientations. Because it depends on coincidences, such a system gives the impression of having taken a snapshot as it detected a hub. There was no global template that detected the positions, angles or numbers of spokes, a circle of a given size or a right angle as a whole. Instead, there was a distributed administration that would identify any incomplete or partially obscured symmetrical pattern and find the position of its hub. The mechanism is similar to that summing orientation.

Figure 9.19 The distributed mechanism with no fixed template for locating and identifying a hub as radial or circular. a) The three orientations of the axes in the array of feature detectors for edge orientation were inferred from the retinal arrangement of the ommatidia on the retina. b) The coincidences of the radial vectors and the tangential vectors at the position of the hub. c) Parallel orientations are summed.

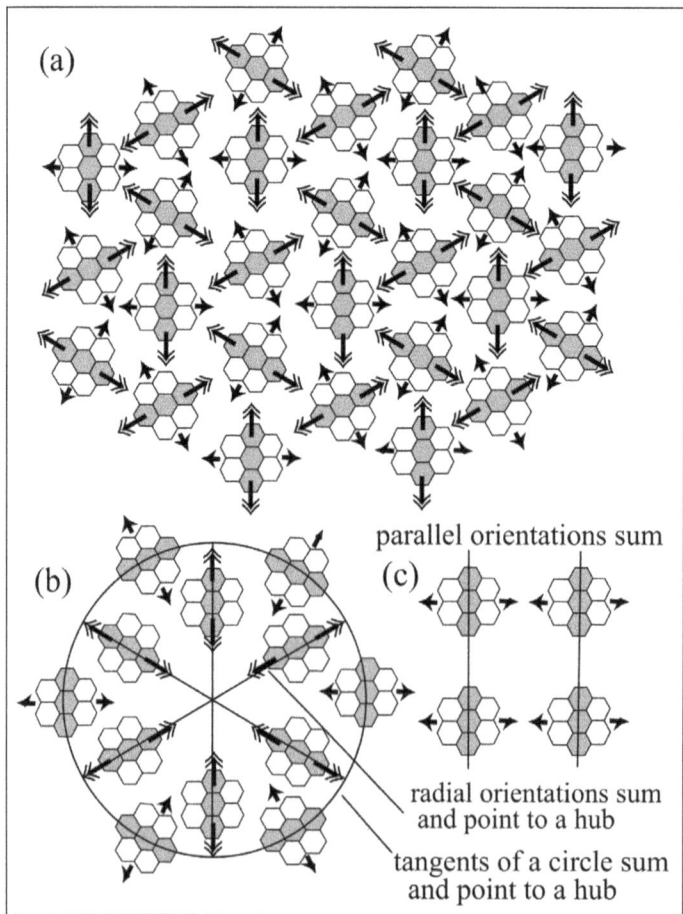

In conclusion, the bees identified radial and circular patterns by the regional coincidences and convergence of local vectors extending from edge orientation detectors, and the position of the hub was also a cue. We saw that every possible use was made of the various ways that the coincidences of the positions and vectors of the edge detectors could be counted in a local region, or their absence noted, but there was no mechanism to reassemble the pattern.

Figure 9.20 Detection of a vertical axis of bilateral symmetry irrespective of pattern. a–c) The bees were trained on seven bilaterally symmetrical patterns simultaneously, taken successively in pairs for 10 minutes on each side in the choice maze. Only three of the patterns are shown here. The pattern with the vertical axis in each pair was not rewarded, so the training was against the preference. Training scores from each pattern were collected separately. d–f) The trained bees were tested on the same seven pairs of patterns rotated through 180°. The same three are shown. The tests were done in random order between periods of continued training, which improved the performance.

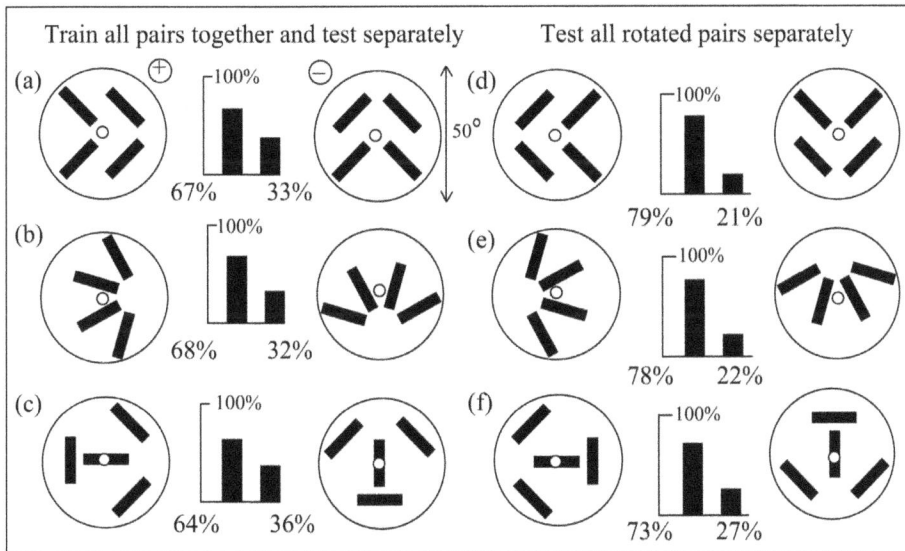

Bilateral symmetry

Bumblebees prefer to settle on flowers that have a more perfect bilateral symmetry than their neighbours. They measure the perfection of the symmetry. Untrained honeybees spontaneously prefer a vertical axis of bilateral symmetry in arbitrary unfamiliar targets irrespective of pattern (Figure 9.11e). The special properties of bilateral symmetry are illustrated by the chevron pattern (Figure 9.15), which, as a whole, displays no average orientation, radial or tangential cue. Bees readily discriminate it from itself rotated by 90° when one of the patterns has a vertical axis of bilateral symmetry, but the resolution of angle is poor—the same as for orientation differences.

The detection of the axis of bilateral symmetry irrespective of pattern was demonstrated with patterns of two pairs of orthogonal bars in the following way. The bees were trained with seven different bilaterally symmetrical patterns of four bars taken in succession, to train the bees to ignore the real pattern (only three patterns are shown in Figure 9.20). The positive (rewarded) target was the pattern placed on its side and the negative targets all had a vertical axis of bilateral symmetry (Figures 9.20a–c). The training was therefore against the innate preference. The bees learn to discriminate the symmetry irrespective of pattern in a few hours of training. On successive days, with continued training, they were tested on the same seven patterns rotated through 180° (only three patterns are shown in Figures 9.20d–f). Although all the patterns in the tests were unfamiliar, the trained bees still picked out the asymmetrical one of each pair, and with an improved performance, because they had more training. These trained bees also discriminated the axis of bilateral symmetry in completely different unfamiliar patterns with different numbers of bars. It does not matter whether the various patterns can be discriminated from each other by other cues, because only the orientation of the axis of symmetry was learned in this experiment, all other cues being inconsistent or the same on both targets.

The mechanism is not such a puzzle as it is in humans, because there is no evidence that the bees really see the patterns. My expectation is that the bees do not recognise the abstract property of bilateral symmetry about an axis in general any more than they recognise pattern or shape in general. As with shape, they find a way to use their feature detectors for the task in hand, irrespective of the pattern. Some time ago, Jones and Buchmann (1974) found that bees landed on zygomorphic flowers in line with the axis of symmetry, even when tipped away from the vertical. Therefore, as the bee scans from side to side, she detects the same sequence from her feature detectors with either direction of the scan, whereas an asymmetrical flower sends back a signal that is different in the two directions. This would be sufficient to identify and measure the bilateral symmetry with a vertical axis.

As another mechanism with distributed administration, it is known that some cues are detected separately on the two sides, probably by the two eyes. If the bee compares the colour, area, height of the centre, radial or tangential, and perhaps other cues, by the two eyes, then a measure of bilateral symmetry can be made in a large number of patterns. In fact, any filter that has two spatially separated pass bands can detect some bilateral symmetry about an axis drawn between them and it is possible that symmetry is detected by several sets of coincidences in the overlapping forward parts of the two eyes.

Other cues

The measurement of size and area

Larger objects project to more facets on the eye and size is measured as the solid angle subtended. The total photon flux within the area is a separate measure and bees can be trained to discriminate a large grey spot that is 50 per cent black from a smaller spot displaying the same amount of black. Bees spontaneously prefer a large black spot to a small one, but the preference can be reversed by training. When there are two or more spots or bars in the same local region of the eye, the bees lump them together and cannot distinguish their separate sizes, but when they are more than 15° apart, bees distinguish them separately, like landmarks.

The feature detectors for areas are probably related directly to the photoreceptors of the retina. The measure of size might be as simple as the summation of responses in a local region of the eye, not necessarily the same size as for edges.

The angle at the eye is combined with the range to give the bee a measure of the absolute size of a black or coloured area, as shown by randomising the angle subtended and the range while keeping the absolute size constant (Figure 7.5).

The reward hole

As the bees were familiarised with the apparatus, they flew into the reward hole many times in the days before the training began. They detected it from the baffle and looked towards the point where they had previously landed on its lower lip. Bees easily remember whether the place just below the reward hole is dark or light. Many years ago, Baumgärtner (1928) found that bees discriminated the relative positions of small coloured rectangles only when displayed close to this point of landing (Figure 1.3). Friedlaender (1931) found that when bees had learned to discriminate an area of black near the reward hole, they lost it when it was moved up (Figures 1.6e and 1.6f).

A filter for the height of the centre

Let us consider what we know about how we locate things in space. Simple tasks that humans take for granted, such as grasping, stepping and tracking, are dependent on an ability to locate objects. Several studies have found that humans locate either the centroid or the midpoint between opposite edges and that the least change in position that can be detected is proportional to the linear separation of the objects. This is Weber's law of separation (Whitaker et al. 2002). The experimental results for human vision can be explained by two-dimensional spatial filters with fixed coordinates on the eye, which detect intensity and operate at several different field sizes to locate position.

Figure 9.21 In these examples, the edge orientation cancelled to zero and the most preferred cue was the location of the centre of black within the local region. a) Training with two separate spots on each target. b) Test with the small spots only; the bees preferred the black at the top. c) There was no preference in a test with the common centres of gravity at the same level. d) Training with T patterns. e) Failure with the centres of gravity moved to the same level, showing that the pattern was not the cue. f) Failure when black was exchanged for white, because the bees had learned the positions of the centres of black.

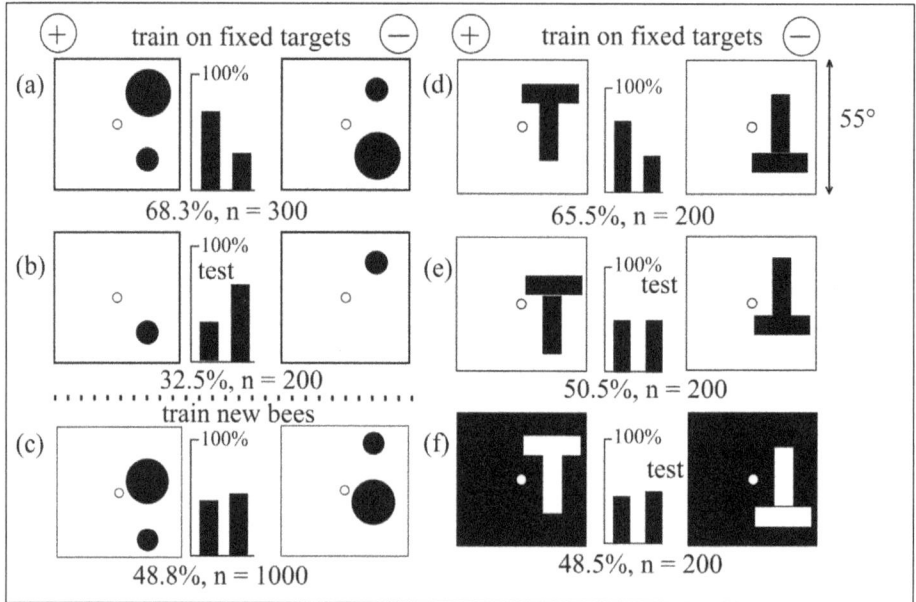

Source: After Horridge (2003b).

The position of the centroid is a feature of any pattern and is used in machine vision as a uniquely defined point. In a lucky discovery, it was found that bees could not discriminate the rotation of an equilateral triangle when the centroid remained at the same position in the vertical direction (Horridge 1997a). The difference in the vertical direction in the positions of the centres of two otherwise equal patterns is a sufficient cue for discrimination—a very small part of the pattern indeed. When there are two separate spots or bars on the target within the local region of the eye, they are not detected separately, but the position of their common centre can be learned (Figure 9.20). Spots or bars further apart on the target (in adjacent local regions of the eye) are discriminated separately. Therefore there is only one filter for position of an area in the vertical direction in each local region of the eye.

Figure 9.22 a–d) Patterns that bees easily discriminate when trained alternately on pairs 1 and 2. a) Orientation at –45° versus +45° and spiral versus sectors. b) Orientation at –45° versus +45°, and the rotation of six spokes. c) Orientation at –45° versus +45° and two spots, one at the top versus one at the bottom. d) Spiral versus sectors (randomly rotated) and two spots, one at the top versus one at the bottom. e–h) Patterns that bees do not discriminate when trained alternately on pairs 1 and 2. e) Horizontal versus vertical gratings, and +45° versus –45°; these patterns were fixed in position during the training. f) A spiral of period 8° versus 12 sectors, and six radial versus six tangential bars; these patterns were shuffled by rotation during the training. g) Radial patterns based on symmetry of three and two other positions of the same patterns; these patterns were fixed in position during the training. h) A spot (sub-tense 16°) at the top of the pattern versus the same spot at the bottom, and the same spot at right versus at left.

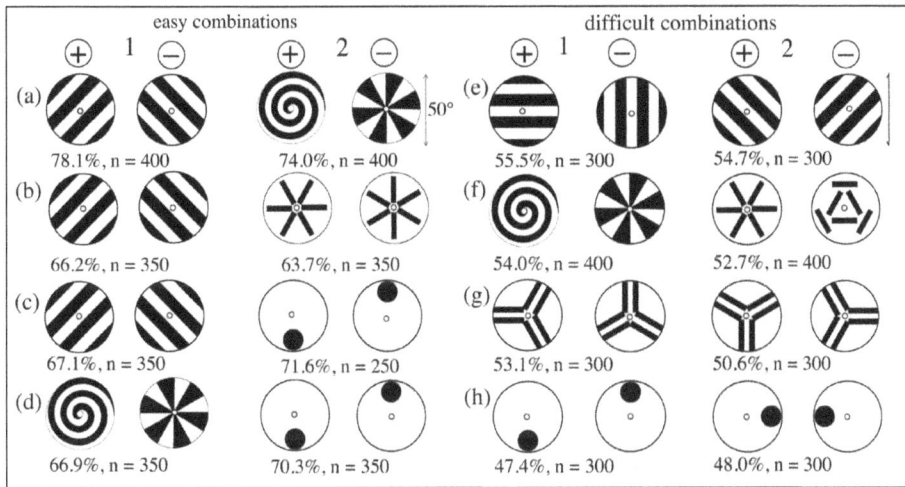

Bees learn one cue of each type in each local region

Bees were trained, first on one pair of patterns for 10 minutes, then on a second pair for 10 minutes, and then back to the first, which was repeated for two hours (Figure 9.22). The pairs of patterns were selected to test the hypothesis that, in each local region of the eye, there was only one channel of each kind and each processed its own cue. If so, two different pairs of patterns that display different states of the same cue would interfere with each other during the learning process. The cues tested were: average orientation of patterns of sectors, edges, radial and tangential edges based on a symmetry of three or six, the position of a black spot and the exchange of black and white.

In Figures 9.22e–h, the patterns all look different to human vision, but the cues in each pair excite the same set of filters in the bee. In Figure 9.22f, the positive cue is tangential for the spiral/sector patterns but is radial for the patterns of six bars, so the bees learn not to use the tangential and radial cues and are left with nothing else, although the pairs of patterns are quite different. The positions of the spots in Figure 9.22h are all different, but the bees are unable to learn that they are positive in two positions and negative in the other two, when they are seen in the same context. The bees do not find a consistent cue in either the positive or the negative targets.

On the other hand, bees easily learn to discriminate when one alternated pair of patterns is of one type and the other pair is of a quite different type (Figures 9.22a–d). The inference is that there is one processing channel for each type of cue in each local region of the eye (Figure 10.1).

The bees failed when they were faced with two simultaneous tasks involving the same type of cue in different states, although all sixteen of the pairs were readily learned individually. They do not learn one of the pairs and ignore the other, which would improve their chance of a reward. Instead, they start to learn afresh each time the patterns are changed, as if each cue channel cannot learn two tasks at the same place. Of course, in a different context, at a different place, the bees might be using the same cues for a different choice because other landmarks are different.

Detection with and without memory of it

In the experiments in which the bees discriminated gratings, they detected the modulation and orientation differences but not the grating pattern. Similarly, they detected spokes or parts of circles as radial or tangential and located the centres but did not recognise the patterns. This helps to clarify the difference between discrimination of patterns and memory of them.

The essential first step is the simultaneous detection of all edges in the local region, but there is no memory at this stage. In each region, the number of excited edge detectors gives the total modulation and their orientations are integrated to give the average orientation (Figure 9.5) and to identify spokes and circles and locate their centres (Figure 9.19b). Other cues, such as the position of the centre of black, colour and the area or size, are abstracted by other pathways in parallel and remembered according to a scale of preferences. All these cues are remembered separately if rewarded, but processing of the image stops short of reassembly.

The preferences for the cues

In Hertz's earliest efforts, preferences were observed when untrained bees selected one pattern from a variety and they learned most readily the patterns that they spontaneously preferred. Preferences were also revealed in differences in the rate of learning and the maximum score achieved. In most discrimination tests in the past decades, the bees had no control over the choice of the images and it was often not clear whether they learned the rewarded pattern, the difference between the rewarded and the unrewarded patterns, to avoid the unrewarded target, or to avoid unexpected cues that were not in the training patterns at all. More to the point, it was not clear how many cues were learned in parallel and in what order. This situation was due to the lack of sufficient tests of trained bees. The preferences were not explored because most of the cues had not been described. Because the preferences were ignored, it was not possible to understand how patterns displaying several cues were discriminated.

Figure 9.23 A way to demonstrate preferences for cues during the learning process; bees were trained to discriminate between a rewarded target displaying two various cues versus a neutral pattern. a) Train with modulation and a large spot. b) Test against the mirror image, with a poor score, showing that these cues are both preferred. c) Test showing that the bees used the spot as a cue. d) Test showing that they also used the modulation. e) Train with parallel bars and a large spot. f) Test against the mirror image, with a high score showing that the position of something was remembered. g) The bees used the spot as a cue. h) The bars were poorly remembered.

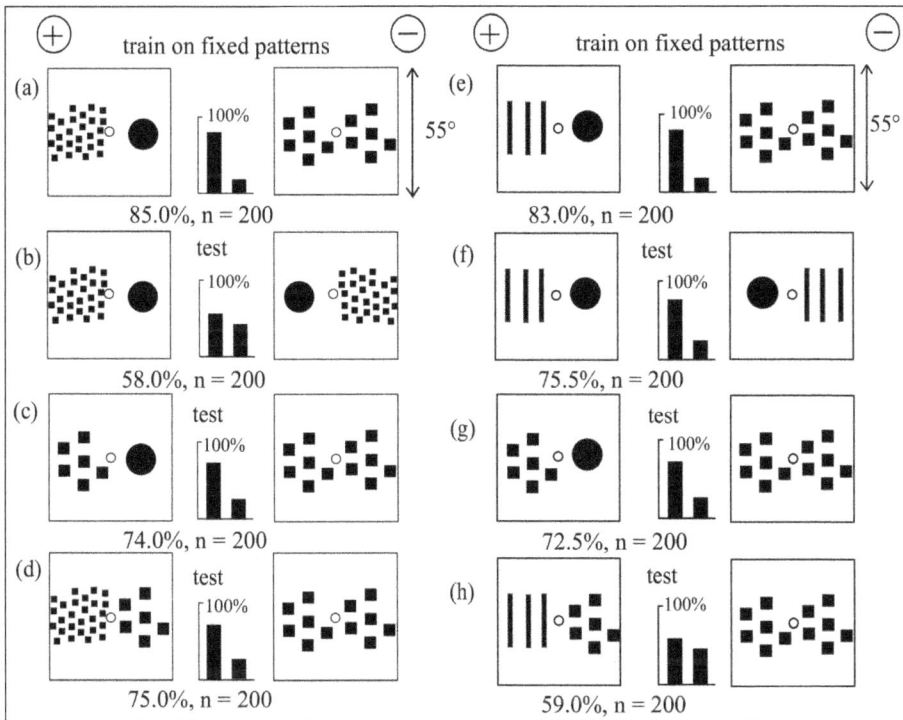

Source: Horridge (2007).

To list the preferences, bees were trained to discriminate between a rewarded target with one pattern on its left side and a different one on the right, versus a neutral pattern (Figure 9.23). This arrangement gave the bees a choice of what to learn on a single target. Tests showed that in some cases they learned two or three cues simultaneously; in other cases, the bees learned one or they preferred to avoid the unrewarded target.

By working with different combinations of patterns, it was possible to put the cues into an order of preference. When two or more were displayed at the same time, the bees learned one first and more strongly than the others. The order of preference during learning was: 1) total area, 2) position of the centre of area, 3) total modulation, 4) radial edges, 5) average local orientation, 6) positions of hubs, and 7) tangential edges. Single black spots and strongly modulated patterns were easily learned. Large black spots were preferred over small ones. Radial spokes and parallel edges were weak cues. Symmetry in a pattern of bars was preferred as a cue over the edge orientations that generated the symmetry. When a weak and a strong cue were presented together, the weak one was scarcely noticed. Various patterns such as a spot, a square cross, a group of small squares and many complex patterns, provided as cues only the area of black, modulation and position of the centre of black. The bees could learn not to avoid circular patterns. When two colours were presented side by side on the rewarded target or on separate targets, the bees had difficulty learning both at the same time. They learned blue in preference to fawn or yellow, even if they had to learn to avoid the blue.

When presented with a pattern on each target, they ignored the cues that were displayed on both targets. When no preferred cue was associated with the reward, they learned to avoid the unrewarded target even if stuffed with cues or blank. In general, they learned to avoid the negative target when the most preferred cues were displayed there.

Salience versus retinotopic cues

In previous work, when a broad black bar or spot was moved more than 10° after training, the bees did not recognise it in its new place, showing that the bees had learned the place. In the choice above, with three parallel bars or a large black spot (Figure 9.23f), the training pattern was distinguished from its mirror image. This was explained by the change in the attraction of the spot in an unexpected position. In other cases, even though the training was successful, the preference was equal on the two sides, so mirror images were confused (for example, Figure 9.23b).

The sensitivity to displacement was related to the field size of the cue. Cues of orientation and the position of a black area were more retinotopic and therefore detected in smaller fields and were not salient. Modulation was detected over larger displacements and therefore in larger fields, so that it was more salient. Small fields implied some failures to detect; large fields implied some failures to localise, but improved the salience. Each cue had its own compromise field size. The most salient cue was a small black spot, and was detected in a large field.

Resolution of the feature detectors

In general, the sizes of the fields of the different cues have not yet been measured. The bees estimated the cues quantitatively and learned absolute size, relative size or angular size of a spot, depending on how they were trained. The minimum detected difference in modulation between two textured patterns was about 30 per cent (Horridge 1997a). When trained to a particular orientation of an edge, and then offered a choice between two others, they preferred the orientation that was nearest to the rewarded one, with minimum detectable difference about 30°.

The widths of the angular sensitivity curves for the orientation of an edge (Srinivasan et al. 1994) or of the axis of bilateral symmetry (Horridge 1996c) were about 90° at the 50 per cent level, because the edge orientation detectors were short. The minimum detected difference in positions of an area of black or the centre of a radial hub in the vertical direction was about 10° (Horridge 2006b). Long training improved the precision of discrimination. It is probable that the field sizes, minimum difference and resolution limit depend also on the pattern. Bees are particularly effective at discriminating the transposition of two coloured panels in the vertical direction on the target, even with no green contrast, and easily discriminate differences of 6° (Gould 1986; Horridge 1999a, 1999b, 2000b). Probably more than two can be learned, but tests of the trained bees with the individual colours in their separate positions have not gone beyond two. Bees make simultaneous use of landmarks in different directions but the minimum angle between them for each of the cues is not known.

Avoidance of a parameter not in the training

It was accidentally discovered that in a test, discrimination is lost when a black spot is added at the same place on both targets. Later, it was noticed that the addition of any parameters that were not displayed in the training caused the bees to act as though they were in the wrong place when they detected a cue that should not have been there. As a result, a small black spot is characterised mainly by parameters that are absent. The decision process makes full use of the available options provided by the repertoire of feature detectors. On arrival at a new place, the bees behave as though they have a list of cues marked as familiar or not, so increasing the variety of labels and useful landmarks. There is less effect when a parameter is duplicated or when an expected parameter is omitted from a test, because they had learned several. All these conclusions were logical inferences from a variety of tests.

Why do they learn more than one cue?

The very high scores obtained when training on a single pattern versus a white target are due partly to the fact that the bees easily detect black (and yet they still make some errors). This soft option was not, however, validated by the subsequent tests, which showed in every case that the bees learned three preferred cues in parallel—notably, the area, the modulation and the position of the centre. High training scores are misleading at the start of the training when the bees first learn to go to anything black. The high scores show that the bees have an easy choice, not that they see the patterns.

Learning several cues in order of preference, with their positions, has two advantages. First, in a natural situation, the coincidence of several cues makes it less likely that they mistake the place. Second, the more cues they learn, the more likely they are to find the reward although some part of the scene is changed.

It was cues all the way

By the 1960s, feature detectors were the best explanation for image processing in vertebrates, as vindicated by work on computer vision. By 1994, the idea had reached bee vision. A different type of experiment, however, was required to show that there was no additional memory of the pattern as distinct from the cues. Instead of shuffling the patterns to eliminate unwanted cues, the patterns now had to be fixed on the targets to give the bees an opportunity to form an eidetic image (or retinotopic memory). The trained bees were then given a large variety of tests to see what they had learned.

A rewarded black square was easily discriminated from an unrewarded oblique bar of the same area (Figure 2.1). This was exactly the pair of patterns used by Wehner (1981:Fig. 86) to illustrate the detection of areas of overlap and non-overlap in the theory of the eidetic image. Tests showed that the real cue was the position of the orientation of the edges of the bar on the unrewarded target and there was no indication that the bees noticed the rewarded black square at all.

Even more significant results emerged after training on a single black bar versus a plain white target (Figure 9.5). The trained bees showed equal preference when the alternative choice was a square, a rotated bar or a line of small squares of the same area, centred on the same place on the target. They could not recognise the training bar when it had been moved to a new place on the target. The cue was anything black of the expected size at the expected place. The bees were sensitive to an additional parameter or change in the magnitude of the cue, but not to shape or pattern.

In numerous further examples, there was no evidence that the bees had remembered or even detected the patterns, only the cues. The cues had won the day by default, but they were limited when alone. Each feature detector is interested only in a field of 3°, but the cues, like the orientation-detector neuron

(Figure 9.4g), have been summed over a field up to 20° wide in a local eye region. The local eye regions collaborate to recognise a familiar place, as described in the next chapter.

This analysis has been exposed with all the tedious detail that has to be explored to reach the simple model (Figure 9.1). It was made possible by the use of patterns that subtended less than 40–50° as seen from the point of choice of the bees, so that the bees could not make use of the configurational layout of widely separated parameters. It is significant that the type of system is not like wax, which moulds to any shape to make a memory, but is a varied collection of innate boxes that collects running totals of a few types of units of data from their local region. The mechanism illustrates how a picture or panorama can be recognised by a simple mechanism although the information in it is greatly reduced.

Endnotes

1. One way to understand this chapter is to read the illustrations, starting with the training patterns at the top, and then consider yourself in the position of the trained bee in the tests, looking in the expected place for the cues learned in the training. It then becomes apparent why the bee succeeded or failed in the test.

10 RECOGNITION OF THE GOAL[1]

We can now return to the topic that caused Forel, Lubbock and Plateau so much trouble in the late nineteenth century: how bees locate and then recognise their destination. The roadblock to progress at that time was that the bees had already arrived at their destination, so they were as confused by the changes in the flowers and rewards as the researchers were about what the bees expected to find. Research since then has shown that bees navigate in the right direction for the correct distance using a variety of flexible mechanisms. They care little about exact appearances along the way, unless they have to search, but they care a great deal about recognition of the exact place of the reward or the hive. They persist in searching because they have not learned an alternative strategy. To study recognition, we need to know exactly what the bees have learned.

Wasps that dug out nests among the sand dunes of the Netherlands provided early indications of the mechanism. During the 1940s, Baerends, van Beusekom, Tinbergen and their Dutch colleagues placed artificial markers around nests and removed other obvious landmarks. When the wasps had learned the layout, the configuration of the markers was changed. Arriving near the nest with food for the young, the wasps made the best match they could between the altered configuration and what they remembered of the previous one. They preferred to rely on distant rather than nearer objects of the same apparent size, and at first used the whole configuration to guide them to the nest hole. With increasing experience, the wasps relied more on a few selected landmarks. Their responses showed that they approached progressively towards the place as they detected the expected landmarks at the expected angles (Figures 10.1a and 10.1b).

Bees approaching the goal along their usual track detect first the most preferred cues on the nearest landmark. This reminds them of the direction to the next landmark, and so on. They orient themselves by reference to their own body coordinates and move in the direction that increases the angles between familiar landmarks. This strategy improves the fit between the scene surrounding them and their memory of it. The whole panoramic context must be appropriate for the recognition, as described over the years by many researchers on landmarks (Rabaud 1928; Thorpe 1956:258; Anderson 1977b; Collett 1992; Collett et al. 2002; Fry and Wehner 2002), but they do not remember a copy of the whole scene. The phrase 'whole panoramic context' means that a number of expected landmarks must be recognised and no unexpected ones, otherwise something is wrong. For the bee, however, there is no 'whole scene', only labels on landmarks, which are all recognised independently of one another.[2] Although odours are significant, bees can rely solely on visual landmarks (Dyer and Gould 1981b; Geiger et al. 1995).

Figure 10.1 Recognition of the goal. a) A ring of pinecones was placed around the nest entrance. b) When the wasps were familiar with this, they were tested on their return with a choice between the original and a modified ring. c) and d) They would also use smaller numbers of cones. e) With two landmarks close together relative to the size of the wasp, the image difference function has a sharp minimum that indicates the position of the nest. f) With landmarks that are far apart, the image difference function is broad and shallow, allowing the insect to circle and wander without getting lost.

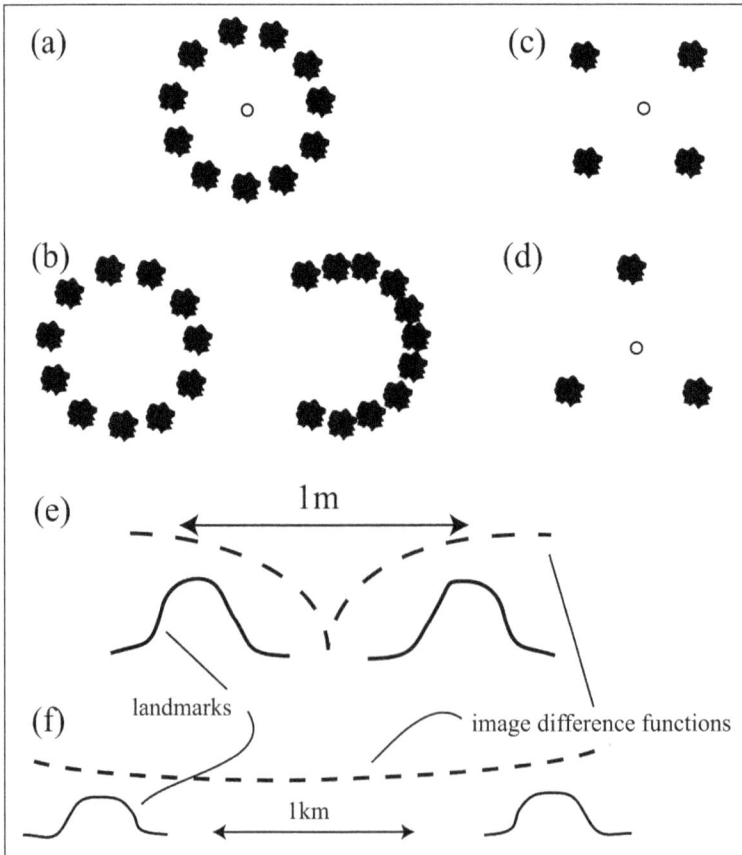

Source: (a–d) partly after van Beusekom (1948).

Figure 10.2 The arrangement of the channels in parallel behind each local region of the eye, as inferred from a wide variety of data. This local system detects one cue of each type in parallel, together forming one landmark label. These local regions are arranged around the head, as illustrated in Figure 10.7.

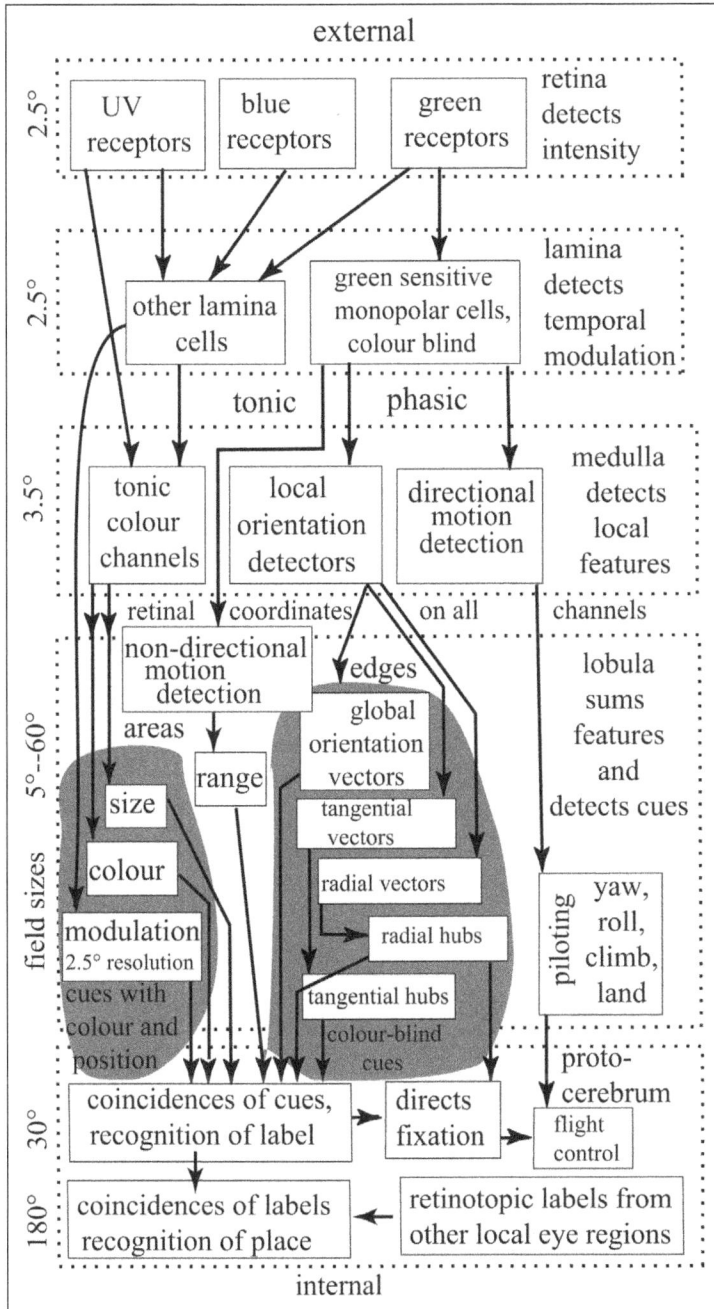

Source: Revised from Horridge (2000a).

The most important aspect of this homing strategy is the scale of the playing field. For the bee or wasp using distant landmarks to head towards home, the directions of cues and their heights above the horizontal change slowly, as though the insect is sliding down a gentle energy slope on which its position at any moment is not very critical (Figure 10.1f). The energy slope is that of the three-dimensional image difference function, which can be calculated from the total change of range, position and height of all the landmarks as the insect makes an incremental movement among them. When bees are using landmarks nearby, the image difference function changes direction quickly, as though the bee is sliding down a steeper energy slope with a sharp indicator of position at the lowest point (Figure 10.1e). The most effective strategy is to be able to switch between distant and close landmarks. This ability to switch between landmarks, and the use of several cues, caused some confusion for early investigators. Exactly the same principles apply to mobile machines with computer vision that recognise a place with a panoramic camera.

From the work described in the previous chapter, we have a list of cues that bees recognise, and if there are signs that further cues exist, we have methods of discovering them. The time has come to put the whole mechanism together.

Parallel pathways in each local region of the eye

Our knowledge of the eye and optic lobe (Chapters 5 and 6), together with the research that lists the cues, can be summarised by a formal diagram with a separate channel for each type of cue in a local region of the eye (Figure 10.2). In the periphery, green receptors connect with the large lamina cells that detect temporal modulation in individual ommatidia. These cells in several separate channels connect with the feature detectors, which detect the direction of local motion from the *sequential* modulation caused by a *moving* edge, and also with local orientation detectors, which detect the *simultaneous* modulation caused by a suitably *oriented* edge. Other lamina neurons detect the modulation in individual receptors (Figure 9.4e) with maximum resolution, and with blue and green-sensitive pathways.

These feature detectors span a group of seven retinula cells in bright light (Figures 9.4 and 9.19). They respond as independent units, so there is no improved detection of the modulation or the orientation angle with increased edge length. Large numbers of local orientation detectors with parallel axes feed into large-field orientation detectors, which therefore have the same axis of orientation as the edge detectors (Figure 9.4f). There is improved detection of large or parallel edges because the summation increases the signal but smoothes the noise. Within the local region of the eye, the summation of different orientations destroys the discrimination of shape and measures the average orientations in local areas of patterns and textures. The vectors of the local edge orientation detectors also feed into other detectors with large fields for the positions of the hubs of circular and radial arrangements (Figure 9.19b). Radial and circular patterns are identified separately but not visualised or reassembled in their layout. All three receptor colour types feed into tonic channels that separately detect colour, size and pattern disruption (on the left in Figure 10.2).

These local regions of the eye subtend about 20°, depending on the cue, and each local region sends a localised label, consisting of one cue of each type, towards the memory. There is no provision for detecting two separate sets of coincidences of cues within the local region. Whether the locally coincident cues are stored in memory depends on the context, the reward or the time of day.

The arrangement of channels has further consequences. There is no path for a transfer between green and blue pathways, otherwise orientation discrimination would not be restricted to green receptors and colour discrimination would be impossible. There is no provision for discrimination of orientation of edges from parallax and when this point was recently tested, no evidence was found (Horridge 2003a). The summation in the local eye region rejects non-coincident excitation and smoothes out the local features. The bee cannot detect two orientations, radials, tangentials, areas, positions or colours *at the same place*. All processing is done by the coincidences of responses in each array of feature detectors of each kind, all of which function independently of each other irrespective of the layout of the pattern. There is nothing special about this universally occurring mechanism of sensory processing. At the level of the local region, discrimination is like tasting a pudding that has a coincidence of flavours, or detecting an odour containing a number of different molecules, irrespective of their spatial pattern in the mixture.

Large and small patterns are differently discriminated

The different processing of large and small patterns has been a troublesome issue, but it provides the key to understanding how bees use the whole panorama. When Wehner (1967) trained bees to discriminate the rotation of a square cross subtending 130° at the bee's eye, he correctly inferred that they located the areas of black on the targets (Figures 4.2a and 4.5). With the same patterns subtending about 45° at the bee's eye, however, Srinivasan et al. (1994) found that the bees could not discriminate a difference in the orientation of the cross at all and they inferred that only the edge orientation could act as a cue while orthogonal orientations were cancelled by summation (Figure 4.2d). It would be some years before it was clear that both observations were correct.

Figure 10.3 The Y-choice apparatus modified by the addition of a transparent baffle in each arm. The targets can be placed at 27cm or 9cm from the baffles to control the angle subtended by the patterns. The decisions of the bees are scored when they pass the baffles.

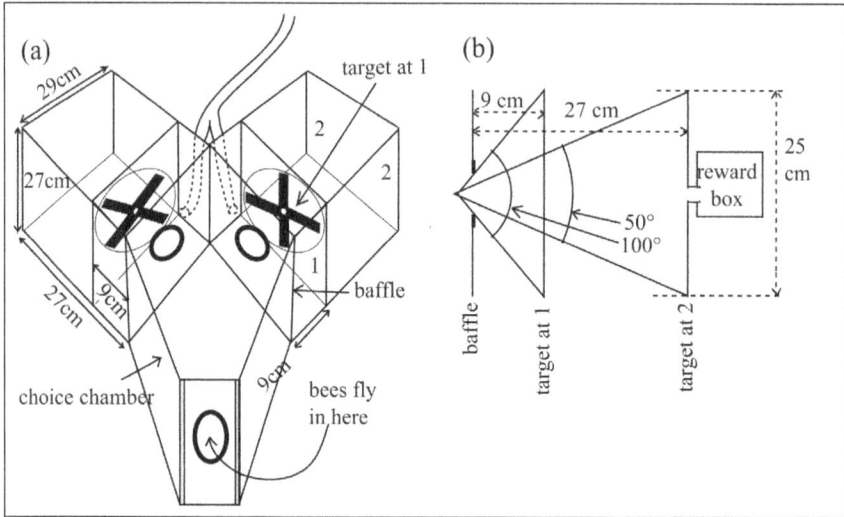

Source: From Horridge (1996c).

Figure 10.4 In the Y-choice apparatus, the bees pass through one of two training tunnels, each of which has four horizontal bars but one is rotated 45° relative to the other. a) The bees learn this situation very well. b) They then transfer their discrimination to targets of two crosses at 45° to each other, with the baffle at 9cm, as in Figure 10.3. c) With the baffle at 27cm, the bees cannot discriminate these targets, but this is not due to a lack of resolution of the eye, as shown by (d), a grating of period 4°.

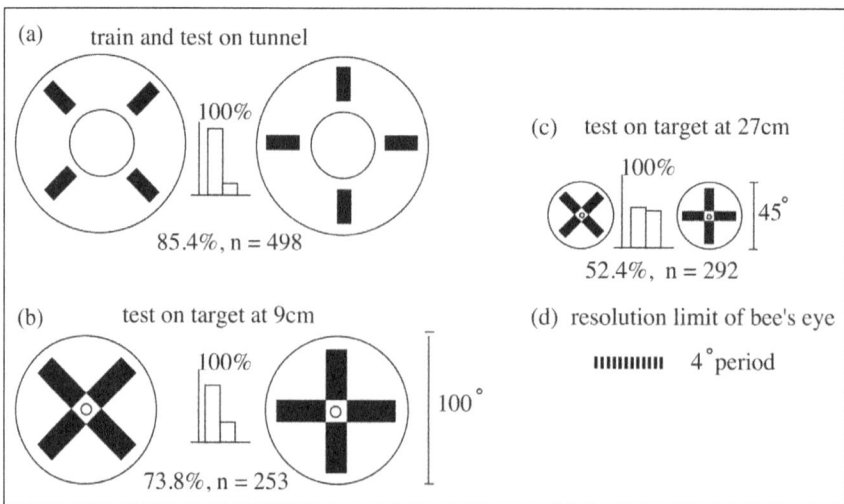

Source: From Horridge (1996b).

The discrepancy between the two results was due to the difference in angular scale, as shown by many similar experiments and by training on patterns of one size and testing on other sizes (Figures 10.4 and 10.5). The size of the local area for the summation of the orientation cue has been measured as 15–25° across by training bees on the orientation cue and then testing them with two orthogonal bars at various distances apart. Also, within the local area, two black spots within 12° of each other were not separated (Figure 9.19c), but further apart they were separate (Horridge 2003b).

Figure 10.5 a) This pattern is not discriminated at 27cm (subtending 45°) from the same pattern rotated by 180°, because there is no difference in cues and there is no eidetic image in a local region of the eye. b) When the criterion of success is landing on the reward hole, or (c) at a range of 9cm, this pattern is easily discriminated.

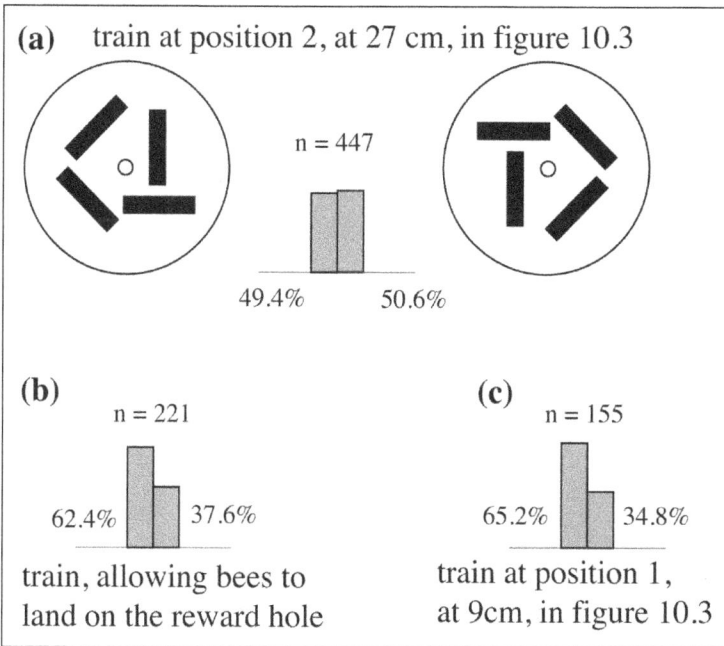

(a) train at position 2, at 27 cm, in figure 10.3

n = 447

49.4%   50.6%

(b) n = 221

62.4%   37.6%

train, allowing bees to land on the reward hole

(c) n = 155

65.2%   34.8%

train at position 1, at 9cm, in figure 10.3

Source: After Horridge and Zhang (1995).

The sizes and separations of local regions on the eye can be measured by comparing discriminations of the same pairs of patterns at different scales (Figure 10.3b). A pattern of four bars that subtended 45° at the point of choice was not discriminated from itself rotated by 180° (Figure 10.5a) because the orientations cancelled and the only cues were the area, the position of the centre of black and modulation, and these were the same on each target. It was easily discriminated, however, when it subtended 100° or when the criterion was landing on the central reward hole (Figures 10.5b and 10.5c). In very large checkerboard patterns, the positions of squares greater than about 10° were discriminated separately (Figure 4.2f). A pattern of plaids subtending 100° with bars 20° long and separated by 10° was easily discriminated from the same rotated by 45° (Horridge 1996b). As another example, a pattern of four gratings at 90° to each other on a target subtending 50° was discriminated from the same pattern rotated by 45° (Zhang and Horridge 1992). Thin black bars at an angle to each other were not discriminated separately and only one averaged orientation could be detected on each side of the target (Horridge 1996a, 2000b). Data such as these showed that the size of the local region for summation of orientation was smaller for gratings than for single bars. These measurements suggested that there were 10 to 15 local regions in the horizontal direction around the eye—more than enough to identify a place. The map of the local areas of the eye is not necessarily the same for each type of cue.

Figure 10.6 Discrimination of the rotation of a pattern of four black and four white equal sectors, with the pattern subtending 100° at the point of choice. a) Training produces a high score. b) The excellent performance persists when the bees see only the peripheral rims 4° wide. c) With new bees and the sectors at 27cm, the performance is still good. d) Rims only 3° wide again provide a sufficient cue. e) With only the central part of the pattern at 27cm, the bees choose at random although the patterns are well above the resolution limit of the eye, as shown by the grating (f) of period 4° at bottom right.

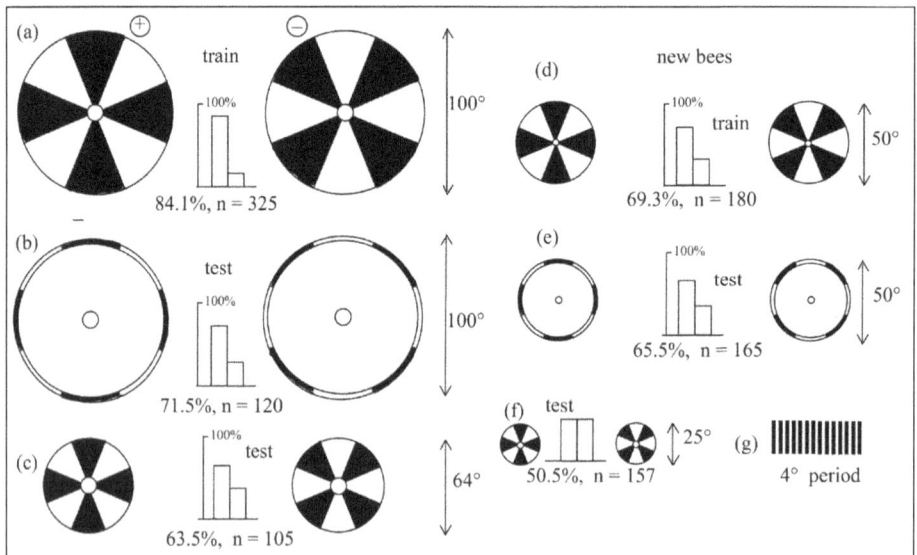

Source: After Horridge (1996c).

An example of how the cue can be a small part of the pattern is illustrated by the discrimination between two very large sector patterns (Figure 10.6). The bees learn the position of black only in the periphery because that is where the black areas fall on different eye regions,[3] as detected from the point of choice.

In experiments with large angles between cues, bees readily detected the position of the correct reward hole by use of a cue at the side of the eye (Figure 8.2); they learned two separate cues of orientation or colour with the two eyes (Giger 1996) or on the two sides of the target (Horridge 1997b). In many experiments over the years, they learned to distinguish two or three artificial landmarks at large angles to each other by colour, orientation or height of the centre.

The same few cues in the landmark labels

In recent experiments, bees were trained in a situation that resembled the natural task of a bee arriving at a foraging site. A black pattern on a white background was displayed in one arm of the Y-choice apparatus at a range of 27cm versus a plain white target in the other arm. The bees were obliged to use this one useful landmark. In the Y-choice maze, the pattern was nothing more than a landmark about the size of a local eye region, so the bees detected only one cue of each type. Various patterns displayed in the training experiments included the previously identified cues. They were: an oblique bar, three parallel oblique bars, an oblique grating, a square cross, six radial spokes, a large or a small spot, a spotty modulation or a ring (Figure 10.2).

The trained bees were given a large number of interleaved tests to discover the order of preference for cues in the learning situation. They preferred to learn first the black area at the expected place, and second, modulation caused by edges at the expected place. These cues were quantified and always available. Next in preference, the orientation cue was learned from a grating that covered the target, but was ignored in a single bar. Next, the bees remembered the existence of a radial pattern and the positions of the centres of black and of radial symmetry. They preferred a blank paper to a circle. All the feature detectors behave as though they are always switched on and in tests the bees recognise and avoid unfamiliar cues that are not displayed in the training.

The cues that bees use in identification of landmarks in the local eye region turn out to be the same as those used for discriminating between fixed patterns on experimental targets. What we thought was pattern perception turned out to be the recognition of the cues displayed on a landmark. The bees in the Y-choice maze were learning the label for the correct place, not a pattern.

There were several advantages of learning several cues. First, in the natural situation, the bees are less likely to mistake the place. Second, the more cues they learn, the more likely they are to find the reward although some part of the scene is changed. Third, the redundancy improves the ability to switch from one cue to another.

It was an accident of the design, in 1987, that the Y-choice maze was approximately the size of the local region of the eye. On the other hand, in the natural panorama, the bees learn the separate labels of landmarks over a much wider range of larger angles.

Figure 10.7 The visual fields of both eyes are divided into an array of local regions around the head. Each of these regions detects any of the cues, including a smoothed measure of modulation (dotted line) and a measure of nearness = 1/ range (dashed line). Cues, as shown by the symbols, are expected in retinotopic directions relative to the midline. The bees recognise a place by the conjunction of expected cues in the expected directions.

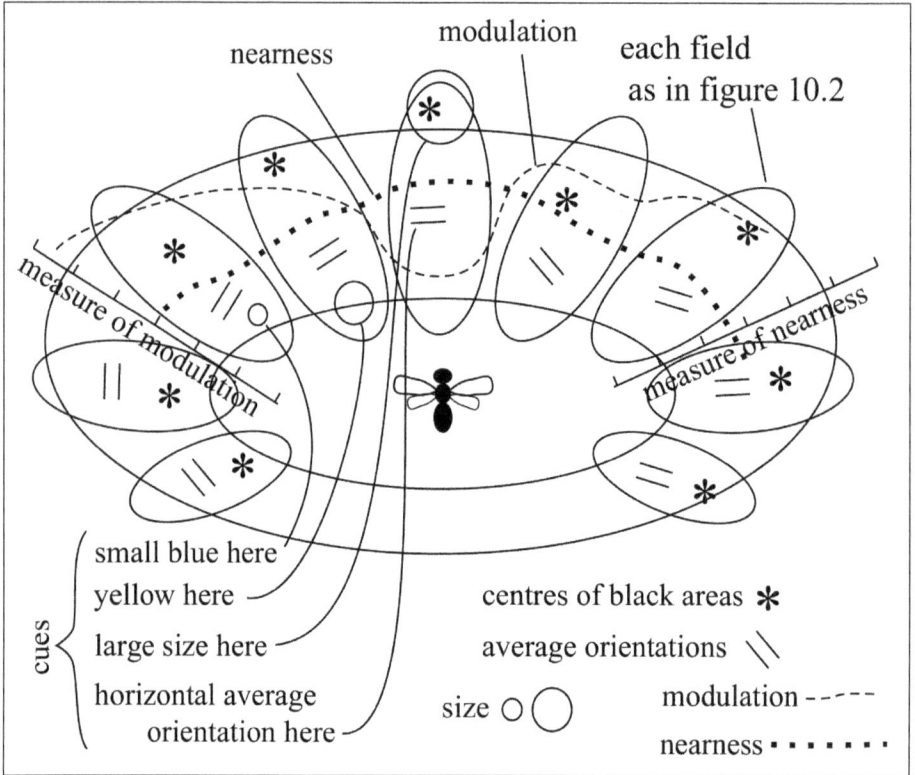

Source: After Horridge (2005b).

The global coincidences of cues

To understand the whole mechanism in the bee, we now assemble an array of local regions side by side around the head to form a whole-eye detector of place. If we repeat the local region (Figure 10.2) about 15 times around the head, we generate an array up to 300° wide that detects up to 10 cues in each local region (Figure 10.7). In all directions around the head, the bee measures the modulation and the nearness of contrasts. For a bee, a place was recognised by a sparse but unique arrangement of landmarks and the angles between them, kept separate in the local regions around the eye, and a small fraction of this array was sufficient to define a place. All experiments revealed that each task was a separate learning experience, there was no evidence that they learned anything more than the cues and in tests they accepted other places that displayed the same cues. By equating the cues from landmark and pattern recognition experiments, we have arrived at a synthesis.

Generalisation after training with patterns

Although few cues were involved at each landmark, the memorised coincidence of the directions of well-separated landmarks ensured that the bees would not accept the wrong place. Conversely, if the reward was moved, they would not accept that the place was wrong. Quite a different behaviour followed training with patterns. We can now explain this distinction.

The patterns were regularly moved on the target or on the flat table to make the bees look at them. This trained the bees to ignore everything outside the targets including local landmarks. Because the responses of the feature detectors were summed to form one cue of each type in each local region of the eye, and the pattern subtended about the same size as the local region of the eye, they learned only one cue in each channel. Because they were limited to one cue of each type in one local region, they would then accept quite different patterns that displayed the same cues. They generalised—that is, made errors—because they had been trained to ignore cues outside a single local eye region.

In nature, the memorised label was the only way that a bee recognised a landmark. The configurational layout of the whole wide panorama around the eye could be detected because it was divided into regions (Figure 10.7). The labels on different landmarks could be similar or not. Because there were several cues commonly available, and many different labels could be distinguished, recognition was much more precise than with a single pattern. Vision for a bee was a succession of landmark labels in different directions—some familiar, some not.

The bee moved about like a blind man navigating by a succession of familiar touch, odour and sound cues. The memory held information about only the coincidences of cues, with poor resolution of positions within the local regions. The rest of the visual input did not pass the cue detectors. There was no reassembly of pattern. In fact, the bees were not interested in patterns. Bees have no pattern perception.

Why patterns were difficult to learn

Bees made several visits before they associated a black and white pattern with a food source. The task was to select one pattern from among several displayed on the front of reward boxes (von Frisch 1914), on a flat table (Hertz 1933) or in a Y-choice maze (Figure 1.1). In all of the experiments with patterns, the rewarded pattern was moved around together with the reward to make the bees look at it, rather than where to go. In recent experiments, the pattern was moved every five minutes. When that is done, the bees must learn to ignore everything outside the target instead of following their natural inclination to pick up several local landmarks in different eye regions. They alternate between learning to go to cues displayed on the rewarded target and to avoid the cues displayed on the unrewarded target. When most of the cues are the same on the two targets, it takes the small brain of the bee some time to grasp the difference. On the other hand, if they find a stationary food source, they make an orientation manoeuvre and immediately learn its location in relation to several convenient landmarks at wide angles, then return for more in a few minutes.

In retrospect, for the whole of the twentieth century, there was a conceptual block to understanding the relation between patterns and landmarks but no lack of experimental data. Bee vision is anti-intuitive, so it is hard to imagine that the mechanism is so simple—and even more difficult to design the right experiments. The bees did not remember the patterns or the landmarks as objects; they remembered the directions of the labels that marked the right place. In each label, the bees learned first a coincidence of modulation, area and position, then the less preferred cues, and they recognised and avoided added cues that were not in the training, but nothing more. The artificial Y-maze apparatus offered only one attractive landmark and one to avoid.

Because the bees were quick to learn to recognise a place but slow to learn a difference in the experimental training, and because it was generally believed that the bees in fact saw the patterns, bee pattern perception became a subject in its own right. For the whole of the twentieth century, however, it was anthropomorphic delusion to accept that bees perceived and discriminated patterns.

The behaviour helps explain the neuron properties

Since the early days of insect visual electrophysiology, many researchers have wondered why the image on the retina is funnelled into relatively few neurons with large fields that make no sense in terms of vision. They were unaware of the total subservience of the bees' visual processing system to a panorama of sparse retinotopic cues averaged over large fields. The fields of about 20° are large only in the context of a bench experiment, not in a compound eye with a panoramic view up to 360°. Large fields throw away detail and all chance of pattern perception within a local eye region, in favour of a few smoothed data points derived from coincidences in an array of extremely simple retinotopic feature detectors.

Endnotes

1. Bees detect something about the configurational or spatial layout of a pattern or shape when several local eye regions overlap it—for example, when the bee examines the target closely or the criterion of success is landing on the reward. If the angle subtended by the target is unknown at the point of choice, it is impossible to analyse the mechanism of discrimination. The solution to this impasse was the accidental use of the Y-choice maze, which limited the field of view to a manageable size.

2. For convergence of ideas, see Lehrer and Campan (2006).

3. Ibid.

11

DO BEES SEE SHAPES?[1]

When the human eye looks at an object, it is almost impossible to avoid seeing its shape. We cannot imagine how we would not see the shape. So it might be difficult for readers to accept the conclusions so far reached—that bees detect cues in simple patterns that they do not see, and remember the directions of cues that enable them to identify places.

It is usually assumed that bees probably remember vague, crude or fuzzy shapes, and that, because they have eyes, the onus of proof lies with those who would show that they do not. The opposite, however, is the case. It is very difficult—in fact, impossible—to show that they remember shapes and quite easy to show that they do not.

Bees can be trained to distinguish between a can of lager and a can of ale, or between a bottle of claret and a burgundy, even with the corks in, as long as the labels differ enough. They easily distinguish between photographs of two different human faces (especially if one has black hair, as in the published example), but it is impossible to show that they see or even detect the whole shape as a shape. In all cases, they might be detecting a small cue that has been learned especially for the occasion, and when tested, this proves to be so.

Going back to Mill's rules (Chapter 2), we see that the general statement that bees see shapes is not true, as shown by numerous examples of pairs of simple shapes or patterns that bees cannot distinguish. When compared with other pairs that were discriminated (Figure 9.14), these examples helped to reveal the cues (Chapter 9).

Having explained much of bee vision by the cues, however, we might have overlooked other ways to detect shape. The only way to find out is to search for them and investigate them one by one.

We have at least three ways to demonstrate that bees do not remember shapes as whole shapes. First, we can show that, when they appear to discriminate between shapes, they in fact use simple cues that can be demonstrated. This is an alternative explanation, but it says nothing about an additional memory of shape unless every example is thoroughly examined. Second, as previously described (Figures 9.14f–j), there are many examples of patterns between which bees cannot discriminate and we can investigate why they fail. This is Mill's method of 'agreement in absence' (of discrimination), which requires numbers of examples to be secure. Third, we can show that although they discriminate between two shapes, bees cannot remember the rewarded or the unrewarded shape that they were trained on when tested versus a different pattern that displays the same cues (as in Figures 11.1f and 11.1g). This positive evidence of absence of recognition of shape is the principal topic of this chapter.

One or several local regions of the eye

The division of the whole eye into local regions that detect separate cues complicates the situation. Therefore, very large patterns that overlap more than one eye region might be discriminated by the spatial layout of their parts, giving the impression that the whole shape is detected. This was the cause of the difficulties in Chapter 4 and the analysis of results with very large patterns was presented in Chapter 10. Therefore, the present discussion refers to small patterns that are covered by one local eye region. As will be seen, this is the whole story when we are discussing results of recent training with the Y-choice maze. The bees were trained to discriminate the patterns and not to look beyond them because the positions of the targets were changed every five minutes to make the bees look at them to identify the place.

The balance of preferences between two targets

When presented with a choice between two patterns, one of which is rewarded and the other not, the bees ignore most of the cues because they are displayed on both targets. The bees learn them on one target and unlearn them on the other. So, cues commonly displayed equally on both, such as average position, modulation, area or blackness, are not used.

The situation is also complicated by the different preferences for the cues. The bees learn first the most preferred cue, even if they learn to avoid it. This can give the false impression that they learn to prefer the rewarded pattern—for example, when all the cues displayed on the rewarded pattern are identical to those on the unrewarded one, but the unrewarded one displays a unique cue that the bees learn to avoid.

Figure 11.1 The colour cue is detected but not the pattern. a) Training patterns.
b) The score was higher when the attraction of the blue spot was removed. c) The
blue spot alone was not recognised because the preferences were balanced on the
two targets. d) The position of the rewarded spot was not remembered because
the training spots were in corresponding positions. e) The trained bees distinguish
between the colours of groups of small spots of the same total area. f) and g) They
cannot distinguish the large training spots from a scattering of small spots of the
same colour; + = rewarded training pattern; – = training pattern without reward.

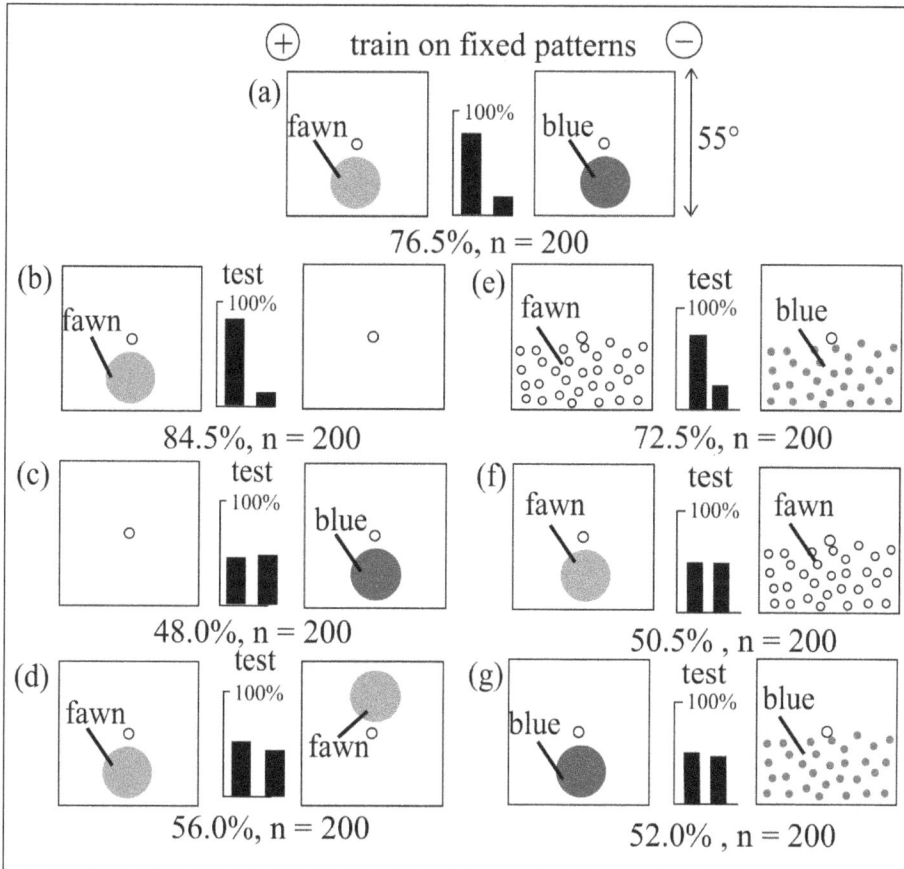

For example, before training, bees preferred a blue spot to a fawn spot, each
subtending 20° at the choice point. When trained to go to the fawn spot
(Figure 11.1a) then tested against a plain white target, they in fact preferred the
fawn spot more than in the original training (Figure 11.1b), but they could not
tell the difference between the plain white target and a blue spot (Figure 11.1c).
At first sight, this seems strange, until we realise that the innate preference for
the blue spot was not completely removed, but only reduced by the training to
the same attraction as a white sheet. The bees' training for the fawn spot was
fully revealed when some residual attraction for the blue was removed. What
mattered was the balance of preferences.

Because the two spots were in the same place on the targets, the trained bees had not learned their position, as shown by a test (Figure 11.1d). They also had not learned their shape. When tested with the original spot versus a scattering of 40 small spots of the same colour with a total area the same as the area of the large spot (Figures 11.1f and 11.1g), they did not distinguish the targets, showing that they did not remember that they had been rewarded on the large spot. They had learned only the colour cue, not its shape, position or modulation.

Figure 11.2 The landmark indicating the reward displays two different colours, versus a blank white target, but the bees learn only that something lies at the right place. a) Training patterns. b) The trained bees failed to distinguish the training pattern from a group of black spots. c) The black spots are sufficient. d) The trained bees scarcely notice if the colours are reversed. e) and f) The fawn and blue spots alone are adequate. g) In a forced choice between the two, the innate preference for the blue spot was unchanged by the training.

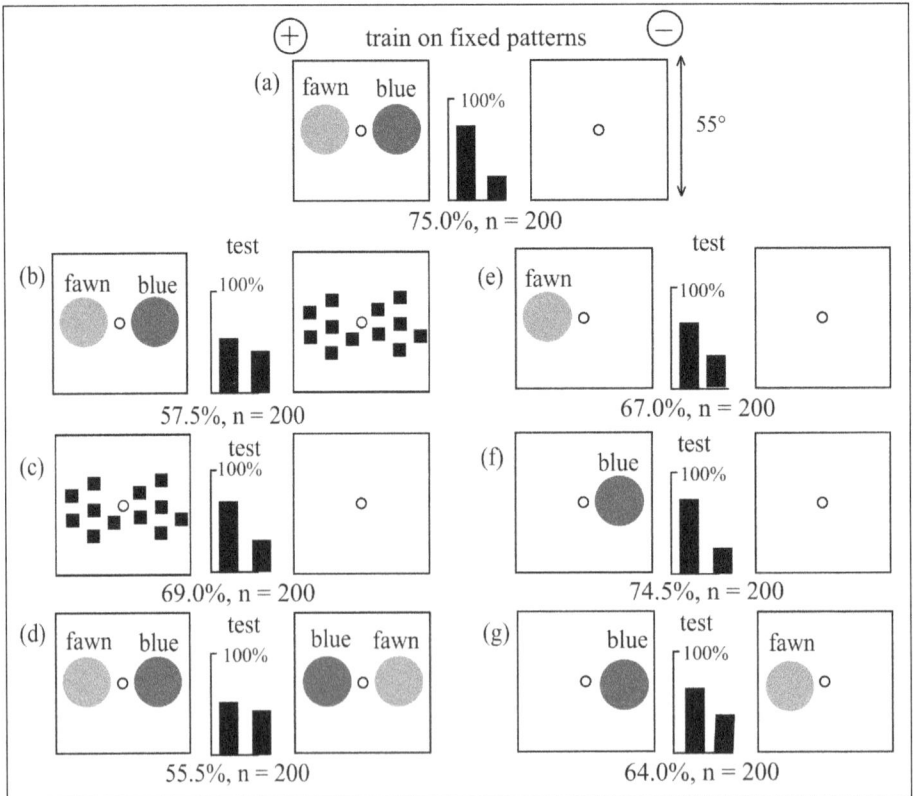

Source: After Horridge (2007), with (g) corrected.

Similarly, when the trained bees were tested with a scattering of 40 small fawn spots versus 40 small blue spots on white backgrounds, the score was as high as in the training (Figure 11.1e). The cue did not have to be in a large spot, implying that a colour learned from a flower nearby could be transferred to the same colour in many scattered flowers further away. There are obvious implications for the evolution of flowering bushes.

In the above training with two coloured spots, the bees clearly detected the fawn and the blue cues separately, because one was rewarded. On the other hand, when bees were trained with the same two spots on the rewarded target versus a blank white target (Figure 11.2a), they appeared to have learned very well. In a test, however, they could scarcely tell the difference between the coloured spots and a black and white neutral pattern (Figure 11.2b) and they responded just as well to the black spots versus a white target (Figure 11.2c). They could scarcely distinguish the training target from its mirror image (Figure 11.2d), so they had not learned much at all. When tested with the fawn and blue spots separately against the white target (Figures 11.2e and 11.2f), the bees' results tell us only that they preferred something to nothing and the blue more than the fawn. When the blue spot was tested versus the fawn one (Figure 11.2g), they preferred the blue, as expected from their innate preference, as though they had learned nothing about the colours in the training.

This example reveals very well our initial mistaken trust in the intuitive conclusion that the bees had learned the shapes and the colours. When you first read the early results illustrated in Chapter 1, no doubt you concluded that bees saw the entire pattern and the colours in their places (Figures 1.2a, 1.2b, 1.3)— as indeed was believed and taught for almost a century. There were, however, no tests of what the bees detected and the uncritical conclusions were based on the intuition that was, according to Mill, 'the intellectual support of false doctrines' that he predicted would ruin society. Well, it ruined the science of bee vision—and no doubt a lot else besides.

So, was all the fine talk about scientific method in Chapter 2 just pub talk and *post hoc et ergo propter hoc* claptrap? Did understanding of bee vision advance in fits and starts by myopic steps guided by anthropomorphism and prejudice? Yes, sometimes it did.

In the past, there have been other examples of training with two or more colours on a rewarded target in which researchers have concluded that the bees learned the colours, although no tests were done. They should all be revisited and tested, in case the conclusions were nonsense. Please don't conclude, however, that bees cannot learn two colours. In Figure 11.1, it was shown that the bees could learn a difference between the same two colours. If trained on a pattern displaying two colours versus the same with the colours interchanged (Figure 11.2d), they would learn the difference in the positions of blue, but perhaps no more. The point is that the task in Figure 11.2a did not require them to learn anything about the pattern or colour, which is the common situation when bees learn to detect a solitary landmark.

More clearly than with black and white shapes, these experiments with colours show that the bees learn very little, sometimes as little as learning to avoid one preferred cue, or that anything is preferred to nothing, or everything else is avoided when they are trained to prefer a blank white sheet. We are now in a sufficiently critical mood to doubt whether bees really see shapes, symmetry or anything else that could be graced by an abstract noun. Fortunately, not many examples have so far been proposed.

Figure 11.3 Position is discriminated but not the shape of the triangle or square. a) Bees discriminate the inversion of the black triangle if the centres are at different positions in the vertical direction. b) They fail in tests or fail to learn when the centres are at the same height. c) They also fail when black and white are interchanged. d) They have difficulty discriminating the rotation of a square.

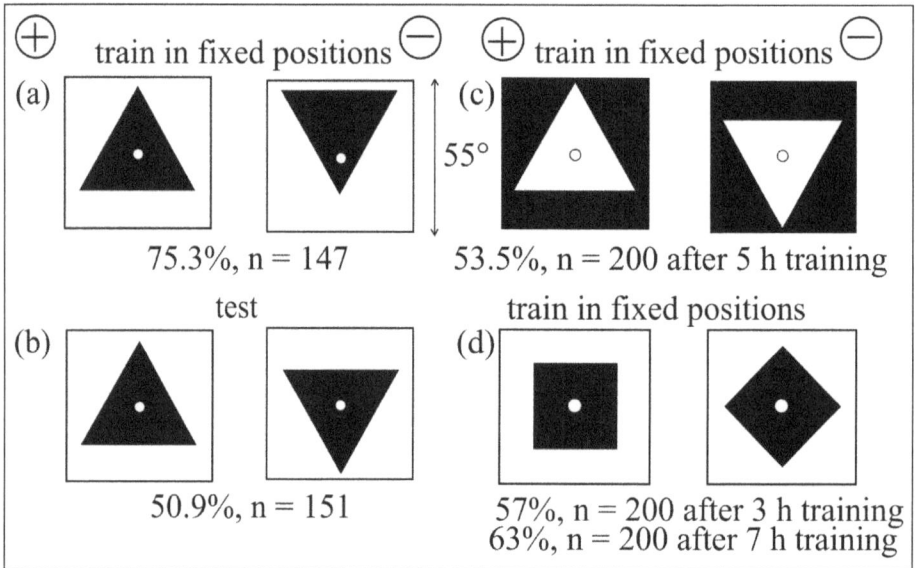

Source: (b) Horridge (1999a).

Bilaterally symmetrical triangles and squares

In 1997, I found that an equilateral black triangle subtending up to 40° at the point of choice was easily discriminated from the same triangle inverted (Figure 11.3a) unless their centres were at the same height (Figure 11.3b). This result made me suspicious of claims of discrimination that required pre-training on other patterns. Even bees that were trained to discriminate the triangles in different positions failed when tested with the centres carefully placed at the same height. The same applied to white triangles on a black background

(Figure 11.3c). The bees detected a difference of about 8° vertically, but not a horizontal shift. The averaged edge orientation on each side of these triangles is vertical, so cannot be a cue.

Bees are very slow to learn to discriminate between a black square subtending 40° at the point of choice and the same rotated by 45° (Figure 11.3d) when the centres are at the same height. Eventually, they learn but the cue is not obvious because edge orientations at right angles cancel out. Of course, very large squares are more easily discriminated.

Mirror-image triangles

In the past, there have been claims that mirror images are variously favoured in training or confused in discriminations, but the trained bees were not tested to reveal what they had learned. In cue theory, mirror images are nothing special.

Bees very readily learn to discriminate between the black triangle when one side is vertical, versus the mirror image of the same (Figure 11.4a). The trained bees were given a variety of tests. They distinguished the triangles when white on a black background (Figure 11.4b) and with edges only (Figure 11.4c), so the cue was probably in the edge orientation. This cannot be the whole story, or else the triangles in Figure 11.3b would be easily discriminated.

The trained bees easily discriminated smaller versions of the two triangles (Figures 11.4d and 11.4e), unless they were moved in the vertical direction (Figure 11.4f), so positions of parts of the areas were not likely cues—as confirmed by testing with the corners only (Figure 11.4g).

A test of the trained bees with the isolated vertical edges revealed some discrimination (Figure 11.4h). In the training, however, there was a vertical edge on both targets, so the bees must have discriminated the difference between their positions. Following the same idea, the trained bees were tested with isolated horizontal edges (Figure 11.4i) and also with vertical edges on one side of the target and horizontal edges on the other side (Figure 11.4j), with surprising success. Clearly, the difference in the positions of the average orientations of the edges on corresponding sides of the two targets was a cue. Finally, in the crucial test, the trained bees failed to discriminate between the original training triangle versus the horizontal and vertical edges without the area of black (Figure 11.4k). They could not recognise the target they were trained on when it was presented versus a different pattern displaying the same cues in the positions where they had been trained to look for them.

Figure 11.4 In contrast with Figure 11.3, the rotation of the same triangle is easily discriminated when one edge is vertical. a) Training patterns. b–e) The trained bees discriminate when black and white are interchanged, when only the edges are displayed or when the triangles are smaller. f) They fail when the small triangles are moved 12° upwards. g) The differing positions of the areas at the corners have little effect. h–j) The average edge orientation on the two sides of the target is a good cue. k) The trained bees fail to distinguish between the rewarded black triangle and the edge orientations alone, so its shape or black area was of no consequence.

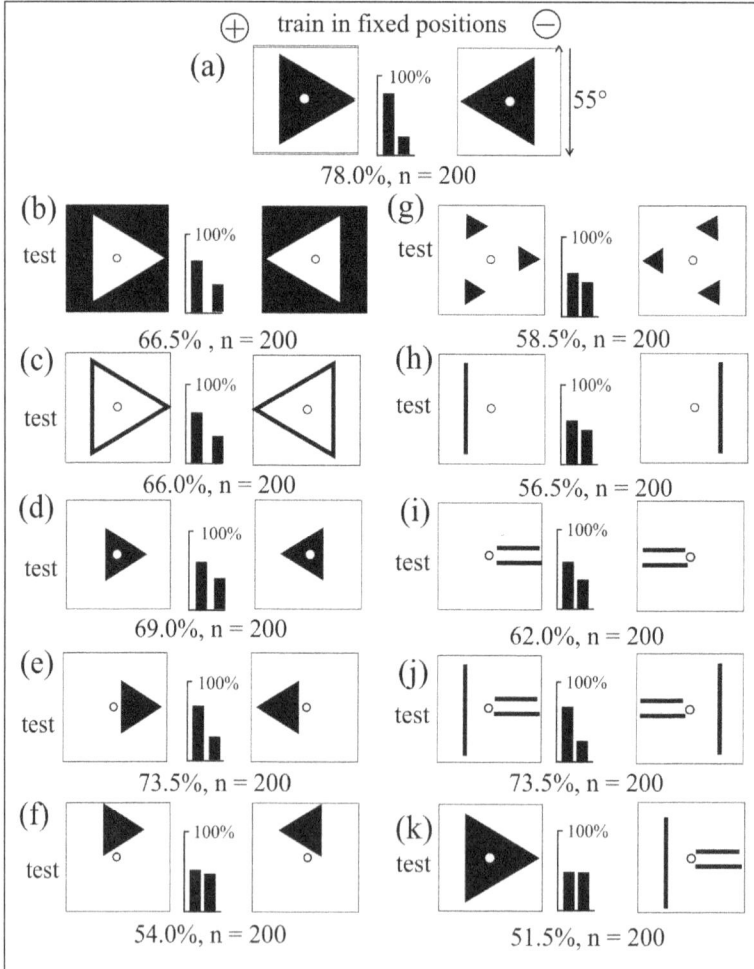

A disc and a triangle

Bees learned to discriminate between a black disc and a triangle of similar area (Figure 11.5a) presented in the Y-choice maze. When the trained bees were tested with the disc versus a random pattern of spots, however, they scarcely recognised the difference (Figure 11.5b). Clearly, they had not learned to go to the disc, as one might suppose from the performance in training. When tested with the spots versus the triangle, they avoided the triangle (Figure 11.5c). Therefore, they did not recognise the disc but they had learned to avoid the triangle.

Figure 11.5 An identification of the cue after training to discriminate between two black shapes. a) Training patterns. b) No preference for the disc versus the spots. c) A similar test reveals an avoidance of the triangle. d) Discrimination does not depend on size or area of black. e) The cue is related to the edges of the empty shapes. f) The black inverted triangle and the triangle at the same centre are equally preferred, so the exact layout of edges is not relevant (compare with Figure 11.3). g) When two oblique lines are added to the disc, the equal preference shows that the cue is the oblique edges on the unrewarded target.

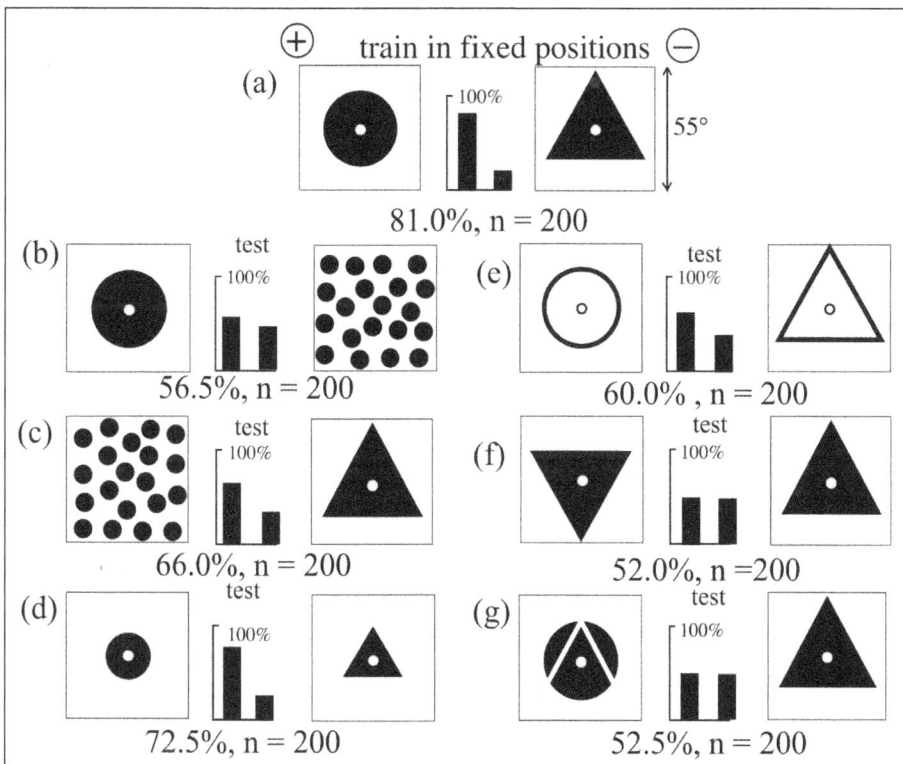

The trained bees discriminated a smaller disc and triangle (Figure 11.5d) and also the isolated edges (Figure 11.5e), but could not distinguish the triangle from the same inverted (Figure 11.5f). This result was similar to that in Figure 11.3b, showing that the bees had not learned the positions of the edge orientations. When white oblique lines were drawn on the disc, however, the bees could not distinguish it from the triangle (Figure 11.5g). Vertical lines serve equally well. Therefore the cue was the average edge orientation on each side. In this case, they did not need to learn the positions of the oriented edges. There was 'absence of proof' that they had learned anything besides the cue, but more importantly, there was a direct demonstration in the tests that they had not learned the shape of either the disc or the triangle.

Figure 11.6 Identification of another cue on the unrewarded target in the discrimination between a ring and a square cross. a) Training patterns. b) and c) The trained bees fail to distinguish the ring from a pattern of spots or a hollow cross. d) A solid black disc is not distinguished from the cross. e–g) The cue is the black around the centre on one target but not the other, irrespective of the pattern; there is clearly no discrimination of shape as assumed by Zhang et al. (1995).

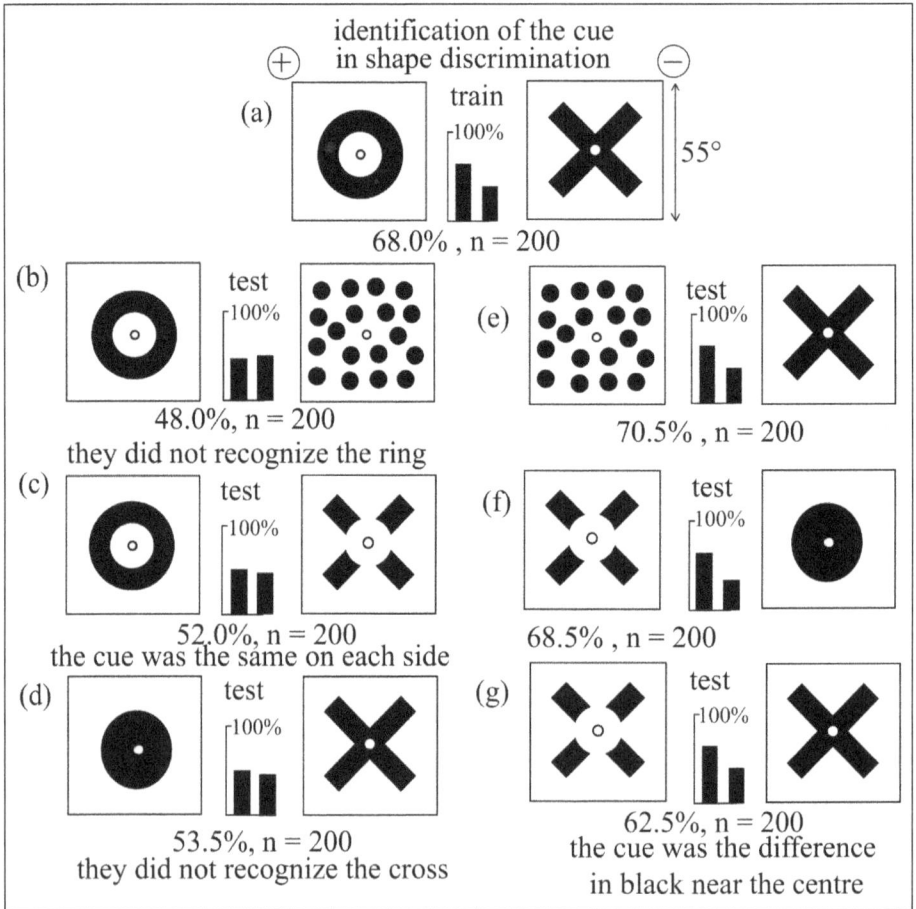

Source: After Horridge (2006a).

A ring versus a cross

This example has an interesting relation to some of the conclusions of previous authors, who assumed that bees discriminated these two shapes. Bees were trained to recognise a large fixed, broad black ring (ID 18°, OD 33.4°) on the rewarded target versus a black cross of similar area (Figure 11.6a). Initially, they avoided the ring innately, but then slowly learned the task. When the trained bees were tested with the ring versus a pattern of spots (Figure 11.6b), however, they could not tell the difference, which suggested that they had learned nothing about the ring.

The trained bees were tested with the ring versus the cross with the centre removed (Figure 11.6c). Now, neither of the test patterns displayed a black area near to the reward hole and the trained bees failed to discriminate. When a black disc (D = 28°) of similar area was tested versus the black cross (Figure 11.6d), there was black around the reward holes on both targets, and again the trained bees failed to discriminate. This result showed that they did not recognise the cross. With the pattern of spots versus the cross (Figure 11.6e), with the cross minus its centre versus the black disc (Figure 11.6f), and with the cross minus its centre versus the intact black cross (Figure 11.6g), they performed as well as in the training, showing that cues were available although the patterns were so altered. Therefore, the necessary and sufficient cue was the white versus black around the reward hole and the real shapes were of no significance at all.

The bees demonstrated quite a subtle way to distinguish between the two patterns. Although we can correct the error of thought that they detected the shape, we cannot say that the bees did not have any other ability to detect something abstract about the patterns. In fact, they picked out the difference in the amount of black around the reward hole, which was a salient detail in the most important direction for them to look (see Figure 1.3).

A ring and a cross versus a white target

The previous experiment showed that when given a ring versus a cross, the bees used the difference in black at the centre as the cue, but learned neither pattern. In the next experiment, the bees were trained with both patterns together versus a white target (Figure 11.7a), as was the normal situation for an isolated landmark. The trained bees performed just as well when tested with a pattern of spots versus a blank (Figure 11.7b), but they failed when tested with the ring and cross versus the spots (Figure 11.7c), so they cared little for the training pattern. They had taken some notice of the position of black on the target, however, as shown by testing with either the ring or the cross in a different position (Figures 11.7d and 11.7e). In a separate experiment, they could not be trained to discriminate between the ring/cross pattern and a pattern of spots (Figure 11.7f).

This result illustrates that the training score is high because the task is easy. The bees detected little more than something on one target and nothing on the other.

Figure 11.7 Excellent recognition of the place, but failure of the bees to recognise either of two simple patterns—one with circular and the other with radial symmetry—when they were presented together on a landmark. a) The bees readily learned the task. b) The trained bees discriminated 12 squares equally well from the white target. c) They failed to discriminate the ring and the cross from the 12 squares. d) and e) The ring and the cross were discriminated separately from the same moved upwards on the target, so something had been learned about the position of the black areas or the radial hubs. f) In a new experiment, the bees could not learn to discriminate the ring and the cross from the 12 squares.

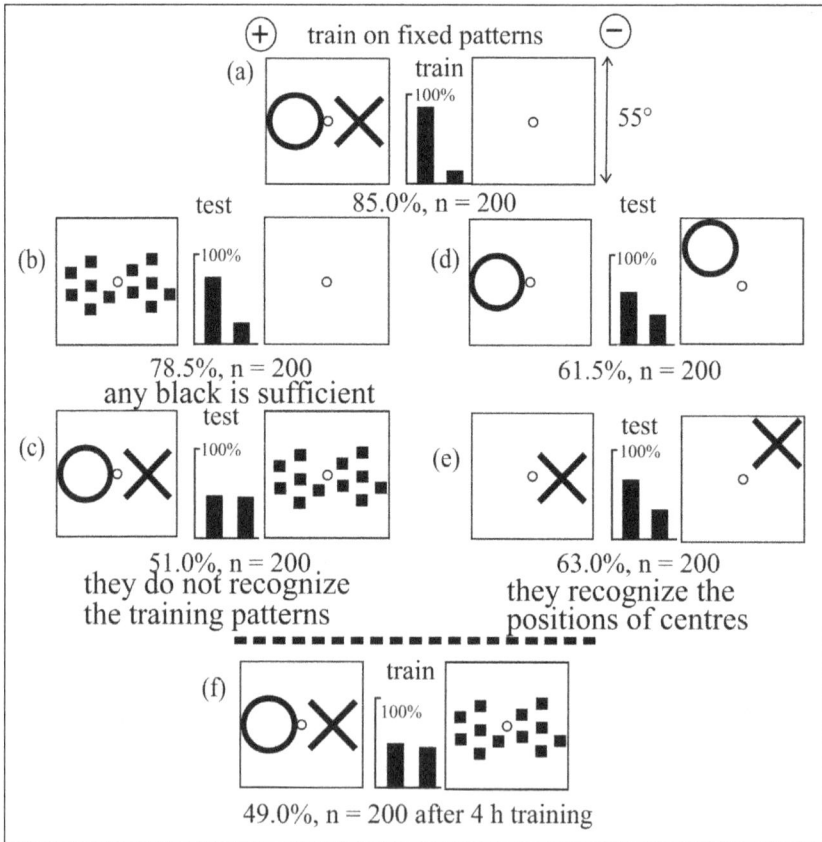

A ring versus a large spot

This is a pair of shapes like the ring and the cross (Figure 11.6), with a large difference in pattern to the human eye but not for bees. In an earlier study that assumed the discrimination of shape, it was claimed that bees trained on a rewarded ring versus a large round spot could transfer the discrimination to patterned targets raised over a patterned background without further training and could discriminate the shapes by the parallax as the eye moved (Zhang et al. 1995).

Figure 11.8 With a ring versus a spot, the cue is the black near the centre.
a) Training patterns. b–d) As long as the cue is present, the shape is of no
consequence. e) and f) Failure to distinguish between the training ring versus
a hollow square or a hollow cross because the cue is lacking. g) The poor
discrimination of a ring from a large spot when both are offset from the centre,
even when the ring is on the unrewarded target. h) Discrimination was excellent
between a small offset spot and ring.

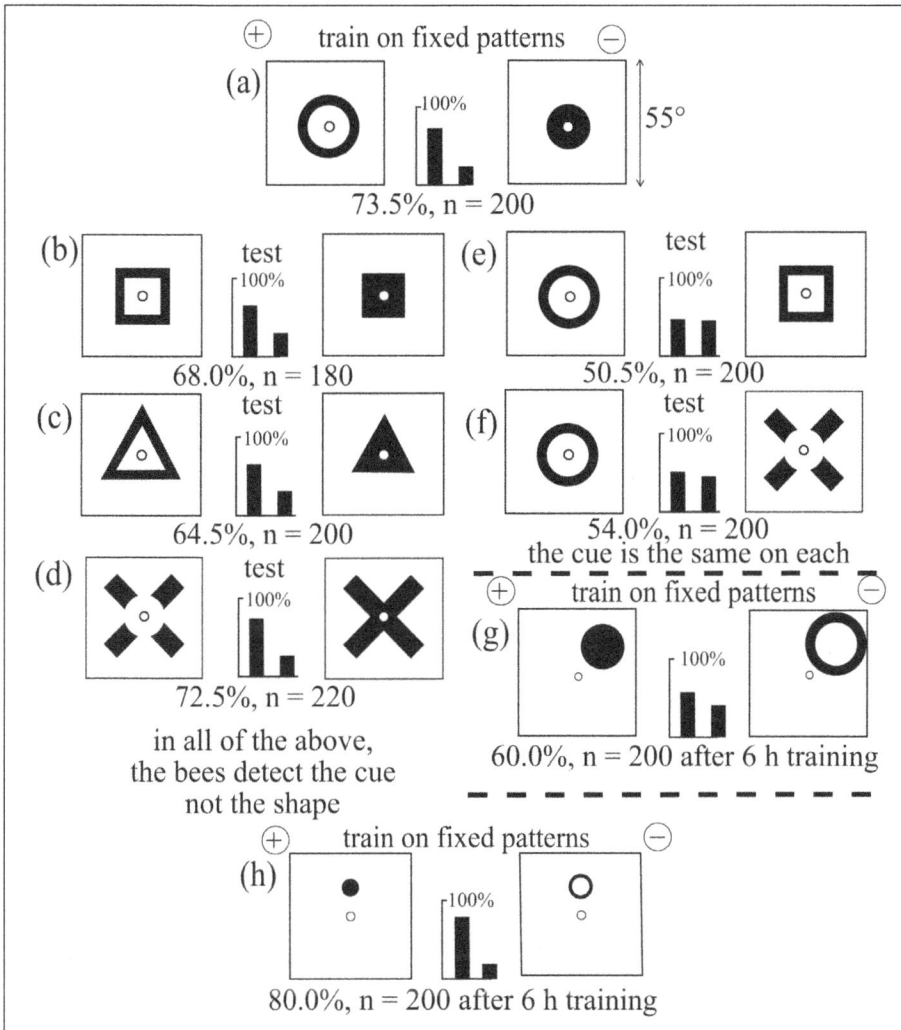

In the new experiments, learning was slow because at first the bees avoided the
ring, but the score reached 70 per cent after two hours of training (Figure 11.8a).
In tests, the trained bees cared nothing about the patterns presented as long as
there was a difference in black around the reward hole (Figures 11.8b, 11.8c,
11.8d). When this cue was lacking, the bees failed, irrespective of the test shapes
(Figures 11.8e and 11.8f).

The cue works only near the reward hole and bees do not detect it in other positions on the targets. They bees will not learn when trained with the ring and spot offset and the ring rewarded. They learn very slowly with the spot and ring offset and the spot rewarded (Figure 11.8g). They probably detect the modulation difference, which in the spot is half that in the ring. Interestingly, they learn the off-axis task better when the spot is very small (Figure 11.8h), but the cues have not been investigated.

Figure 11.9 A discrimination task in which the bees remembered three cues. a) The training patterns—one inverted relative to the other. b) Moving the patterns in the vertical direction had little effect (compare with Figure 11.4f). c) Removing the orientation of the straight edge reduced the score. d–f) The positions of the straight edges, the positions of the centres and the directions of the curvature were all adequate cues. The modulation, the area of black and symmetry were not learned because they were the same on both targets.

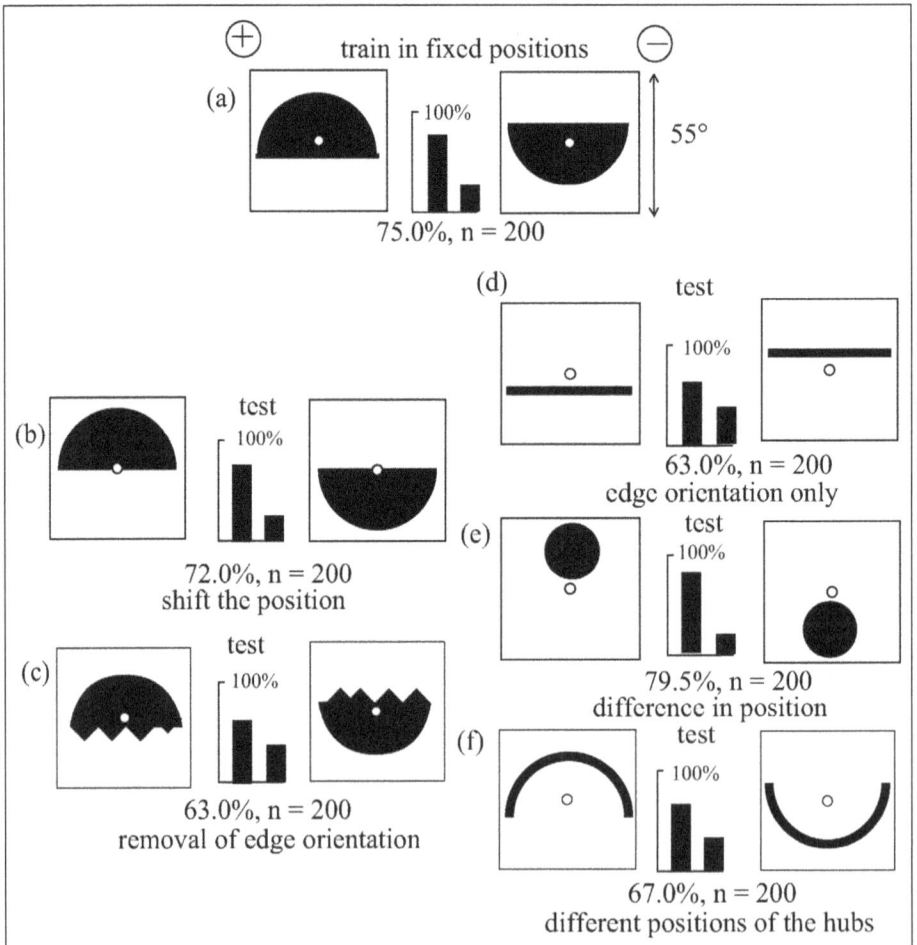

The D shape versus the same inverted

The bees learn this task easily (Figure 11.9a). A small shift of the pattern vertically has little effect (Figure 11.9b), but removing the orientation cue from the straight edge has a greater effect (Figure 11.9c). Three types of cue were easily demonstrated with the trained bees: the difference in position of the horizontally oriented edge (Figure 11.9d), the difference in position of the area of black (Figure 11.9e) and the difference in the direction of the curvature of the curved edge (Figure 11.9f). Each of these differences functioned separately, irrespective of the pattern in the test. In these tests, the bees accepted patterns that were different from those in the training as long as they displayed no unfamiliar cues.

A thick black O versus a large letter S

In this experiment, the bees were trained with a large black O (as in Figure 11.8) subtending OD = 33.4° and ID = 18° at the point of choice, versus a large black letter S of the same area (Figure 11.10a). The patterns were the same as those used by Chen et al. (2003). The naive bees detected the O and at first avoided it. As a result, they learned very slowly. After two hours' training, the score was 65 per cent. On subsequent days, tests were done only when the training score was more than 70 per cent.

The bees learned the unrewarded pattern. When the trained bees were tested with the training O versus a pattern of randomly arranged spots of the same total area (Figure 11.10b), the result was 53.5 per cent at a time when the average training score was 78 per cent, so clearly the bees had not learned to go to the O. When the trained bees were tested with the pattern of spots versus the S (Figure 11.10c), the result was 66 per cent in favour of the spots when the average training score was 78 per cent, so the bees had learned to avoid the S more than the O.

The cue was not related to the topology of the shapes. Two gaps were made in the O, each subtending angles of 40° at its centre, and the broken O was tested versus a figure of eight (Figure 11.10d), with a score of 65.5 per cent. When the figure of eight was tested against the S (Figure 11.10e), discrimination was very poor (55 per cent). Finally, the mirror image of the S was weakly discriminated from the S (Figure 11.10f), with a score of 61 per cent, which suggested that there was an additional cue beside the black near the centre. The topology was not a factor because the trained bees responded similarly to the eight and the S, and it was irrelevant whether the O was open or closed.

Figure 11.10 The cues were unrelated to the topology. a) The training task. b) Failure with the O versus a pattern of spots. c) Test with the pattern of spots versus the S. d) Test with a broken O versus a figure of eight of similar area. e) Test with the figure of eight versus the S. f) Test with the mirror image of the S versus the S; in each of these tests the bees discriminated irrespective of the topology. g–k) The identification of the cues. g) Discrimination of a broken S versus a black disc. h) Discrimination of a broken S versus an oblique bar. i) Failure to discriminate the oblique bar from the S. j) The O was discriminated from the thin bars. k) Failure to discriminate the O from the thin bars rotated through 90°. The cues were therefore the black near the centre and the orientation of the central bar of the S—both in the unrewarded target.

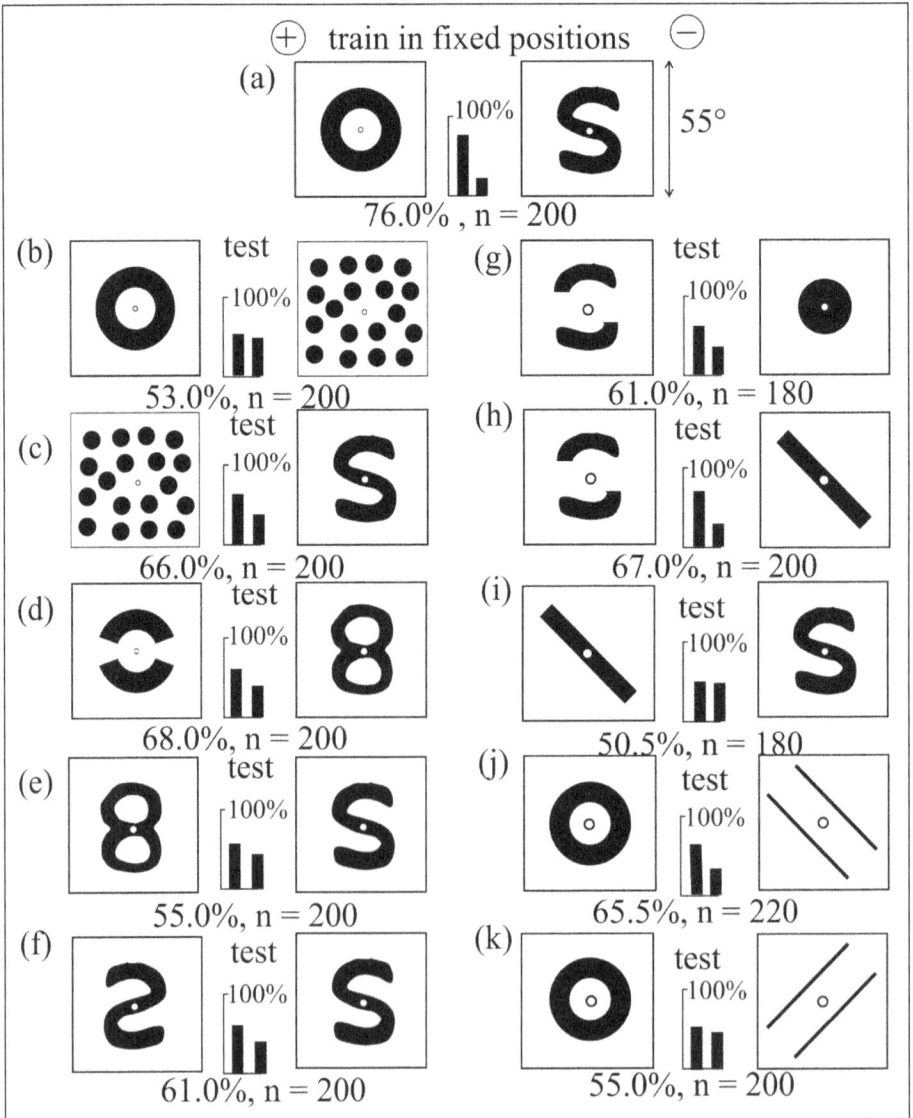

Quite different tests were required to demonstrate the real cues. The same trained bees were tested with the S with the middle section deleted versus a black disc of similar area (Figure 11.10g). They were also tested with the S with its middle section deleted versus an oblique bar with the same orientation as the middle section of the S (Figure 11.10h). These tests, taken together with the tests in Figures 11.10d–f, showed that white near the reward hole was preferred to black near the reward hole irrespective of the rest of the pattern.

This idea was corroborated by showing that the trained bees failed to discriminate the oblique bar versus the S (Figure 11.10i). They had certainly not learned the shape of the S. The O was then tested versus two thin bars with no black near the centre (Figure 11.10j). There was excellent discrimination, showing that a cue was detected, but discrimination was lost when the thin bars were turned through 90° (Figure 11.10k). Taken all together, the results were consistent with the detection of two cues already familiar from earlier work: the black near the reward hole and the average edge orientation at a certain position. The bees did not learn the difference in the topology of the O and the S, although, in the absence of appropriate tests, that was an earlier conclusion (Chen et al. 2003).

Discrimination of the rotation of a sector pattern

Until quite recently, it was accepted that bees could be trained to remember the layout or the global aspects of a pattern. For example, with reference to a proposed eidetic image of a sector pattern (as in Figure 11.11a), 'insects are able to compare a stored neural image…with a current neural image…has directly been shown in honeybees…The only factor that can account for the bees' ability to discriminate…is the exact retinal position of the black and white sectors' (Wehner 1981:476). In fact , for 25 years, no factors were tested.

To analyse the situation, bees were trained on two patterns of six sectors—one rotated by half a period relative to the other (Figure 11.11a). It was most interesting to discover that the trained bees failed to recognise the rewarded pattern versus the same pattern that was seriously rearranged (Figure 11.11b). The bees had not learned the position of the hub because this cue was the same on both training targets (Figure 11.11c). The trained bees failed when the horizontal sectors were removed from the training patterns (Figure 11.11d), but they discriminated very well when only the horizontal sectors were displayed (Figure 11.11e). This test gave the game away.

So, after 25 years of support for eidetic vision, when the tests were done the positions of the horizontal sectors on the negative target were a sufficient cue (Figure 11.11).

Figure 11.11 The curious discrimination of patterns of sectors. a) Training patterns, one rotated by half a period relative to the other. b) The trained bees failed to recognise the rewarded pattern versus the rearranged pattern. c) The bees had not learned the position of the hub because it was the same on both training targets. d) The trained bees failed when the horizontal sectors were removed. e) They discriminated with the horizontal sectors displayed. The cue in the training was therefore the position of the horizontal sectors on the unrewarded target.

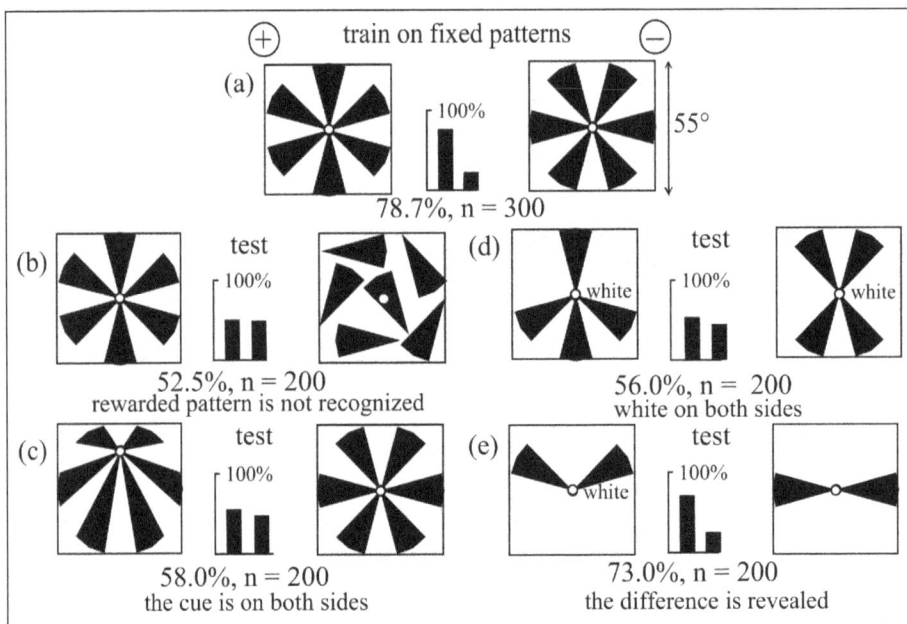

Source: After Horridge (2006a).

Spots

Lest it be thought that the bees or I favour solitary shapes rather than patterns, I have searched for evidence that bees can count or remember regularities or patterns displayed in groups of spots. Black spots are suitable units because individually they display few cues—namely, area, modulation and position.

First, when the training spots are fixed in position during the training, the bees learn to distinguish between two spots and three of the same total area and something about their positions. The performance depends strongly on the size of the spots but is reduced as the number of spots is increased (Figure 11.12, left side). The performance also depends on the size of the targets and is improved when the spots fall into different local regions of the eye.

When the training patterns are rotated randomly during the training, the bees cannot even learn to distinguish between two spots and three of the same total area, no matter how large the targets or how long they are trained (Figure 11.12, right side).

Figure 11.12 Failure to learn to discriminate the number or spatial layout of patterns of spots. Bees were trained on each pair of patterns separately. Those on the left were fixed in position during the training; the pairs on the right were rotated at intervals during the training. On the left, the positions of a few large spots were learned better than more spots with the same total area. On the right, even the difference between two and three spots was not learned. Previous claims that bees could count had no controls for the position effect.

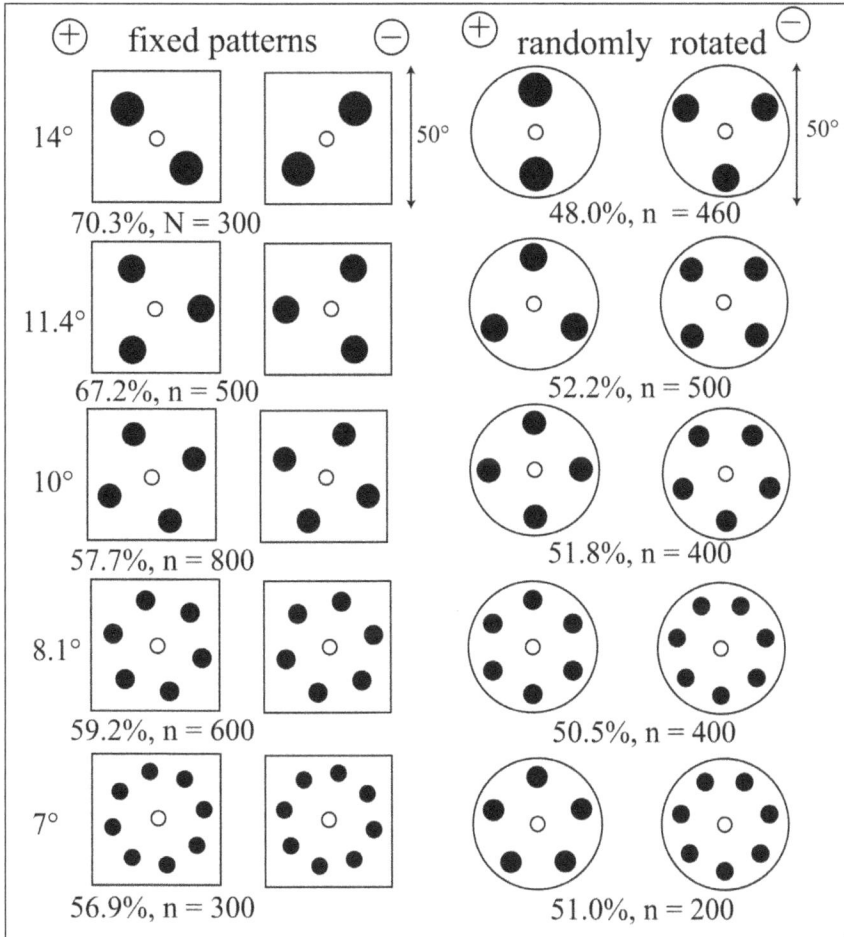

The same few cues are used every time

The choice of tests was the result of a long history of progressive understanding of the way that bee vision worked. Once a way was found for defining the test set for each pair of patterns that was discriminated, it was possible to discover exactly what the bees had learned in each case. Each example yielded the same general conclusions. They learned to ignore cues that were the same on both targets and they remembered one or more simple cues in order of preference, but nothing about the layout or shape.

This implies that for each pair of patterns that is detected in each local eye region, the bees learn a selection from the same small repertoire of cues. When a new pair of patterns was substituted, the bees were obliged to learn the new situation, using the same order of preference of a few cues. In each context, therefore, they could learn only one task, but in a different context, there would be other cues in other local regions of the eye.

Conclusion

The strategy was to present two fixed training shapes that differed in a simple way. The trained bees were given numerous tests, which progressively identified and refined the cues that they used. Tests that resulted in failures to discriminate were an essential part of the analysis. This process was not a test of a theory of vision; it was a logical investigation of what the bees really detected.

The trained bees did not learn shape in general; they learned to discriminate by detecting and learning the position of one or more simple cues. There are only a few of these cues and they are used over and over again. Different pairs of shapes displayed the cues in different combinations in different strengths. Discrimination of shapes involved the coincidences of cues that were detected together in a local region of the eye, not the reassembly of the layout.

Endnotes

1. It follows from the results in this chapter that when bees discriminate between two shapes they learn something for that occasion only, not the recognition of shapes in general, and also that the performance does not imply cognition.

12

GENERALISATION AND COGNITIVE ABILITIES IN BEE VISION[1]

"if that truth involves the putziness of other people or events, so be it,
but if it involves the narrator's own schmuckiness, limitations, prejudices,
foibles, screw-ups at the event, etc, then these get told about too"
(David Foster Wallace, 2008).

For a century, there have been claims of something in bee vision more subtle than the coincidences of feature detectors and cues. Anthropomorphism—that is, the tendency to put human capabilities into the brains of the bees—was not openly supported, but cognition trickled down from work on higher animals.

Under the general heading of cognition in vision, the oldest belief was that the bees really saw and remembered the spatial layout of patterns. Also, it was thought that bees generalised patterns that looked similar to them. More recently, it was proposed that bees recognised patterns as a whole, that they detected patterned shapes over a patterned background and detected abstract features such as symmetry, topology and other pattern qualities irrespective of the real pattern. Indeed, they do distinguish certain global features such as size, total length of edge or modulation, average edge orientation and the presence of circles or spokes, but with only a limited repertoire of cues.

Some strange conclusions can be found in high places. For example, Giurfa et al. (2001) state that bees 'interpolate visual information', 'categorize visual information' and 'learn contextual information'. They 'form sameness and difference concepts', 'transfer to the same or a different sensory modality', perform 'delayed matching' or 'non-matching to sample tasks', 'learn specific objects and their physical parameters' and 'master abstract inter-relationships' such as 'sameness and difference'. These claims of cognitive abilities were based on the performance of bees that were not tested in a way that would easily have eliminated those conclusions.

Of course, anyone is free to persist with the idea that bees recognise things, rather than places. Of course, bee behaviour can be described in the terminology of the cognitive sciences, with no reference to the analytical work since that of Hertz, which showed that when presented with unfamiliar patterns, trained bees chose according to the cues that they learned in the training. Bees, however, do not recognise the patterns they were trained on when these are tested against other patterns that display the same cues (Chapter 11). This final chapter further shows that even our own experiments that supported the so-called cognition of bees rested on very shaky ground.

Generalisation

In the experiments of Mathilde Hertz (summarised, 1933), bees were trained to come for a reward of sugar solution at a flat white table where a group of patterns of similar size were shuffled in position at intervals. One of the patterns was consistently rewarded and the others were not. The bees learned to go to the rewarded one if it differed from the others in length of edge or certain other features (Figures 1.2d, 1.2e, 1.4). When tested with unfamiliar patterns, the trained bees accepted some but not others. For example, when trained on circles, the bees treated them as equal to a pattern of spots (Figure 1.5). This does not look like recognition of similarity. Hertz inferred that, although the patterns were indeed different, the bees recognised certain cues or parameters, such as a measure of the modulation or total length of edge, the area of black and the presence of symmetry. The acceptance of unfamiliar patterns was called generalisation and was attributed to two factors: the low-level recognition of parameters held in common and the existence of higher-level categories, such as a similarity detected by the bees.

Also, bees could learn to generalise when some features were shared in common between a number of training patterns, and the trained bees then recognised the same features in other patterns. When wasps (*Vespa germanica*) were trained simultaneously with different kinds of equilateral triangles, they distinguished unfamiliar triangles from squares or other shapes (Verlaine 1927). This example of generalisation during the training was said to be a remarkable performance that suggested a higher cognitive function, but there was no consideration of simple cues as the explanation.

There was not universal acceptance of generalisation. In his useful (but usually ignored) review of the topic, Carthy (1958) was equivocal. He accepted that patterns were preferred or discriminated by differences in edge length irrespective of pattern, but also gave examples of patterns of similar edge length that were discriminated and others that were not. Carthy assumed that the bees had a limited repertoire and poor recognition and he made the telling remark that 'the bees might be reacting to only parts of the pattern and not to the whole'. Almost certainly, he had in mind the pioneering work of Lashley (1938), who showed that rats learned only a token part of the pattern that indicated the reward. Later bee researchers also ignored Lashley.

Categorisation

The term 'generalisation' comes from the way humans generalise many different shapes such as different chairs or the letter 'a' in different fonts, and cannot prevent themselves from unconsciously categorising everything that they consciously see. The members of a category can be substituted for each other without loss of understanding. Human language and vision depend on a long process of learning the useful categories detected by all the senses.

In other animals, intermediate between bees and primates, some patterns can be substituted for each other. For example, a rival male can successfully substitute almost any patch of red for the red breast of the robin or the stickleback and still initiate an attack. Because there are numerous levels of complexity and different kinds of visual systems, generalisation is hard to pin down. Bees fly about, visit flowers and navigate with landmarks, so it has been assumed that they also categorise things. This was summarised succinctly as 'patterns have to be grouped into invariance classes' (Wehner 1975). In the light of recent experiments, perhaps this should have read 'patterns are naturally grouped into invariance classes by the cues abstracted from them'.

From the earliest training experiments to the present time, there were therefore two extreme explanations of generalisation—almost opposites in their mechanism. In one, the general properties were related to categories that classified things or qualities and within which there was generalisation. On the other hand, substitutes are accepted because there are insufficient cues to distinguish them from the genuine article. Neither of these explanations was validated by tests on trained bees.

Spatial memory

In the early twentieth century, there was a variety of theories that memories were represented spatially in the mammalian brain—some even by analogy with magnetic fields. For example, following the ideas of Pavlov, 'neuron paths are established between parts of the brain'. 'We use Semon's term "engrams" to denote these physiological paths and Head's term "neural schema" as a permissible synonym' (Campion and Elliot Smith 1934). The engrams could be in or out of consciousness. The neural schema were hypothetical reassemblies of patterns in the brain.

The Gestalt theory, popular in the first half of the twentieth century and still influential today, was based on the idea that the visual image was laid out as a spatial field that would be preferred or remembered when its neural organisation matched the previously established image in memory. Another principle of Gestalt theory was that parameters such as symmetry, roundness, star shape, coarse or fine texture and size were detected as generalised features because the visual system was adapted to detect and remember them. For most of the twentieth century, this was the dominant conceptual scaffolding and many experiments with bees were designed with this theory in mind. With improved techniques, however, modern neuroscience has not found any reassembled schema or images, even in humans, although there are neurons that look like cue detectors.

Bees certainly generalise

When Hertz shuffled the positions of patterns on a flat white table to make the bees look for them, the bees did not use information about their flight directions in relation to the orientations of the patterns. The parameters that were described were those that could be used despite the training strategy—notably, the colour, edge length, circle versus spoke, area or size, irrespective of the pattern. When trained to a pattern of a particular total edge length or modulation, versus a variety of other patterns, the bees looked for the training cue in entirely unfamiliar targets and were not interested in the real patterns (Chapter 1). In the vertical plane also, bees trained to one pattern readily accepted some unfamiliar patterns (Baumgärtner 1928; Friedlaender 1931; Wiechert 1938). From detailed experiments (Chapter 4), it was inferred that the bees simply totalled the areas of overlap of black and a measure of the edge length in a global comparison of the training and the test patterns (Cruse 1972: Anderson 1977a). This was very low-level stuff. There remained, however, a belief in something more than quantifying the parameters. As a separate mechanism, Hertz thought that radial and circular symmetry were detected as a whole, irrespective of the detail, and inferred high-level cognitive mechanisms. In contrast, from similar data, I infer a distributed low-level mechanism (Figure 9.19).

The observed generalisations of bees fell into categories that Wittgenstein would call 'natural families'—in this case based on clearly definable simple parameters that did not overlap or merge into each other, so providing some indication of their validity. Other possible parameters, such as angles between edges or counting the corners, spots or bars, did not yield data of the same kind.

In a more suspect example, Mazokhin-Porshnyakov (1969) trained bees to discriminate between a large, hollow triangular pattern (rewarded) versus a number of ring-shaped patterns of different sizes, all presented on a horizontal surface (with the orientation randomised). The large triangles were composed of many smaller triangles and the large rings of many smaller rings, so that the bees might distinguish triangles and rings from a distance as well as from close up. The trained bees were then able to discriminate between triangles and rings of unfamiliar sizes or orientation or with different background and form of outline. Because generalisation implied cognitive behaviour, and because memories of images were believed to be laid out spatially in the brain, Mazokhin-Porshnyakov inferred that the bees had learned the generalised concept of 'triangularity'.

This example illustrates the flaw in all work designed to test a theory. The data were compatible with the theory, but the theory was not corroborated by further tests that could have disproved it. The bees obviously learned something from the training—possibly a small part of the pattern—but it was concluded that they learned 'triangularity'. This faulty logic persists to this day. Later, it was shown that the bees indeed learned a few cues, but not a triangle (Anderson 1972).

Similar data; different conclusions

Following similar work with ants, Jander et al. (1970) trained wasps to discriminate an oblique black bar (Figure 4.6a) and showed that they detected the orientation when black and white were interchanged (Figure 4.6c). The trained wasps, however, confused the training bar with the white bar on a black background (Figure 4.6d). This result was interpreted in terms of rows of symmetrical detectors of modulation (Figures 4.1b and 4.1c). Generalisation was not mentioned.

At the same institute, Wehner (1971) trained bees to come to a huge oblique black bar (subtending 130° long) on a white background versus a plain white target (Figure 4.6e). Unlike the wasps trained by Jander (Figure 4.6d), the trained bees easily distinguished between the black bar on white versus a white bar on black (Figure 4.6h). Wasps and bees had learned sufficiently to respond to the edge orientation, but Wehner's bees had learned the position of black as well. Wehner (1971) inferred that 'the information about the direction of a visual stimulus is laid down in the central nervous system as an invariant information irrespective of the actual contrast condition'. This was in fact the experimental result expressed in different words, not an explanation. Local feature detectors were not mentioned.

Wehner then proposed that the bees must be able to distinguish between the patterns that they were observed to generalise, to exclude the possibility that they simply could not detect the differences. When several patterns are generalised, however, it does not imply that they are separately distinguishable. Indeed, they could be identical. Categories are based on usage and vary with the agent. For example, sheep distinguish between each other but humans do not distinguish between sheep. Bees distinguish between larvae that need feeding and those that do not, but probably not between individual larvae. Although illogical, the proviso that the trained bees must be able to distinguish between the patterns that they are supposed to generalise has persisted in the literature to the present time (Benard et al. 2006).

Later, the patterns in the brain became rather volatile. For example, 'generalized information can be transferred later on to other stimulus configurations, which never occurred during the training' (Wehner 1975). This apparently destroys the idea that the memory is a shape in the brain. The inclination of a bar was discriminated 'even if the contrast was completely reversed'. 'Therefore a two-dimensional matching…has to be followed by a sampling mechanism according to invariance classes…Preprocessing of the pictorial input has to be studied first if one wants to solve the classification problem.' All this mental gymnastics, based on few results, assumed the image in the brain before recognition. In my view, however, the engram was unsupported by experiment and it was a 'devoted attempt to force nature into the conceptual boxes supplied by professional education' (Kuhn 1970:5).

Later, categories and spatial images dropped out: 'even an "experienced" bee does not seem to build up in its mind abstract search images consisting of pure geometrical forms that are invariant against other visual parameters such as hue of colour, size, contrast, or fine pattern detail' (Wehner 1981). What, then, is the way forward? One way is by more of the same. From 1995 on, several researchers found examples of transfer to unfamiliar patterns by trained bees and concluded that the patterns were generalised. Ignoring numerous examples of unlike patterns that were interchangeable and published testable explanations in terms of cues, and making no critical tests of their own, they said that the bees had cognitive abilities (Giurfa et al. 2003; Stach et al. 2004; Benard et al. 2006).

Generalisation within the training regime

Bees in flight have a very good appreciation of the sizes and ranges of contrasting objects around them. When the rewarded parameter was kept constant during the training while the other parameters were randomised, the bees could be trained to choose a black disc at a certain range irrespective of the angular size of the disc (Lehrer et al. 1988). They could also remember a disc of a certain absolute size irrespective of the apparent angular size (Horridge et al. 1992). The bees learned to generalise from the randomisation during the training. The angular size, the absolute size and the range all turned out to be parameters that could be learned.

The same strategy was used with a pattern of vertical parallel bars on one target versus a similar but horizontal pattern on the other (van Hateren et al. 1990). The positions and widths of the bars were randomised during the training, so that the bees 'made their decision on the basis of orientation only'. For a time, these results suggested that the orientation was detected irrespective of position and that 'specific features of the pattern, such as bars and edges, are extracted and their orientation analysed as in the mammalian cortex' (Srinivasan et al. 1993b). As shown later, however, with a vertical versus a horizontal bar, the bees ignored the orientation and preferred to learn the modulation difference. The parameters were recognised in tests only in the places on the target where they occurred during the training (Horridge 2003a, 2007).

Figure 12.1 An error of interpretation, shown within the square. a) In the modified Y-choice apparatus (Figure 10.3), bees were trained on horizontal versus vertical random gratings, so they learned the orientation cue. b) The trained bees were tested on the composite bars at various distances. c–e) At the 9cm range, the trained bees preferred the small horizontal bars, but at 27cm, they preferred the large composite bar. Memory of local and global orientation was inferred. f–h) The illustrations are now drawn at the relative sizes detected by the bees. The bees preferred bars similar in size to those in the training patterns and the horizontal edge orientation within the small area where they had learned the modulation or orientation cue in the training, as shown by the dashed circles in (h).

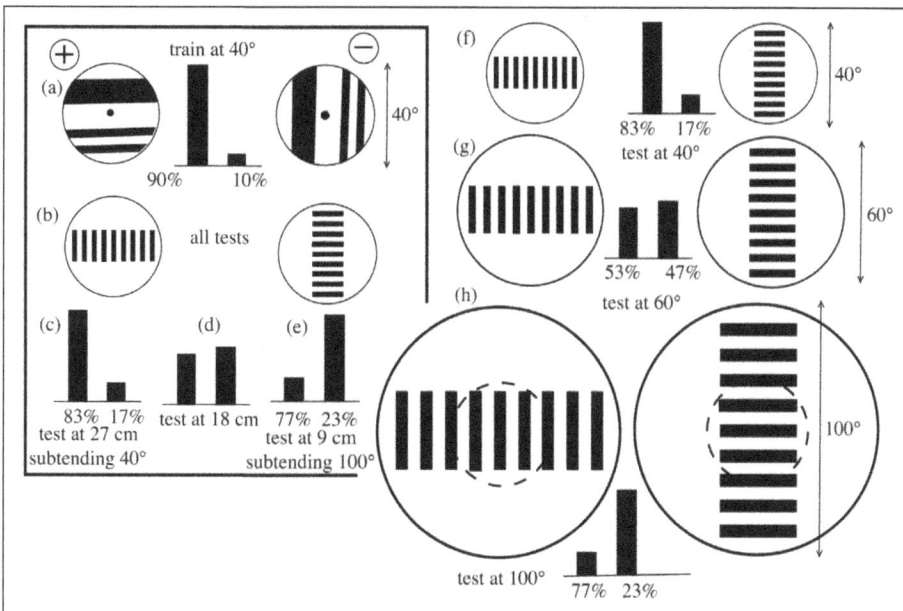

Source: After Zhang et al. (1992).

Other inferences of cognition

Global versus local perception: a dog's breakfast again

In our paper (Zhang et al. 1992) that claimed to be the first attempt to examine 'whether bees analyse patterns in terms of their local properties, global properties, or both', our introduction was based on our reading of human psychophysics. In the experiments, bees were trained to prefer horizontal edges by using gratings of random period (Figure 12.1a) and then tested on two large bars, each composed of many small bars at right angles to the axis of the main bar. Seen globally, there was therefore one large bar on each test target, but locally there were many smaller bars at right angles to them (Figure 12.1b).

When the trained bees made their choice at a range of 9cm, they chose the horizontal orientation of the small bars, but from 27cm, they chose the horizontal orientation of the large bar. At 18cm, the effects of the global and local orientations were supposedly cancelled (Figures 12.1c–e). In other experiments, bees trained on bars composed of smaller bars could use either the global orientation or the local orientation in tests where only one was available.

The result was not queried at the time although there were severe faults in the experiments. In fact, before baffles were introduced in 1995, the bees could have detected the global orientation from a distance and then the opposite orientation of the small bars at a later point in the flight path. Although at a range of 27cm the small bars were separated by spaces of 4°, from a greater distance, they were not separately resolved. Conversely, the bees probably detected little of the global pattern from a range of 10cm because they had been trained to expect the orientation cue within a target subtending 45°. Moreover, it has since been shown that the perceived orientation is a sum over each local region of the eye. The illustration has now been revised to clarify the situation faced by the bees (Figures 12.1f–h), but there are other problems.

In these experiments, the bees were allowed two visits on each side of the apparatus in each test, so they could have improved their score at the second visit. This is relevant only to the marginal successes. Also, vertical edges generated more modulation than horizontal edges because bees in flight scan in the horizontal plane. Luckily, our conclusions were cautious: 'Although our experiments demonstrate the existence of local and global analysis, they do not shed light on the underlying processes' (Zhang et al. 1992). How could they, without numerous tests of greater variety?

We in fact suggested modulation as a cue: 'the coarse and fine gratings are detected and analysed in terms of the different temporal signatures that they produce' (Zhang et al. 1992). Indeed, it was later found that bees preferred to learn the modulation cue rather than an orientation cue (Horridge 2007), and untrained bees and wasps preferred patterns rich in modulation to those rich in orientation (Jander et al. 1970; Lehrer et al. 1995). Other work showed that the detectors of edge orientation were only local and that they did not span gaps to detect global orientation (Horridge 2003c). 'Global perception' was simply a cover for ignorance, but for years we knew no better.

If we had known more at that time about detection of cues in fixed patterns, we would have tested for modulation and locations of black and orientation cues. Finally, some of the data were suspect because there was a limited variety of 10-minute tests. In hindsight, our suggestion of global detectors was no explanation at all. It was a form of words that was consistent with human impressions of what the bees detected. This is exactly what science is supposed to eliminate.

Figure 12.2 The scores in training experiments with pairs of patterns, each with four different orientations in the four quadrants. a) With a difference in average edge orientations on the two sides of the targets. b) As before, but with the right target rotated. c) An example with no average orientation. d) With a difference in average edge orientations on the two sides of the targets. e) With radial versus tangential cues and also orientation differences on the two sides. f) With radial versus tangential cues. g) Mirror images of (f), with the same cues. h) Patterns with no detectable difference in cues.

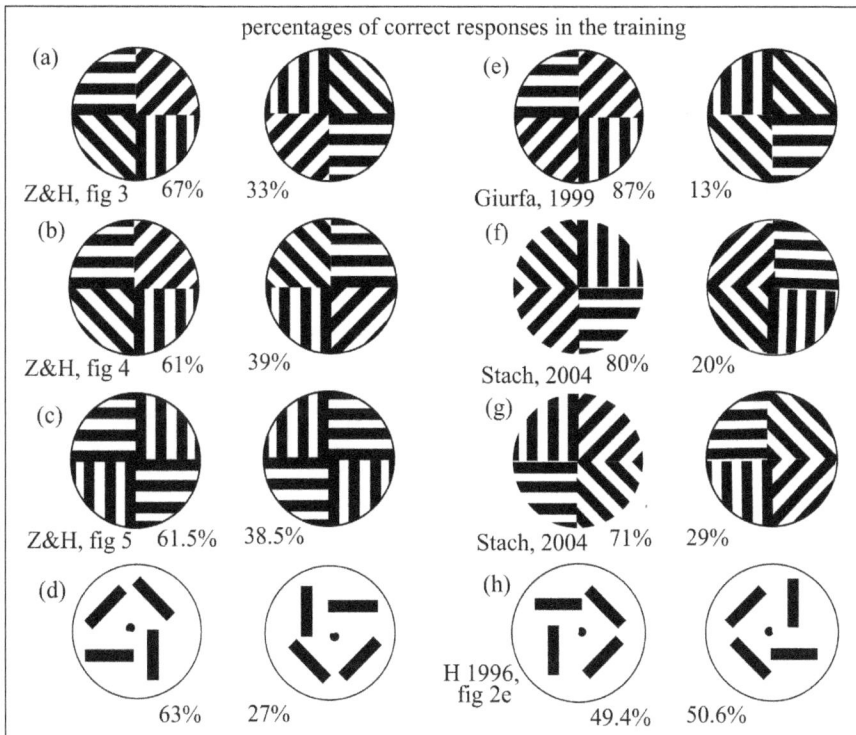

Sources: (a–c) from Zhang and Horridge (1992); (e) from Giurfa et al. (1999); (f, g) from Stach et al. (2004); (h) from Horridge and Zhang (1995:Fig. 6a).

Separate regions of the target

In our next experiment, we planned 'to see how many parts of a pattern could be discriminated separately, and whether discrimination was lost on rotation or inversion of the parts' (Zhang et al. 1992). A target was divided into four quadrants with a differently oriented grating of period 8° in each quadrant.

These patterns confused subsequent researchers but not the bees. In the Y-choice maze, bees discriminated the rewarded training pattern from a similar pattern with the quadrants rearranged (Figure 12.2a). Increasing the number of sectors to eight or 16 showed that the smallest effective sectors subtended about 22° at the eye, which was more than 100 facets or a similar number of unit orientation detectors. This calculation gave 'some idea of how an array of numerous templates, each individually ineffective, can collaborate together to make specific ensembles that fit the pattern sufficiently well' (Zhang and Horridge 1992).

In fact, this was all rubbish because the design of our experiment and the data were faulty. First, we were unaware at the time that one side of our training targets had more horizontal edge and the other side more vertical edge, and that the bees processed the average orientation separately on each side of the target. This cue was there for all to see (Figures 12.2a and 12.2b). In another pair of similar patterns (Figure 12.2c), the orientation cues were more likely to cancel out but something was apparently discriminated. Also, we did not test what the bees really detected or even whether they remembered an ensemble at all. Third, from 1990 to 1996, the bees were allowed two visits (10 minutes) on each side of the apparatus in the tests, which was sufficient for them to add a few points to the borderline scores.

With similar naivety, and similar patterns with orientation cues in four quadrants, Giurfa et al. (1999) allowed the bees to approach close to the targets, which therefore subtended very large angles at the final choice point, so the configurational layout of areas of black could be discriminated. They concluded that when trained with a pattern of four quadrants versus a blank target, the bees learned mainly the lower half of the rewarded pattern, but when trained with one pattern versus another (Figure 12.2e), they learned all the pattern—and to avoid the unrewarded pattern. In their training pattern, however, there were radial versus tangential edges and also differing average orientations on the two sides, which the authors did not mention, providing obvious parameters for the bees. Either one or both of these parameters was also displayed in their test patterns, so the results threw no light at all on global vision.

More recently, bees were trained with similar patterns but with shuffled thickness and positions of the bars, versus a similar unrewarded group with a different pattern of orientations (Stach et al. 2004; Stach and Giurfa 2005). This time, the targets subtended 37° at the point of choice. Discrimination depended on green contrast and therefore edges were involved. In the training targets and tests, there were opposite average orientation cues on one side of the targets (Figures 12.2f and 12.2g) and on the other side there were radial versus tangential edges that the authors did not mention. In some tests, the pattern was reduced to one bar in each quadrant while retaining the difference in average orientation on the two sides (as in Figure 12.2d); in others, the details were shuffled within each quadrant, but the parameters remained for all to see. In tests, the trained bees discriminated with black and white reversed, as would be expected because the feature detectors for edges were symmetrical (Figure 9.4).

The ability of the bees to discriminate the unfamiliar patterns was described as a generalisation and 'after a long training with a single pair of patterns, bees built a simplified holistic pattern representation that included all four edge orientations in their appropriate spatial relationship and that allowed transfer to novel stimuli preserving such a positive layout' (Stach and Giurfa 2005). This conclusion was a guess for which there was no evidence and no test of global vision. Moreover, obvious radial parameters were displayed. There was no evidence that the bees also generalise their response to patterns with fewer correct orientations, depending on their match with the trained layout because the parameters in the training remained in the tests. There were no tests of what the bees really detected. There was certainly no evidence for the claim that the bees responded to 'the perceived lay-out' in patterns of this size.

The same data supposedly demonstrated 'categorization based on sets of multiple features' and the bees 'were shown to assemble different features to build a generic pattern representation which could be used to respond appropriately to novel stimuli sharing the same basic layout' (Benard et al. 2006), revealing the persistence of unsubstantiated ideas about spatial reassembly in the brains of targets that subtended 37° at the point of choice. The authors say the 'results show that honeybees can recognize visual patterns on the basis of the global layout made from four different orientations, common to a series of different patterns'. In fact, there were no tests of whether the layout of quadrants was noticed at all by the bees and the test data were compatible with the recognition of the obvious rad/tan or orientation cues. Moreover, there were abundant published data to show that the individual bars, the separate quadrants and the whole patterns could not be discriminated if the rad/tan and orientation cues cancelled out in patterns subtending 37° (for example, in Figures 12.2h and 9.14j).

Illusory contours

By 1993, it was possible to 'suggest, perhaps for the first time, the existence of feature-extracting mechanisms in the insect visual system that might be comparable, functionally, to those known to exist in the mammalian cortex' (Srinivasan et al. 1993). This lyric was inspired by an inference that insects perceived illusory contours. When they had been trained to discriminate between the orientations of shuffled orthogonal gratings, bees apparently saw the contours of the Kanizsa rectangle illusion (van Hateren et al. 1990:Fig. 4). It was supposed that, as in the human cortex, lines of edge detectors with similar orientation were strung together. Bees, like humans, also responded as though they saw an illusory orientation at a fault line across a regularly striped pattern

(Horridge et al. 1992). There seemed to be nothing wrong with the idea of illusory contours, but at the time we did not know that different edge orientations in close proximity cancelled each other or that edge detectors did not span across gaps that were resolved.

When the experiments were repeated, they failed. There had been two changes to the design of the experiments. Until 1996, there were no baffles in the apparatus so the bees could enter at full speed and make a fast decision from further away. Also, they were allowed 10 minutes and two visits on each side in the tests, which allowed them to improve their success rate. After 1996, however, the baffles halted them and they took longer to peruse the targets from a fixed distance. They also had only five minutes on each side and many varied tests were intercalated, so they saw the same test at long intervals between other tests. With these precautions, the bees did not detect illusory edges or the edges at fault lines (Horridge 2003a). Also, David O'Carroll told me that he could not repeat the detection of illusory contours when recording from single neurons of the dragonfly lobula.

Transfer of shape between green and blue channels

In a brief paper, Zhang et al. (1995) trained bees to discriminate between a wide horizontal bar (rewarded) and a similar vertical bar (subtending about 36° by 8° at the choice point). To prevent input via the motion-detecting system, the edges of the bars displayed contrast against background only to the green receptors. The trained bees could immediately distinguish between such bars when they were presented in blue contrast.

The observations were not in doubt, but there was no evidence for the conclusion that 'shape is memorized in a generic form regardless of whether it is initially sensed by green-contrast, blue-contrast, luminance-contrast or motion-contrast signals' Zhang et al. (1995). The shapes of the bars or the orientations at the edges were not even probable cues for stationary bars. In the light of later findings (Giger and Srinivasan 1996), it was impossible for the bees to detect orientation with the blue channel alone, and in any case, the probable cue was the modulation difference. In another experiment, the authors in fact showed that the cue was the difference in modulation between horizontal and vertical bars and modulation was detected by both green and blue receptors. When the bars were oblique, the bees learned the orientation cue and could transfer to similar targets with green contrast but not to ones with blue contrast, because there was no difference in modulation with the oblique bars. In the first experiment, the bees did not transfer between green and blue channels; they had learned the modulation cue, which was not colourblind. The conclusion, however, has been frequently quoted as evidence of cognitive transfer of shape discrimination.

The Dalmatian dog; shape from parallax

Continuing the same saga:

> To investigate whether bees encode shape in a generic form, regardless of input channel, we began by asking if bees that have learned a shape defined in terms of luminance contrast can recognize the same shape when it is defined in terms of motion contrast. (Zhang et al. 1995)

Accordingly, bees were trained to discriminate a thick black ring (rewarded) from a large black spot of the same area, both centred on the central reward hole. The 'trained bees can immediately distinguish between the same shapes when they are presented as black-and-white textures, of pixel size 4 mm square, 6 cm in front of a similarly textured background'. The same trained bees then learned to discriminate between black and white random-pixel textured oblique bars, a task that they could not do before they learned 'that motion contrast is the relevant cue' (Zhang et al. 1995). Even more remarkable, having learned to discriminate the two shapes with motion cues, the trained bees recognised them in blue contrast. In the earlier version (Zhang and Srinivasan 1994), a textured Dalmatian dog on a textured background was illustrated, upside-down, to make recognition of it more difficult. Miriam Lehrer used a textured elephant in one of her illustrations.

First, let us look at the internal evidence for misplaced conclusions. Pixels of 4mm square on the background would subtend 0.8° and even the pixels raised 6cm in front would subtend 1° at the point of choice, and would not be resolved. Second, the discovery that equal lengths of edges at right angles cancelled out the orientation cue (Srinivasan et al. 1994) implied that when the pixels were large enough to be resolved, the orientation cues were cancelled. Third, discrimination of orientation required only edge detectors (Figure 9.4), not motion detectors (Srinivasan et al. 1993). Fourth, the bees were allowed 10 minutes at each arm of the Y-maze before the patterns were changed, giving an average of two choices at each test, so they could more easily reach the relatively weak borderline scores that were recorded. Finally, there was no test for whether the bees saw the shapes at all. Furthermore, when I repeated the original training with exactly the same patterns, the bees learned to avoid the spot, the cue was the absence of black near the reward hole and the trained bees had no memory at all of the shapes (Figure 11.8).

When this experiment was repeated with larger pixels that were resolved, the bees failed to discriminate. The bees trained on plain black patterns would not discriminate textured patterns raised 6cm above a textured background. Furthermore, the same bees readily discriminated between two orthogonal bars of plain white paper that were raised 6cm above plain white targets, showing that weak shadows provided sufficient cues (Horridge 2003a).

In this example, the experiment was based on a good idea and the experimental data were compatible with the premise. The patterns, however, were inappropriate, the data were suspect, alternative explanations were available, there were several reasons to reject shape perception in general and shape from parallax in particular and the results could not be repeated.

Bilateral symmetry about an axis

Untrained bees have a preference for patterns with a vertical axis of bilateral symmetry, irrespective of the pattern (Lehrer et al. 1995). Bees learned to discriminate the vertical axis in patterns of two pairs of bars at right angles to each other (Figure 9.15). In agreement with the contemporary ideas about global templates, it was proposed that 'bilateral symmetry assists discrimination' and 'if there is a global filter for this pattern, it has broad angular tuning' (Horridge 1996a).

Then, to demonstrate the cue of symmetry about an axis, with the newly introduced baffles in place, bees were trained all day on seven quite different bilaterally symmetrical patterns that were taken successively for 10 minutes each. The patterns all displayed the same four black bars in various arrangements. The rewarded ones had a vertical axis of symmetry and the unrewarded one in each pair was the same pattern rotated through 90° (Figure 9.20). Bees readily detected the orientation of the axis in tests with unfamiliar patterns.

From these results, I inferred global filters that

> perhaps work in the same way as the face detectors in human vision…
> It is difficult for us to appreciate that the bees are sensitive to the pattern
> as a whole and discriminate a global feature of it without remembering
> the locations or orientations of individual bars, but in our own vision
> we are familiar with our discrimination of colours without being able
> to identify their constituent wavelengths. In this respect, bee vision
> of form resembles our vision of colour; the components of it are not
> separately discriminated.

> Like the smile of the Cheshire cat in Lewis Carroll's *Alice Through the
> Looking Glass*, the abstract feature, the smile, persists although the cat
> is no longer distinguished. Generalization of this type is the essence of
> vision, in that whole objects and complex relationships are recognized
> irrespective of local variables. (Horridge 1996a)

With science like this, who needs poets?

In fact, low-level cues must have been detected in all the symmetrical patterns, but the tests were never done. Scanning in flight of bilateral symmetry yields the same sequence of feature detection in either direction. Alternatively, bees can discriminate the average orientation and the averaged positions of the centres of black, colour or other cues, separately on the two sides of the target. When these averages, together with radial and tangential cues, are at equal heights on the two sides, they could be sufficient to distinguish bilateral symmetry about a vertical axis in many patterns. The topic needs further investigation.

Topology as a cue

Having found that bees discriminate between a rewarded black O and an unrewarded black S (Figure 11.10a), Chen et al. (2003) proposed that the 'topological properties constitute a formal description of fundamental perceptual organizations, such as distinguishing [a] figure from [the] background, parsing visual scenes into potential objects, and performing other global, Gestalt-like operations'. This was typical gobbledygook borrowed from the cognitive sciences.

Chen et al. made four tests of the trained bees that neither proved nor disproved topology as a cue, but they tried none of the possible tests that would have disproved it, nor did they demonstrate what cues the bees had really learned. They also found that discrimination of the S (rewarded) from the O was learned extremely rapidly, which is now explained by the innate avoidance of the O, so probably it was not learned at all.

The choice of the broad black ring as the rewarded target was most unfortunate. In a repetition of the same experiments, the cues were the presence of black near the reward hole *on the unrewarded target* and the orientation of the middle section of the S (Figure 11.10). The topology was irrelevant.

Even without the critical tests, the discrimination between a closed and an open shape in no way demonstrated that bees recognised the topology, any more than a discrimination between two pictures of human faces showed that bees recognised faces as faces or as individuals.

Preference for radial symmetry irrespective of pattern

About 1994, attention was drawn to the evolutionary advantages of symmetry in a variety of animals and plants. Animals are intrinsically asymmetrical but symmetry has been perfected by sexual selection and forward locomotion. Bees preferred to forage from symmetrical flowers (Møller 1995) and it was supposed that flowers adopted and rewarded symmetry to attract bees.

In a circular apparatus, bees were trained to come to neutral targets placed vertically at the back of four out of 12 compartments (Figure 9.11). The training patterns were then replaced with 12 patterns with different levels and kinds of symmetry in equal numbers and the bees' choices were recorded. When the test patterns were of the same kind, the bees preferred larger periods and broader bars or sectors. When the patterns differed in type, the bees preferred radial and avoided circular patterns. Bilaterally symmetrical patterns were preferred when their axes were vertical.

In line with the ideas of the time, it was proposed that 'filters tuned to radiating and circular shape elements…would enable the bee to use global parameters to discriminate numerous patterns with only a small number of specialized neurons' (Lehrer et al. 1995). The discrimination of edges as radial or tangential when they lay at different angles to each other was, however, at odds with the discovery that edges at large angles to each other reduced the orientation cue (Srinivasan et al. 1994). To resolve the discrepancy, it was proposed that 'bees have additional filters, of which the minimum number is two types in polar co-ordinates that resemble radial sectors and concentric circles', and that the 'large field or global detectors of polar symmetry inhibit the orientation detectors' (Horridge 1994). Later, edge detectors in radial or circular directions on the eye were grouped into 'innate global filters for radial and tangential contours in the pattern as a whole' (Horridge 1996c). None of these proposals was tested. They became firmer as time passed and they were quoted by others. In hindsight, it was an illustration of science in progress, naive moonshine or a misleading catastrophe—depending on your standpoint.

Training on radial symmetry

Before the work on preferences, bees were trained simultaneously with radial patterns (rewarded) versus tangential ones (Figure 9.12a), with the positions of black shuffled at intervals. The bees transferred their training to quite different patterns displaying the same cues. Unfortunately, it was found later that bees innately preferred radial patterns and avoided circles, so they might have learned nothing. This error was later corrected (Horridge 2006b, 2007).

A pattern of three or six equally spaced radial bars was readily discriminated from the same target rotated by half the angle between the bars, but rotation of a target with four, five or seven radial bars was poorly discriminated (Horridge 2000b). This result was 'consistent with the proposal that there is a family of global filters at small angles to each other with 3 arms and another family with 6 arms'. 'The early visual processing retains the resolution of the retina, but at a higher level the memory has available only the outputs of large-field filters' (Horridge 1997c). By 1998, 'generalization over a range of certain related images…can be explained by…coarsely tuned filters but not by an eidetic image, or universal learning mechanism' (Horridge 1998a).

Despite these observations, there were other examples where different orientations were remembered separately in the same region of the target. No-one discussed the discrimination of edges at angles in the same pattern

(Zhang et al. 1992), different orientations in concentric circles (Horridge and Zhang 1995) or different numbers of sectors in radial patterns (Wehner 1981:Figs 59, 67). There was obviously more to be found, *dessous des cartes*.

No global filter for radial symmetry

The discriminations of symmetry presented a problem because they were independent of the scale and layout, so that many templates would be required. It was impossible to imagine global filters that fitted the data. The usual tests showed that radial or circular edges were not reassembled. The bees remembered only the radial or tangential character and the position of the centre of symmetry irrespective of the position of black, pattern or scale. Later, it was discovered that bees could be trained to discriminate between the left and the right halves of a symmetrical pattern, either radial or circular. Therefore, symmetrical global filters with a single output were ruled out because either half of the pattern would excite the same filter (Horridge 2006b).

The feature detectors proposed for edge orientation in other experiments did not string together to span across gaps (Figure 9.8). They were short, independent and about 3° long (Figure 9.9). The feature detectors for radial and tangential patterns were demonstrated by the same tests as used for detectors of edge orientation and turned out to be the same (Figure 9.19).

There were now sufficient data for an explanation of symmetry detection by local feature detectors feeding into larger fields, like all sensory processing. Edges anywhere in the pattern were treated as radial when they converged towards a hub or as tangential when lines at right angles to them converged towards a hub (Figure 9.19). The position of the hub and its radial or circular character were remembered, but the original layout of the feature detector responses was lost. This was a distributed, local and flexible mechanism that would find an average centre and identify the pattern as radial or circular by a distributed administration, irrespective of the size or pattern. Hypothetical global filters were excluded and replaced by an evidence-based explanation.

How the nexus between patterns, landmarks and place was broken

In the nineteenth century, many efforts were made to understand how bees returned exactly to the rewarded place. A common technique was to give a reward on a flower, then change the flower for another of a different shape or colour or hide it with a few leaves. Felix Plateau, for example, correctly concluded that the bees ignored the altered shapes and the colours of artificial flowers. Tedious exact repetition of Plateau's experiments showed that the bees went unerringly to the place where they had found the reward irrespective of the shapes, but were lost if the place was moved (Forel 1908:170).

When Turner (1911) trained bees to distinguish between two boxes—one with horizontal black stripes and a reward of sugar inside, the other with vertical stripes and no reward—he interchanged the positions of the boxes at intervals to make the bees look for the rewarded pattern, *irrespective of the place*. By shuffling the positions while keeping the cue constant, Turner had broken the nexus between the recognition of the label on the box, which was the horizontal edges, and the recognition of the place, which required several landmarks at wide angles to each other.

The technique was adopted by von Frisch (1914) and used with various modifications by all later investigators. Hertz (1933) placed black patterns flat on a white table and placed a reward of sugar solution next to one of them. She broke the connection between reward and place by shuffling the patterns on the table, so the bees ignored everything except the cues in the correct pattern, irrespective of the place.

For the first time, it was noticed that the bees took a much longer time to learn. Also, they either became tolerant or liable to mistakes when trained in this way and accepted unfamiliar patterns, which was called generalisation. In contrast, when rewarded at a fixed place, they returned after a single visit and never made an error.

Let's explain. When the positions of the patterns were shuffled or two targets were interchanged, the bees were obliged to look for the familiar cues on the patterns and they were trained to ignore everything outside the patterns. Vision was restricted to one or two forward-looking local regions of the eye by the shuffling or alternation of the patterns because the rest of the eye learned to ignore the surrounding place. So blinkered, they could no longer use the coincidences of landmarks between different eye regions (Figure 10.7). Alternating or shuffling the targets exposed the bees to errors by restricting the memory to a local region, which processed only one of each type of cue. The number and variety of cues that could be learned was inadequate to distinguish every pattern, so recognition was easily fooled. The observed ambiguity, or confusion of the bees, was called *generalisation*.

This bit of history shows how bee trainers were fooled by their own training technique, combined with the small repertoire of cues. For almost a century, they believed that bees generalised patterns because they saw them as similar or they belonged to the same bee category. The error of thought was established in the literature, heels were dug in, territories were defended, referees unjustly rejected papers and contention seriously slowed the advance of understanding.

At the same time, the coincidences of cues, the total area, the position of the centre of area, total modulation, average local orientation, the tangential or radial nature of edges and positions and types of hubs were each summed over a local region. This removed the detailed distributions of contrasts within the experimental patterns. It was called *global vision*, but it was in fact an artefact of the training and testing technique that was restricted to one local region of the eye.

On the other hand, bees were not interested in the training pattern in an unfamiliar position and generalisation was not observed in the identification of a place. The recognition of a very large target, or a pattern that the bees were allowed to examine closely or a natural situation, involved the retinotopic detection of a variety of features over very wide angles in the whole scene (Figure 10.7) and the bees made the best fit with cues in their expected places at large angles to each other (Thorpe 1956; Collett et al. 2002). In a similar way, blind people identify a place by sound, smell and touch all around. Despite the great number of publications with 'pattern perception' in the title, bees detect coincidences of cues, not patterns.

Coincidences in neuron responses and learning

The explanations of visual recognition offered here have been in terms of the coincidences between the cues and expected positions of landmark labels. This explanation has a long history. Sherrington (1906) called it 'integration'. Hebb (1949) wrote an influential book with the idea that the coincidences of inputs, including those from reward channels, would strengthen synaptic contacts on a key neuron and trigger the growth of new synapses when learning occurred. Eccles (1957) described in detail the summation or inhibition of coincidences of the inputs at synapses as the key to understanding all nervous systems. Moreover, the immense, new topic of adaptive neural nets in artificial learning systems relies on the idea that the coincidences of different inputs allow the neural net to learn.

Whether or not there is a range of bee behaviour that makes use of something more thoughtful than the learning of rewarded coincidences, or the avoidance of punished ones, seems now to be a matter of opinion.

Cognitive visual behaviour in route finding and navigation

Much of this discussion depends on the education and life experiences of the contestants. An education in the Napoleonic system of Continental Europe, or as an ethologist, will lean you towards accepting intuitive explanations of performance and reliance on definitions of terms that are usually simply taken from cognitive psychology. In contrast, English empiricists or American comparative physiologists will lean towards mechanistic analysis.

The least justified, most dogmatic or fundamentalist opinion that I can find comes from Professor Randy Gallistel, of Rutgers University, who does not work primarily on bees but has just written a book on cognition. Gallistel would say that the word 'cognitive' implies computation, so if the bee computes, it has cognition—simple as that! The bee does path integration and optic flow summation, therefore it computes, therefore it has cognition. The problem with this is easily discerned when I point out that a bindweed stem describes an excellent spiral as it winds up a stick and my slide rule computes without needing a battery, therefore they have cognition. Even my watch computes.

A more reasonable view comes from Adrian Dyer, who in fact works on bees (but allows the bees to land on the targets and thereby blocks his own progress). Dyer would say that bees will reveal cognition if they can use memory to solve a novel or abstract task, and he can point to several published accounts of performance. For example, bees that have learned mazes are faster at solving an unfamiliar maze. Trained bees accept some unfamiliar targets in place of the learned one. Bees familiar with a foraging ground adjust their foraging method according to the place (and time of day). Moreover, there is some evidence that bees count. Again, the problem is that these accounts are of the *performance*, and the bees have not been tested thoroughly to see what they have really learned. The idea of cognition was an intuitive inference that was not deduced from experimental results, but was a word taken from the cognitive sciences, put into the title of the paper and then claimed to be a causative agent.

Even worse, we might already have a mechanistic explanation available for these performances. For example, Hertz showed that bees discriminated between targets on the basis of more or less modulation (Figure 1.4), so four objects would generate more modulation than three of the same objects, enabling the bees to pass the test of counting. Perhaps 'cognition' equates to that which we do not yet understand.

It is clear from this example that empirical experimental data about parameters or landmark labels can replace cognition as a causative agent. It is not so clear that cognition can ever be demonstrated as a causative necessity, because there might always be an undiscovered mechanistic explanation.

Coming closer to home, Srinivasan would grant 'cognition' to any animal that can do something, such as an ability to categorise, navigate complex mazes or other tasks that might require thought in a human. Srinivasan would say, 'If it looks like cognition, sounds like cognition, acts like cognition, then it is cognition.' There are more problems here than the requirement to test unsupported inferences of cognition. For example, robots perform tasks more difficult than bees and there are distinguished psychologists who would allow cognition for robots but not for bees. Performance that looks like cognition is a feature of computer programs that look ahead and predict moves in chess games and also of systems with feedback loops that counteract unexpected forces and stabilise our posture. The performance is just the beginning. We look for the mechanism, not for a word that tells us that the bee does something interesting.

Finally, another hardworking experimentalist, Randolf Menzel, has been involved many times in the discussion of cognition in bees, mainly over the question of whether bees remember maps of their territory. Menzel has again shown that bees remember the territory that they have explored, either voluntarily or in search of the randomised position of a food source. He would probably allow the term 'cognitive' for novel behaviour that emerges from a combination of memories that creates expectations of outcomes and (contra Dyer and Srinivasan) he would exclude those examples where behaviour is directly controlled by the sequence of stimuli.

From quite a different standpoint, Tye (1997) reviews the most remarkable bee performances and argues that the bees are not aware of what they are doing, and are therefore not cognitive, and also that the localisation of a light by a subject with blind-sight is a response to a stimulus and not cognition.

Need I say that the analysis of the mind of the bee cannot be based on performance alone. Before a book on 'What do bees think?' can be written, there must be some experimental analysis of several kinds, followed by detailed tests and validation, otherwise 'cognition' is just a word in an arbitrary definition.

After all this, what does the bee see?

Of course, we can never know what bees really sense when they see. In human terms, they see nothing. To the experimentalist, the expression 'What do bees see?' is a query about what stimulation they detect, not about the sensations of the bee when the visual system is in action. They detect cues and direction of movement in each local region of the eye, but these stimuli are mixed in the optic lobes with other modalities from other parts of the animal. Their appreciation of their surroundings must be like that of a blind man who uses all available inputs to control his movements.

We can *guess* what bees really detect. For example, some disturbed bees chased me away from their hive, so I am not going back there—they might SEE ME. Alternatively, we can *propose* that the disturbed bees detect and follow any large moving object, even against a textured background. Then we can devise experiments to *test* this proposal. We might *conclude* that the bees follow the largest moving object through a forest of trees and bushes when there is an odour trail generated by spilt honey, a bear or bee pheromone. For every question, we follow the steps: guess, proposal, tests, conclusion, belief and unwarranted extrapolation, then rejection. Given a sensitive imagination, assiduous observation, efficient experimentation and much thought, we slowly analyse the behaviour that the bee presents. *This* is the way that small science advances.

To make an analysis at all, we depend on the repertoire of the animal. If the bee does not respond to the training or the tests, we can go no further. The bee might have detected the stimulus but was not aroused by it. For this reason, there might be a lot that we will never suspect. This is not proof that the bee will never be fully understood. If, however, the bees respond to one group of patterns but not to a related group of patterns that differs in a defined way from the first group, we are on the way to discovering a cue.

It is easy to show that bees detect edges and areas separately, that shapes are not reassembled in their memory and that orientation is cancelled by edges at angles to each other. With that visual mechanism, we can only suggest that bee vision is similar to detecting the separate tasty molecules in coffee or hearing sounds from an orchestra.

Detection and perception

The human visual system has several kinds of lapses from conscious vision that could help us imagine the vision of the bee. One of these is the ability to be aware of our surroundings although not particularly conscious of them.

Subliminal perception is the ability to take in brief or weak signals that are not consciously detected at the time. In humans, they can be recorded by brain imaging or correlated with electrical potentials, so there is no doubt of their existence, even if nothing is reported. One example is subconscious priming, when a word is flashed so briefly on a screen that it is not seen but can still be correctly reported. Other examples are masked perception, inattention blindness and diverted attention, all of which block conscious vision but the stimulus can be correctly reported later. That is all that is required by a bee that remembers a route and a place, but is not interested in pattern perception. Classically, subliminal perception was regarded as an automatic process that was independent of consciousness, and perhaps that is the way we might think about bee vision.

In humans, some brain lesions (not retinal lesions) cause a situation called 'blind-sight', in which the subject has no conscious vision in a part or whole of an eye, but is able to report correctly a strong stimulus such as a colour, a black spot or a large familiar object and its position. Perhaps it means something to suggest that bee vision is all blind-sight and therefore not cognitive by Pye's definition!

Retrospect

The idea, which persisted for 100 years, that pattern perception was based on the reassembly of a central image laid out in the brain served the bee badly. The inferences of cognitive analysis of visual images by bees were compatible with the original data and in line with the general theories of the time, but some results were explained by merely describing the performance in different words and then guessing higher processes, so causing years of confusion. As the experimental testing of trained bees progressed, the local cues offered a low-level mechanistic explanation but the detection of the spatial layout of the pattern was not ruled out because the patterns were huge and overlapped several local eye regions. Contemporary publications, moreover, added new conclusions without cancelling the old ideas of cognition.

We can now infer that the training procedure limited the bee's vision. For the task in hand only, they learned to ignore all except a few cues in a local area of the eye. They generalised because they recognised the few cues they had learned and no unexpected cues were detected. Further discoveries, however, are never ruled out. After all, humans have a sensory processing system that depends entirely on peripheral arrays of simple feature detectors and a distributed administration.

Endnotes

1. To understand the depth of the divide between intuition and empirical methods, or between ethology and the mechanistic analysis of behaviour, this chapter should be read in conjunction with Chapter 2.

AFTERTHOUGHTS

Let us start at the small end, with the nerve cells. Electrophysiology is an attractive technique for mechanistic analysis. When successful, the data flow from the electrodes and amplifiers. To explain even a little of the behaviour, however, electrophysiology requires a thorough knowledge of the neuron anatomy. One can go only so far with these methods and then find that the different kinds of data do not easily fit together. Neurons are relatively simple, but their combinations are devilishly difficult to unravel. The bee has small neurons, as yet obscure neuron anatomy in detail and behaviour with too many interacting causative agents for total analysis.

Next come the bees. Bees are unique for the study of olfactory and visual behaviour because they readily learn to use these stimuli to come for a reward. Analysis of the behaviour alone leads to a map of the formal interactions that explain this particular behaviour of bees. This map can look like the gross structure of the nervous system. This map of interactions then guides the anatomist and electrophysiologist towards the identification of the real map of signal transmission, possibly by using large tropical bees.

Next come the investigators. They need many skills, especially the experience to spot error. Students need to learn fast and thoroughly, but select what they want for themselves; read all the literature and take notes and notice gaps and anomalies that they would like to investigate. They need to think straight and to separate the reported experimental data, which are usually boring but correct, from the fanciful thoughts, discussions, postulates and theories that relate to the data, but might be misleading. Most science in this world is armchair or media science, but real discovery depends on the nuts and bolts, the nitty-gritty, the nose to the grindstone, with an observation every day, on the job.

It is a great shame that the most reliable accounts—the critical scientific reviews by experts—are scattered in expensive subscription journals. They should be available freely because the curriculum lags behind the frontier of knowledge, the textbooks are always out of date and perhaps the teachers are as well.

Turning now to the world of thought, philosophy and theory, it is clear from the preceding chapters that advance is premeditated but particular discoveries usually emerge from a series of experiments in which many interesting facts turn up. Then thought is applied, sometimes for a long time, even years, before there is any conclusion or an idea for a new experiment. At this stage, it is obligatory to think about the problem all the time. Advances originate mostly from projects that have lasted many years with secure funding. After the real work is done, the so-called philosophy of science appears as armchair or media activity. The actual experiments are soon forgotten among the talk and print about the implications and extrapolations by the chattering classes.

You might have noticed that my idea of analysis of behaviour is to aim to find a model or description of the mechanism in terms of the components that are likely to be involved, not in terms of concepts that are invented by intuition for this or similar occasions. At every step, testing the trained bees to discover what they really detected with their eyes has validated the postulated mechanism. So, I did not analyse 'innate' responses in terms of 'drives', but in terms of coincidences of responses of feature detectors, cues and labels that were separated, identified and characterised. Bee vision is unique and borrowed terminology is misleading.

Those who observe and describe the performance then dream up a theory or a mechanism that might attract media excitement, but usually they drive the topic into a morass of untested assumptions. Those who study mammals or birds do not have the luxury of hoping for a mechanistic explanation, so they use the terminology and intuitive concepts of cognitive science, mathematical models or the nerve nets of the connectionists, ignoring the possibility that these are entirely fictional and misleading. That effort leads to much discussion and wastage of time.

This divide between intuitive ethology and mechanistic comparative physiology can be traced back to the conflict between Kantian principles, as taught in countries that were conquered by Napoleon (roughly), and empirical principles, as taught in the English-speaking world (roughly). For obvious reasons, research on bee vision has suffered disproportionately on account of this legacy.

What did the analysis of bee vision teach us, besides the basic principles of a visual system of medium complexity with a small brain? We discovered the kind of system. The memory mechanism is not like wax that takes up any shape, but is a set of independent preformed boxes that can be ticked, and when the same combination recurs, the place is automatically recognised. Bees have no general concepts of texture, shape or topology but only a memory of cues in immediate past experience. Bees remember cues and places, not patterns or shapes.

The historical approach warned us of errors of thought, dangers of language and failures of successive paradigms. Bee vision, however, illustrates how the information in a picture or panorama can be greatly reduced and yet be recognised by a simple mechanism. It shows also that the evolution of bee vision has reached a level where no more processing can be done without the reassembly of the pattern, which requires vast extra processing power. Also, the bee has a seeing system, for large or small visual fields, that has been evolved for the recognition of places on a large or small scale, with minimum components and weight, which might be useful for machines with computer vision.

Finally, what of science as a grand and infallible discipline that is essential for saving the future of humankind? The tracks of those involved in the minute topic of bee vision do not appear to lead in that direction. Their blind alleys explored, adherence to ineffective methods, failure to consider and repeat the works of others, ignorance of the literature in languages other than their own and unwillingness to change their opinions all look like typical conservative acceptance of the status quo and preference for a discussion instead of another experiment. Others persist, however, and yet others become committed. They read more, look again at the natural world and think more deeply. Science follows a circuitous path, highlighted here and there by a flash of brilliance, forever attracting enthusiastic new students and charting an unexpected path, forever inadequate to comprehend the whole because there is so much more to be discovered.

SUMMARY OF THE MODEL OF BEES' VISUAL PROCESSING

For a bee, the **parameters** in the external panorama display two types of features—**areas** and **edges**—which are processed separately and not reassembled. Apart from motion detectors, the peripheral units of vision are feature detectors of two types: **intensity detectors** that respond to areas and **modulation** and **orientation detectors** that respond to passing edges. The feature detector responses are summed by type and position in each local region of the eye to form several types of **cues**, each with the position of its average. This summation destroys the local pattern. The cues are within the bee and must therefore be characterised by testing trained bees. Cues can be remembered. The coincidence of different cues in a local region is the **label on a landmark** in that retinotopic direction. Bees learn landmark labels **to identify a place** and find the reward. To a bee, a pattern is just another landmark.

The receptors and feature detectors

In each ommatidium of the compound eye, bees have three colour types of ordinary photoreceptors, with their spectral sensitivity peaking in the ultraviolet, blue and green. A change in the intensity of light in the receptors, such as that caused by a passing edge, causes a **modulation** of the electrical response propagated to the arrays of neurons below (Figure 1). The receptors feed into lamina neurons that amplify the modulation and cut out the persistent signal from constant illumination. These in turn feed into **feature detectors** of four kinds.

The feature detectors for modulation have balanced excitatory and inhibitory inputs that are arranged so that they detect contrast at edges but are insensitive to changes in brightness (Figure 2b). They are of two kinds: the pure modulation detectors that signal heterochromatic modulation but not the direction of edges, and the detectors of edge orientation that are green sensitive and colourblind. Both kinds detect simultaneous modulation of a small group of seven receptors (Figure 2). The sizes of these feature detectors have been measured as 3 ommatidia wide. There are also green-sensitive colourblind detectors of sequential modulation in adjacent receptors, which detect the direction of motion of a contrast across the eye.

Figure 1 Overview of the stages of visual processing of edges (on the left) and areas (on the right), from receptors to recognition of place. a) Receptor array. b) Lamina cell array. c) Formation of cues by summation. d) Formation of landmark labels by coincidences. e) Recognition of place.

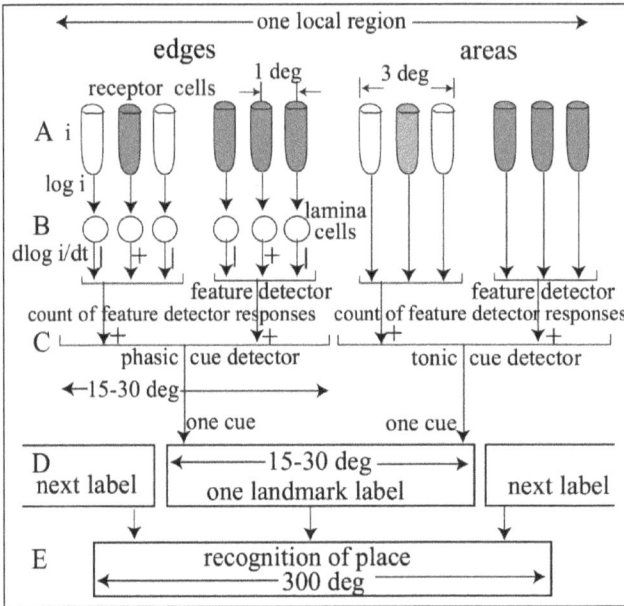

The feature detectors for edge orientation are symmetrical about their axis of orientation (Figures 2c–e), as shown by the inability of the bee to distinguish which side of an edge is dark and which is light. To be able to separate all edge orientations, there must be at least three kinds with 120° between the orientations of their axes, probably in line with the coordinates of the retina. They are green sensitive and colourblind.

The feature detectors for size, colour and brightness of areas of the image are probably the receptors themselves, because these features require a steady signal, not a modulated one (on the right in Figure 1). Between them, the feature detectors together respond to the parameters that the bees detect in the external panorama. The parameters are outside the bee; the responses of the feature detectors are inside.

Figure 2 The convergence of receptors on the four types of feature detectors for edges, all of which are insensitive to intensity changes. a) The receptors. b) The radially symmetrical modulation detector; this detects edges, not just small spots. c–e) The detectors of edge orientation with bilateral symmetry are green sensitive and colourblind; they cannot distinguish between the two sides of an edge. The numbers show the relative excitation and inhibition by light.

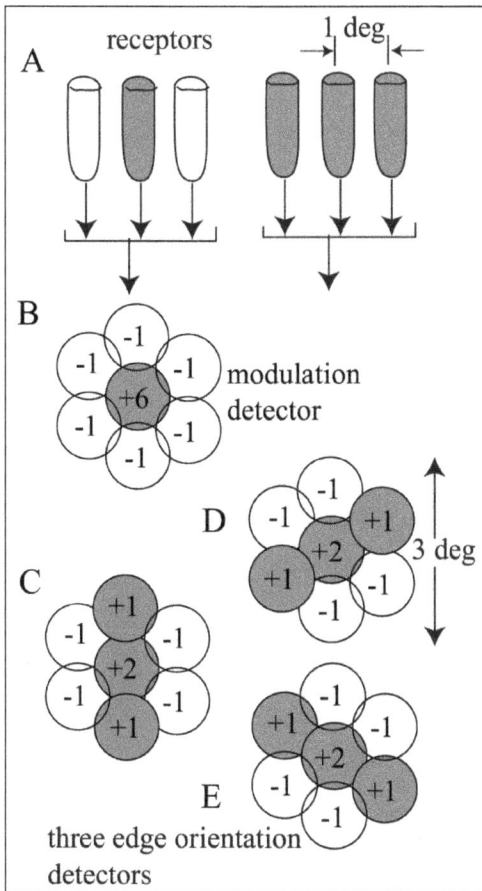

Source: After Horridge (2005a).

The cues relating to edges

The cues are formed by the summation of responses of each kind of feature detector within the local region on each side of the target or pattern, as though the bees look towards the centre and divide the view between their eyes. Just as the receptors count photons, each cue detector totals the coincident responses of its own array of feature detectors in a local region of the eye (Figure 3). There can be several different cues in a local region, but because they are totals, there is only one of each kind. Although simple and sparse, the cues in a local region are usually sufficient to identify a small pattern (<40°) or a landmark. The absence of a cue is itself a cue. If they are rewarded, the cues are learned and remembered in the range of positions where they were displayed during the training.

Figure 3 At least three different orientations are separated by the edge orientation detectors (Figures 2c–e). The responses of these are summed in such a way that edges at right angles cancel, so the pattern is lost but the predominant orientation and its position remain.

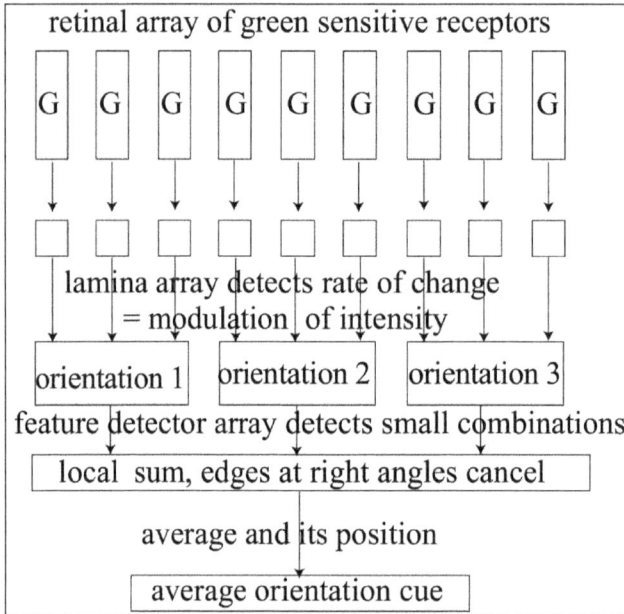

The summation of edge orientation detectors to form cues smoothes out the detail in the local region (Figure 3). The bees detect and learn the cues but they have no information about the distribution of the feature detectors that were summed. Consequently, there are many pairs of different patterns that the bees cannot distinguish. In tests, the trained bees detect familiar cues in unfamiliar patterns but the real patterns are of no interest.

To a bee, the orientation cue with its position is a kind of average orientation of edges in a local region. The responses of the edge orientation detectors are summed in such a way that edges at right angles cancel, so the pattern is lost but the predominant orientation and its position remain (Figure 4d). This is the first counter-intuitive property of bee vision. For example, in a square or a square cross the orientation cue is cancelled by the edges at right angles to each other that are resolved by the feature detectors for edge orientation. The greatest gap that can be spanned in a row of small squares is 3°. This is a measure of the maximum size of the edge orientation detectors. Similarly, orientation is destroyed when a bar is broken up into squares or cut into square steps.

Figure 4 Examples to illustrate the summation of feature detectors for edge orientations in various ways to form cues; pattern is lost but cues emerge.
a) A line of detectors with oblique orientation. b) Detectors with vertical axes. c) Mixed orientations cancel. d) The orientation cue is cancelled in the edges of a square but weak hubs are detected at the corners. e) and f) A tangential and a radial cue and their hubs. The orientation is cancelled but the modulation and position of the hub remain.

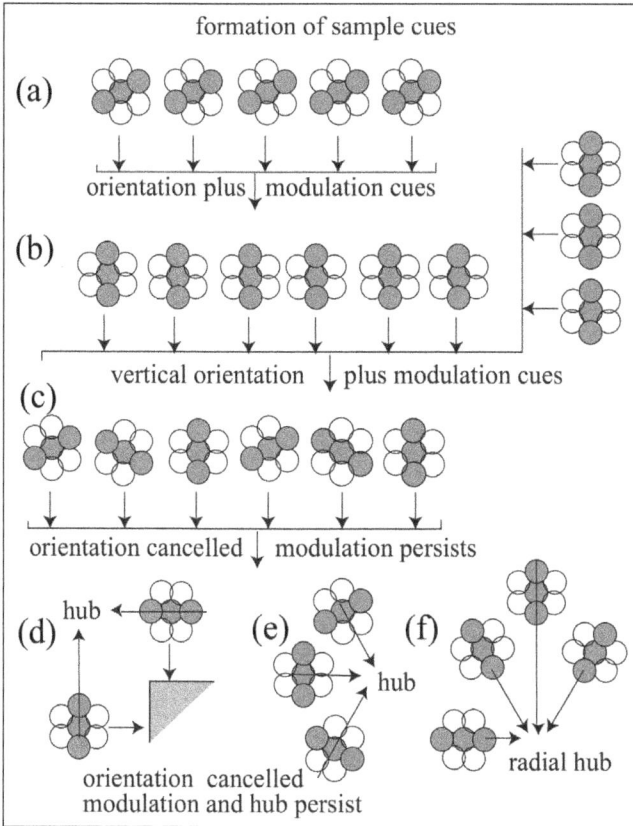

The edge detectors also collaborate to detect the hubs of radial or circular patterns (Figures 4e and 4f). The type of pattern—radial or tangential—and the position of the hub can be learned, but again, the real layout of the pattern is lost in the summation.

There are surprisingly few types of cues. There is an order of preference for learning the cues in the training situation, with total modulation in a local region the most preferred, then area, position of centre, a black spot, colour, radial edges, bilateral symmetry, average orientation and finally tangential or circular edges, which are avoided. Heterochromatic modulation (Figure 2b) and green-sensitive edge modulation (Figures 2c–e) are separate cues. Despite searches, no more cues have been found. This is a small but obviously adequate collection of cues for the varied life of a bee (Figure 5).

315

Figure 5 Representations of the cues in human terms, in order of preference. In the outside world, the parameters look like this to us. The bee detects only excitation that is processed as it passes from neuron to neuron. Each of these cues is represented in the bee as a quality, a quantity and a position on the eye. A neuron's activity can be defined in the same way.

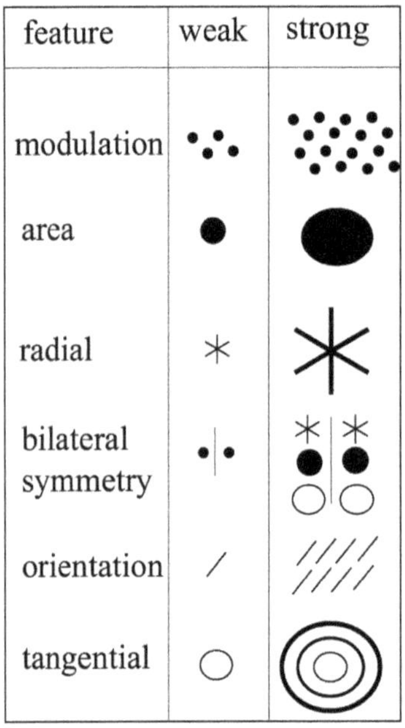

feature	weak	strong
modulation		
area		
radial		
bilateral symmetry		
orientation		
tangential		

Cues related to areas

The responses of blue, green and presumably the unstudied UV receptors are totalled separately in the local areas on each side of the target. They are totalled as the number of excited receptors multiplied by the brightness. The position of the centre is a cue, but areas are not related to edges, so there is nothing about shape. This is the second counter-intuitive property of bee vision.

The positions of blue, green and yellow areas are discriminated separately, but not all the areas are learned separately, blue being the preferred and sometimes the only position learned, even when on the unrewarded target. The positions of the centres of two areas of black or colour can be remembered as cues, but where they are close together, the bees detect their common centre. This merging of the two areas diminishes as the spots move apart, from an angle subtending 5°, until at 15° they are quite separate.

Much of the natural panorama displays a variety of colours and orientations of edges with a strong modulation cue for bees, but within each local region the orientation can be cancelled out and nearby areas summed together, so only the modulation of green and blue receptors and the average colour of the local areas remain.

Landmark labels; place recognition

The group of cues that is detected at the same time in a local region of the eye forms the label of a landmark, irrespective of whether there is a single or several real landmarks in that part of the panorama. The label can be learned. All that matters is that the bees remember the coincidence of responses of cues in that local region of the eye. Landmark labels are therefore retinotopic—that is, at a place on the eye. The group of landmark labels at wide angles to each other that is detected at the same time by the whole eye makes the key to the recognition of a place (Figure 6).

Figure 6 The display in the panorama that is detected by the bee. Each oval subtending about 30° represents a local region in the bee eye. Within each region no more than one cue of each kind is detected. The combination of landmark labels, with their directions, enables the bee to recognise a place.

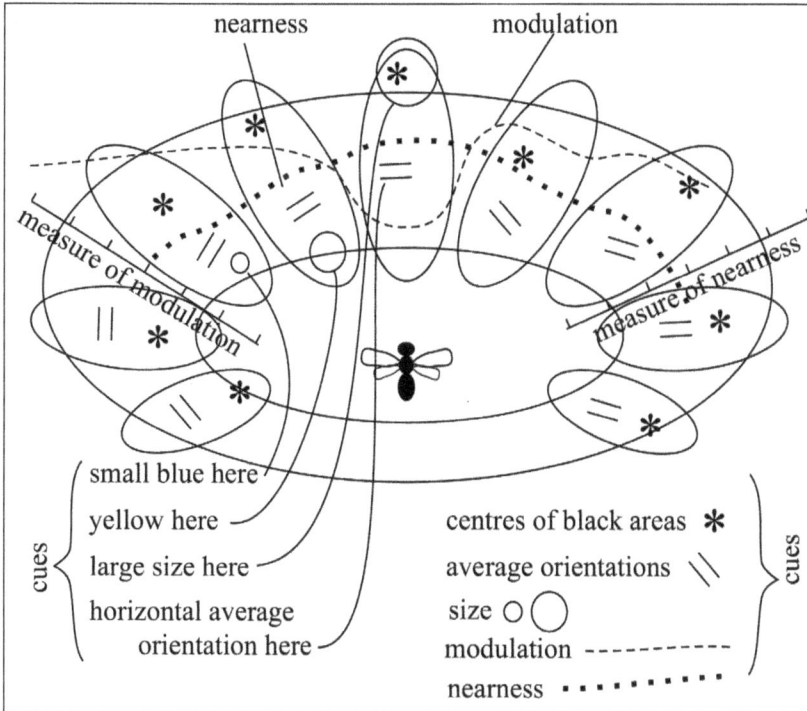

The resolution of orientation of an edge on a vertical surface is poor because the feature detectors are independent and so short. A difference of 45° is the limit for a single bar, 30° for a parallel grating. At each stage in processing, there is a compromise between the resolution, which is better in small summation fields, and the sensitivity or the ability to find the target, which is better in large fields.

317

The mechanism is designed for a very large visual field for the whole eye, and it works for a compound eye of any size. At each level in processing, the coincidence of inputs is the signal to pass the response to the next level. The whole process from receptors through to feature detectors and then to cues and landmark labels (Figure 1) is done region by region on the eye, and therefore in coordinates related to the position of the head and body axis (Figure 6). For this reason, bees scan the scene in the horizontal direction as they fly and orient their head and body to detect landmark labels and identify the place of the reward. In Skinner's terminology, learning the labels to recognise a place must be done by 'operant' conditioning, which is now part of 'active vision'. The control of the bee's active vision is largely unstudied.

Resolution in the processing hierarchy

Parts of the above model were inferred from measurements of resolution of features, cues and landmarks (Figure 1). Resolution depends on the angular sub-tense and shape of the field of the detector and on the separation between detectors, and is not the same for each cue. At the level of receptor responses, electrical recording gives 2.5°. For the feature detectors, we have for modulation a resolution of 2°, which is better than for a single receptor on account of the lateral inhibition (Figure 2b). For directional edge detection, bees have 3° (Figures 2c–e), and for detection of a small black spot, 2–3°. At the level of coincidences of feature detector responses to form cues, we have: modulation in regions of 20° across; orientation in regions of 15–20° across; position of areas of black or colour, 12–16°; for the position of the centre, 5°. At the level of coincidences of cues to form a landmark label, we have areas up to 45° across for the summation and a resolution of 15–20° for the separation between neighbouring landmarks. The three stages of processing have resolutions of approximately 2–3°, 5–20° and >20°. The fields of the cues are two to 10 times the diameter of the fields of the feature detector.

BIBLIOGRAPHY

Most reference lists in published papers aim to cover the most recent literature. This list also aims to bring out the origins of our knowledge in the older literature. To make the best advantage of this list, use the 'Find' capability on your computer to search for authors whose name is not first in the reference and to search for topics indicated by the titles. This list is a beginning only. It is advisable to search also on the Internet and on the search engines of the most appropriate journals, using the topic, name and initials to find the recent bibliography of these authors.

Adelson, E. and Bergen, J. 1985, 'Spatiotemporal energy models for the perception of motion', *Journal of the Optical Society of America*, vol. 2, pp. 284–99.

Ali, M. A. (ed.) 1984, 'Photoreception and vision in invertebrates', Plenum, London and New York.

Aloimonos, Y. (ed.) 1993, *Active Perception*, Erlbaum, Hillsdale, NJ.

Anderson, A. 1972, 'The ability of honeybees to generalize visual stimuli', in R. Wehner (ed.), *Information Processing in the Visual Systems of Arthropods*, Springer, Berlin, pp. 207–12.

Anderson, A. M. 1977a, 'Shape perception in the honeybee', *Animal Behaviour*, vol. 25, pp. 67–79.

Anderson, A. M. 1977b, 'A model for landmark learning in the honey bee', *Journal of Comparative Physiology*, vol. 114, pp. 335–55.

Arnett, D. W. 1972, 'Spatial and temporal integration properties of units in the first optic ganglion of dipterans', *Journal of Neurophysiology*, vol. 35, pp. 429–44.

Arvanitaki, A. 1937–61 [see list of recordings from molluscs in Bullock and Horridge (1965:1372)].

Aung, S., Srinivasan, M. V. and Zhang, S. W. 2003, 'Honeybee navigation, properties of the visually driven "odometer"', *Journal of Experimental Biology*, vol. 206, pp. 1265–73.

Autrum, H. (ed.) 1979–81, *Vision in Invertebrates. Volume VII. Parts 6A, 6B and 6C*, Springer, Berlin.

Autrum, H. and von Zwehl, V. 1962, 'Die spektrale Empfindlichkeit einzelner Sehzellen des Bienenauges', *Zeitschrift für vergleichende Physiologie*, vol. 48, pp. 357–84.

Baader, A., Schäfer, M. and Rowell, C. H. F. 1992, 'The perception of the visual flowfield by flying locusts. A behavioural and neuronal analysis', *Journal of Experimental Biology*, vol. 165, pp. 137–60.

Backhaus, W. 1991, 'Color opponent coding in the visual system of the honeybee', *Vision Research*, vol. 31, pp. 1381–97.

Baerends, G. P. 1941, 'Fortpflanzungsverhalten und Orientierung der Grabwespe *Ammophila campestris*', *Tijdschrift Entomologie*, vol. 84, pp. 68–275.

Baerends, G. P. 1959, 'Ethological studies of insect behaviour', *Annual Review of Entomology*, pp. 207–34.

Barlow, H. B. 1952, 'The size of ommatidia in apposition eyes', *Journal of Experimental Biology*, vol. 29, pp. 675–84.

Barlow, H. B. 1961, 'Possible principles underlying the transformations of sensory messages', in W. A. Rosenblith (ed.), *Sensory Communication*, MIT Press, Cambridge, Mass., pp. 217–34.

Barlow, H. B. 1965, 'Visual resolution and the diffraction limit', *Science*, vol. 149, pp. 553–5.

Barlow, H. B. and Levick, W. R. 1965, 'The mechanism of directionally selective units in rabbit's retina', *Journal of Physiology*, vol. 178, pp. 477–504.

Barlow, H. B., Frisby, J. P., Horridge, A. and Jeeves, M. A. 1993, *Natural and Artificial Low-Level Seeing Systems*, Oxford University Press, Oxford.

Barnett, P. D., Nordström, K. and O'Carroll, D. C. 2007, 'Retinotopic organization of small-field-target-detecting neurons in the insect visual system', *Current Biology*, vol. 17, pp. 1–10.

Baumann, F. 1975, 'Electrophysiological properties of the honey bee retina', in G. A. Horridge (ed.), *The Compound Eye and Vision of Insects*, Oxford University Press, Oxford, pp. 53–74.

Baumgärtner, H. 1928, 'Der Formensinn und der Sehschärfe der Bienen', *Zeitschrift für vergleichende Physiologie*, vol. 7, pp. 56–143.

Benard, J., Stach, S. and Giurfa, M. 2006, 'Categorization of visual stimuli in the honeybee *Apis mellifera*', *Animal Cognition*, vol. 9, pp. 257–70.

Bernard, C. G. (ed.) 1966, *The Functional Organization of the Compound Eye*, Pergamon, Oxford.

Berry, R., Stange, G., Olberg, R. and van Kleef, J. 2006, 'The mapping of visual space by identified large second-order neurons in the dragonfly median ocellus', *Journal of Comparative Physiology*, vol. 192, pp. 1105–23.

Bethe, A. 1898, 'Dürfen wir den Ameisen und Bienen psychische Quälitaten zuschreiben?', *Archiv für gesampte Physiologie*, vol. 70, pp. 15–100.

Beusekom, G. van, 1948, 'Some experiments on the optical orientation in *Philanthus triangulum*', *Behaviour*, vol. 1, pp. 195–226.

Bidwell, N. J. and Goodman, L. J. 1993, 'Possible functions of a population of descending neurons in the honeybee's visuo-motor pathway', *Apidologie*, vol. 24, pp. 333–54.

Bishop, L. G. and Keehn, D. G. 1967, 'Neural correlates of the optomotor response in the fly', *Kybernetic*, vol. 3, pp. 288–95.

Borst, A. 1991, 'Fly visual interneurons responsive to image expansion', *Zoologische Jahrbücher, Physiologie*, vol. 95, pp. 305–13.

Borst, A. and Bahde, S. 1988, 'Visual information processing in the fly's landing system', *Journal of Comparative Physiology*, vol. 163, pp. 167–73.

Boycott, B. B. 1961, 'The functional organization of the brain of the cuttlefish *Sepia officinalis*', *Proceedings of the Royal Society*, vol. 153, pp. 503–34.

Braddick, O. J. and Sleigh, A. C. (eds) 1982, *The Physical and Biological Processing of Images*, Springer, Berlin.

Braitenberg, V. 1967, 'Patterns of projection in the visual system of the fly. I. Retina-lamina projections', *Experimental Brain Research*, vol. 3, pp. 271–98.

Brünnert, U., Kelber, A. and Zeil, J. 1994, 'Ground nesting bees determine the location of their nests relative to a landmark by other than angular size cues', *Journal of Comparative Physiology*, vol. 175, pp. 363–9.

Buchner, E. 1976, 'Elementary movement detectors in an insect visual system', *Biological Cybernetics*, vol. 24, pp. 85–101.

Buchner, E. 1984, 'Behavioural analysis of spatial vision in insects', in M. A. Ali (ed.), *Photoreception and Vision in Invertebrates*, Plenum Press, New York, pp. 561–622.

Buchner, E., Götz, K. G. and Straub, C. 1978, 'Elementary detectors for vertical movement in the visual system of *Drosophila*', *Biological Cybernetics*, vol. 31, pp. 235–42.

Buddenbroch, W. von, 1937, *Grundriss des vergleichende Physiologie*, Borntraeger, Berlin.

Buddenbroch, W. von, 1952, *Vergleichende Physiologie. Volume 1. Sinnesphysiologie*, Birkhäuser, Basel.

Bullock, T. H. and Horridge, G. A. 1965, *Structure and Function in the Nervous Systems of Invertebrates*, Freeman, San Francisco and London.

Burkhardt, D. 1962, 'Spectral sensitivity and other response characteristics of single visual cells in the arthropod eye', *Symposium of the Society of Experimental Biology*, vol. 16, pp. 86–109.

Burkhardt, D. and Streck, P. 1965, 'Das Sehfeld einzelner Sehzellen—eine Richtigstellung', *Zeitschrift für vergleichende Physiologie*, vol. 51, pp. 151–2.

Burtt, E. T. and Catton, W. T. 1962, 'A diffraction theory of insect vision. Part I. An experimental study of visual acuity in certain insects', *Proceedings of the Royal Society of London*, vol. 157, pp. 53–82.

Burtt, E. T. and Catton, W. T. 1969, 'Resolution of the locust eye measured by rotation of radial striped patterns', *Proceedings of the Royal Society of London*, vol. 173, pp. 513–29.

Butel-Repen, H. von, 1900, *Sind die Bienen Reflexmaschinen*, Verlag Arthur Giorgi, Leipsig.

Butler, R. 1971, 'The identification and mapping of spectral cell types in the retina of *Periplaneta americana*', *Zeitschrift für vergleichende Physiologie*, vol. 72, pp. 67–80.

Cajal, S. R. y, 1909, 'Nota sobre la estructura de la rétina de la mosca (*Mosca vomitoria*)', *Trabajos del Laboratorio de Investigaciones Biológicas del Universidad, Madrid*, vol. 16, pp. 109–39.

Cajal, S. R. y, and Sanchez, S. D. 1915, 'Contribución al conocimiento de los centros nerviosos de los insectos. Parte I. Retina y los centros opticos', *Trabajos del Laboratorio de Investigaciones Biológicas del Universidad, Madrid*, vol. 13, pp. 1–168.

Campan, R. and Lehrer, M. 2002, 'Discrimination of closed shapes by two species of bee, *Apis mellifera* and *Megachile rotundata*', *Journal of Experimental Biology*, vol. 205, pp. 559–72.

Campion, G. G. and Elliot Smith, G. 1934, *The Neural Basis of Thought*, Kegan Paul, London.

Carthy, J. D. 1958, *An Introduction to the Behaviour of Invertebrates*, Allen & Unwin, London.

Cartwright, B. A. and Collett, T. S. 1979, 'How honey-bees know their distance from a near-by visual landmark', *Journal of Experimental Biology*, vol. 82, pp. 367–72.

Cartwright, B. A. and Collett, T. S. 1983, 'Landmark learning in bees; experiments and models', *Journal of Comparative Physiology*, vol. 151, pp. 521–43.

Cartright, B. A. and Collett, T. S. 1987, 'Landmark maps for honeybees', *Biological Cybernetics*, vol. 57, pp. 85–93.

Catton, W. T. 1998, 'A test of the visual acuity of the locust eye', *Journal of Insect Physiology*, vol. 44, pp. 1145–8.

Catton, W. T. 1999, 'The effect of target orientation on the visual acuity and the spatial frequency response of the locust eye', *Journal of Insect Physiology*, vol. 45, pp. 191–200.

Chen, L., Zhang, S. W. and Srinivasan, M. 2003, 'Global perception in small brains: topological pattern recognition in honey bees', *Proceedings of the National Academy of Science of the USA*, vol. 100, pp. 6884–9.

Cheng, K., Collett, T. S. and Wehner, R. 1986, 'Honeybees learn the colours of landmarks', *Journal of Comparative Physiology*, vol. 159, pp. 69–73.

Cheng, K., Collett, T. S., Pickhard, A. and Wehner, R. 1987, 'The use of visual landmarks by honeybees; bees weight landmarks according to their distance from the goal', *Journal of Comparative Physiology*, vol. 161, pp. 469–75.

Chittka, L. and Menzel, R. 1992, 'The evolutionary adaptation of flower colours and the insect pollinators' colour vision', *Journal of Comparative Physiology*, vol. 171, pp. 171–81.

Chittka, L., Dyer, A. G., Bock, F. and Dornhaus, A. 2003, 'Bees trade off foraging speed for accuracy', *Nature*, vol. 424, pp. 388.

Collett, M., Harland, D. and Collett, T. S. 2002, 'The use of landmarks and panoramic context in the performance of local vectors by navigating bees', *Journal of Experimental Biology*, vol. 205, pp. 807–14.

Collett, T. S. 1971, 'Visual neurons for tracking moving targets', *Nature*, vol. 232, pp. 127–30.

Collett, T. S. 1992, 'Landmark learning and guidance in insects', *Philosophical Transactions of the Royal Society of London*, vol. 337, pp. 295–303.

Collett, T. S. 1993, 'Route following and the retrieval of memories in insects', *Comparative Physiology and Biochemistry*, vol. 104A, pp. 709–16.

Collett, T. S. and Baron, J. 1994, 'Biological compasses and the coordinate frame of landmark memories in honeybees', *Nature*, vol. 368, pp. 137–40.

Collett, T. S. and Kelber, A. 1988, 'The retrieval of visuospatial memories by honeybees', *Journal of Comparative Physiology*, vol. 163, pp. 145–50.

Collett, T. S. and King, A. J. 1975, 'Vision during flight', in G. A. Horridge (ed.), *The Compound Eye and Vision of Insects*, Oxford University Press, Oxford, pp. 437–66.

Collett, T. S. and Land, M. F. 1975a, 'Visual control of flight behaviour in the hoverfly *Syritta pipiens L.*', *Journal of Comparative Physiology*, vol. 99, pp. 1–66.

Collett, T. S and Land, M. F. 1975b, 'Visual spatial memory in a hoverfly', *Journal of Comparative Physiology*, vol. 99, pp. 59–84.

Collett, T. S. and Land, M. F. 1978, 'How hoverflies compute interception courses', *Journal of Comparative Physiology*, vol. 125, pp. 191–204.

Collett, T. S. and Lehrer, M. 1993, 'Looking and learning: a spatial pattern in the orientation flight of the wasp *Vespa vulgaris*', *Proceedings of the Royal Society of London*, vol. 252, pp. 129–34.

Collett, T. S. and Patterson, C. J. 1991, 'Relative motion parallax and target localization in the locust', *Journal of Comparative Physiology*, vol. 169, pp. 615–21.

Collett, T. S. and Rees, J. A. 1997, 'View-based navigation in Hymenoptera: multiple strategies of landmark guidance in the approach to a feeder', *Journal of Comparative Physiology*, vol. 181, pp. 47–58.

Collett, T. S. and Zeil, J. 1998, 'Places and landmarks: an arthropod perspective', in S. Healy (ed.), *Spatial Representation in Animals*, Clarendon Press, Oxford, pp. 18–53.

Collett, T. S., Dillmann, E., Giger, A. and Wehner, R. 1992, 'Visual landmarks and route following in desert ants', *Journal of Comparative Physiology*, vol. 170, pp. 435–42.

Collett, T. S., Fry, S. N. and Wehner, R. 1993, 'Sequence learning by honeybees', *Journal of Comparative Physiology*, vol. 172, pp. 693–706.

Crozier, W. J. 1928–37 [numerous papers with colleagues in the *Journal of General Physiology*].

Cruse, H. 1972, 'Versuch einer quantitativen Beschreibung des Formensehens der Honigbiene', *Kybernetik*, vol. 11, pp. 185–200.

Dafni, A., Lehrer, M. and Kevan, P. G. 1999, 'Spatial flower parameters and insect spatial vision', *Biological Reviews*, vol. 72, pp. 239–82.

Dahmen, H. 1991, 'Eye specialization in waterstriders: an adaptation to life in a flat world', *Journal of Comparative Physiology*, vol. 169, pp. 623–32.

David, C. T. 1979a, 'Height control by free-flying *Drosophila*', *Physiological Entomology*, vol. 4, pp. 209–16.

David, C. T. 1979b, 'Optomotor control of speed and height by free-flying *Drosophila*', *Journal of Experimental Biology*, vol. 82, pp. 389–92.

David, C. T. 1982, 'Compensation for height in the control of ground speed by *Drosophila* in a new "Barber's Pole" wind tunnel', *Journal of Comparative Physiology*, vol. 147, pp. 485–93.

David, C. T. 1986, 'Mechanisms of directional flight in wind', in T. Payne, M. Birch and J. S. Kennedy (eds), *Mechanisms in Insect Olfaction*, Oxford University Press, Oxford, pp. 53–69.

De Souza, J., Hertel, H., Ventura, D. F. and Menzel, R. 1992, 'Response properties of stained monopolar cells in the honeybee lamina', *Journal of Comparative Physiology*, vol. 170, pp. 267–74.

de Vries, H. 1956, 'Physical aspects of sense organs', *Progress in Biophysics*, vol. 6, pp. 208–64.

Dietrich, W. 1909, 'Die Facettenaugen der Dipteren', *Zeitschrift für Zoologie*, vol. 92, pp. 465–539.

Dill, M. and Heisenberg, M. 1995, 'Visual pattern memory without shape recognition', *Philosophical Transactions of the Royal Society of London*, vol. 349, pp. 143–52.

Dill, M., Wolf, R. and Heisenberg, M. 1993, 'Visual pattern recognition in *Drosophila* involves retinotopic matching', *Nature*, vol. 365, pp. 751–3.

Douglass, J. K. and Strausfeld, N. J. 1995–96, 'Visual motion detection circuits in flies', *Journal of Neuroscience*, vol. 15, pp. 5596–605; vol. 16, pp. 4551–62.

Douglass, J. K. and Strausfeld, N. J. 2005, 'Sign-conserving amacrine neurons in the fly's external plexiform layer', *Visual Neuroscience*, vol. 22, pp. 345–58.

Doujak, F. E. 1984, 'Electrophysiological measurement of photoreceptor membrane dichroism and polarization sensitivity in a Grapsid crab', *Journal of Comparative Physiology*, vol. 154, pp. 597–605.

Doujak, F. E. 1985, 'Can a shore crab see a star?', *Journal of Experimental Biology*, vol. 166, pp. 385–93.

Dubs, A., Laughlin, S. B. and Srinivasan, M. V. 1981, 'Single photon signals in fly photoreceptors and first-order interneurons at behavioural threshold', *Journal of Physiology*, vol. 317, pp. 317–34.

Dyer, A. G. and Gould, J. L. 1981a, 'Honey bee navigation', *American Scientist*, vol. 71, pp. 587–97.

Dyer, A. G. and Gould, J. L. 1981b, 'Honey bee orientation: a backup system for cloudy days', *Science*, vol. 214, pp. 1041–2.

Dyer, A. G., Neumeyer, C. and Chittka, L. 2005, 'Honeybee (*Apis mellifera*) vision can discriminate between and recognise images of human faces', *Journal of Experimental Biology*, vol. 208, pp. 4709–14.

Eccles, J. C. 1957, *The Physiology of Nerve Cells*, Johns Hopkins Press, Baltimore.

Efler, D. and Ronacher, B. 2000, 'Evidence against a retinotopic-template matching in honeybees' pattern recognition', *Vision Research*, vol. 40, pp. 3391–403.

Egelhaaf, M. 1985, 'On the neuronal basis of figure-ground discrimination by relative movement in the visual system of the fly. II. Figure-detection cells: a new class of visual interneurons', *Biological Cybernetics*, vol. 52, pp. 195–209.

Egelhaaf, M. 1987, 'Dynamic properties of two control systems underlying visually guided turning in house flies', *Journal of Comparative Physiology*, vol. 161, pp. 777–83.

Egelhaaf, M. and Borst, A. 1993, 'Movement detection in arthropods', in F. A. Miles and J. Wallman (eds), *Visual Motion and Its Role in the Stabilization of Gaze*, Elsevier, Amsterdam, pp. 53–77.

Erickson, R. P. 1982, 'The across-fiber pattern theory: an organizing principle for molar neural function', *Contributions in Sensory Physiology*, vol. 6, pp. 79–110.

Ernst, R. and Heisenberg, M. 1999, 'The memory template in *Drosophila* pattern vision at the flight stimulator', *Vision Research*, vol. 39, pp. 3920–33.

Esch, H. E. and Burns, J. E. 1995, 'Distance estimation by foraging honeybees', *Journal of Experimental Biology*, vol. 199, pp. 155–62.

Exner, S. 1875, 'Über das Sehen von Bewegungen und die Theorie des zusammengestzten Auges', *Sitzungsberichte Akademische Wissenschaft Wien, Abteilung III*, vol. 72, pp. 156–70.

Exner, S. 1891 [1988], *Die Physiologie der facettirten Augen von Krebsen und Insecten*, Franz Deuticke, Leipsig [Translated by R. C. Hardie as *The Physiology of the Compound Eyes of Insects and Crustaceans*, Springer-Verlag, Berlin].

Exner, S. 1894, *Entwurf zu einer physiologischen Erklärung der psychischen Erscheinungen*, Franz Deuticke, Leipsig.

Fabre, J. H. 1879 [the experiments on navigation are published in the volume of *Souvenirs Entomologiques* for 1879].

Field, D. J. 1987, 'Relations between the statistics of natural images and the response properties of cortical cells', *Journal of the Optical Society of America*, vol. 4, pp. 2379–94.

Fisher, R. A. 1935, *The Design of Experiments*, Oliver and Boyd, London.

Forel, A. 1908, *The Senses of Insects*, Translated by M. Yearsley, Methuen, London.

Fraenkel, G. S. and Gunn, D. O. 1940, *The Orientation of Animals*, Oxford University Press, Oxford.

Free, J. B. 1970, 'Effect of flower shapes and nectar guides on the behaviour of foraging honeybees', *Behaviour*, vol. 37, pp. 269–85.

Friedlaender, M. 1931, 'Zur Bedeutung des Fluglochs im optischen Feld der Biene bei senkrechter Dressuranordnung', *Zeitschrift für vergleichende Physiologie*, vol. 15, pp. 193–260.

Frisch, K. von, 1914, 'Der Farbensinn und Formensinn der Bienen', *Zoologisches Jahrbucher, Physiologie*, vol. 35, pp. 1–188.

Frisch, K. von, 1947, 'The dances of the honey bee', *Bulletin of Animal Behaviour*, vol. 5, pp. 1–32.

Frisch, K. von, 1957, *A Biologist Remembers*, Gombrich, Lisbeth.

Frisch, K. von, 1965 [1967], *Tanzsprache und Orientierung des Bienen*, Springer, Berlin [translated into English as *The Dance Language and Orientation of Bees*, Harvard University Press, Cambridge, Mass.].

Frisch, K. von, 1971, *Bees, Their Vision, Chemical Senses, and Language*, Cornell University Press, Ithaca, NY.

Frisch, K. von, Lindauer, M. and Daumer, K. 1960, 'Über die Wahrnehmung polarisierten Lichtes durch das Bienenauge', *Experientia*, vol. 16, pp. 289–301.

Fry, S. N. and Wehner, R. 2002, 'Honeybees store landmarks in an egocentric frame of reference', *Journal of Comparative Physiology*, vol. 187, pp. 1009–16.

Gallistel, C. R. and King, A. P. 2009, *Memory and the Computational Brain: Why cognitive science will transform neuroscience*, Wiley/Blackwell, New York.

Gavel, L. von, 1939, 'Die kritische Streifenbreite als Mass für die Sehschärfe bei *Drosophila melanogaster*', *Zeitschrift für vergleichende Physiologie*, vol. 27, pp. 80–135.

Geiger, K., Kratzsch, D. and Menzel, R. 1995, 'Target-directed orientation in displaced honeybees', *Ethology*, vol. 101, pp. 335–45.

Gibson, J. J. 1950, *The Perception of the Visual World*, Houghton Mifflin, Boston.

Gibson, J. J. 1979, *The Ecological Approach to Visual Perception*, Houghton Mifflin, Boston.

Giger, A. D. 1996, PhD thesis, The Australian National University, Canberra.

Giger, A. D. and Srinivasan, M. V. 1995, 'Pattern recognition in honeybees: eidetic imagery and orientation discrimination', *Journal of Comparative Physiology*, vol. 176, pp. 791–5.

Giger, A. D. and Srinivasan, M. V. 1996, 'Pattern recognition in honeybees: chromatic properties of orientation analysis', *Journal of Comparative Physiology*, vol. 178, pp. 763–9.

Giger, A. D. and Srinivasan, M. V. 1997, 'Honeybee vision: analysis of orientation and colour in the lateral, dorsal and ventral fields of view', *Journal of Experimental Biology*, vol. 200, pp. 1271–80.

Giulio, L. 1963, 'Elektroretinographische Beweisführung dichroitischer Eigenschaften des Komplexauges bei Zweiflüglern', *Journal of Comparative Physiology*, vol. 46, pp. 491–5.

Giurfa, M. 2003, 'Cognitive neuroethology: dissecting non-elemental learning in a honeybee brain', *Current Opinion in Neurobiology*, vol. 13, pp. 726–35.

Giurfa, M. 2007, 'Behavioral and neural analysis of associative learning in the honeybee: a taste from the magic well', *Journal of Comparative Physiology*, vol. 193, pp. 801–24.

Giurfa, M. and Lehrer, M. 2001, 'Honeybee vision and floral displays: from detection to close-up recognition', in L. Chittka and J. D. Thomson (eds), *Cognitive Ecology of Pollination*, Cambridge University Press, Cambridge, pp. 61–82.

Giurfa, M. and Menzel, R. 1997, 'Insect visual perception: complex abilities of simple nervous systems', *Current Opinion in Neurobiology*, vol. 7, pp. 505–13.

Giurfa, M. and Vorobyev, M. 1998, 'The angular range of achromatic target detection by honey bees', *Journal of Comparative Physiology*, vol. 183, pp. 101–10.

Giurfa, M., Hammer, M., Stach, S., Stollhoff, N., Müller-Deisig, N. and Mizyrycki, C. 1999, 'Pattern learning by honeybees, conditioning procedure and recognition strategy', *Animal Behaviour*, vol. 57, pp. 315–24.

Giurfa, M., Schubert, M., Reisenman, C., Gerber, B. and Lachnit, H. 2003, 'The effect of cumulative experience on the use of elemental and configural visual discrimination strategies in honeybees', *Behavior and Brain Research*, vol. 145, pp. 161–9.

Giurfa, M., Vorobyev, P., Brandt, R., Posner, B. and Menzel, R. 1997, 'Discrimination of coloured stimuli by honeybees, alternative use of achromatic and chromatic signals', *Journal of Comparative Physiology*, vol. 180, pp. 235–43.

Giurfa, M., Vorobyev, P., Kevan, P. and Menzel, R. 1996, 'Detection of coloured stimuli by honeybees: the role of chromatic and achromatic contrast', *Journal of Comparative Physiology*, vol. 178, pp. 699–709.

Giurfa, M., Zhang, S. W., Jenett, A., Menzel, R. and Srinivasan, M. V. 2001, 'The concepts of "sameness" and "difference" in an insect', *Nature*, vol. 410, pp. 930–3.

Goldsmith, T. H. and Bernard, G. D. 1974, 'The visual system of insects', in M. Rockstein (ed.), *The Physiology of Insects. Volume 2*, Academic Press, New York, pp. 165–272.

Goodman, L. J. 1960, 'The landing responses in insects. I. The landing response of the fly *Lucilia sericata* and other Calliphorinae', *Journal of Experimental Biology*, vol. 37, pp. 854–78.

Goodman, L. J. and Fischer, R. C. (eds) 1991, *The Behaviour and Psychology of Bees*, CAB International, Wallingford, Oxford.

Goodman, L. J., Ibbotson M. R. and Bidwell, N. J. 1991, 'Spatial, temporal and directional properties of motion-sensitive visual neurons in the honeybee', in L. J. Goodman and R. C. Fisher (eds), *The Behaviour and Physiology of Bees*, CAB International, Wallingford, Oxford, pp. 203–26.

Götz, K. G. 1965, 'Die optischen Übertragungseigenschaften der Komplexaugen von *Drosophila*', *Kybernetik*, vol. 2, pp. 215–21.

Götz, K. G. and Buchner E. 1978, 'Evidence for one-way movement detection in the visual system of *Drosophila*', *Biological Cybernetics*, vol. 31, pp. 243–8.

Götz, K. G., Hengstenberg, B. and Biesinger, R. 1979, 'Optomotor control of wing beat and body posture in *Drosophila*', *Biological Cybernetics*, vol. 35, pp. 101–12.

Gould, J. L. 1976, 'The honey bee dance–language controversy', *Quarterly Review of Biology*, vol. 51, pp. 211–44.

Gould, J. L. 1982, *Ethology: The mechanisms and evolution of behaviour*, Norton, New York.

Gould, J. L. 1984, 'Natural history of honey bee learning', in P. Marler and H. S. Terrace (eds), *The Biology of Learning*, Springer, Berlin, pp. 149–80.

Gould, J. L. 1985, 'How bees remember flower shapes', *Science*, vol. 227, pp. 1492–4.

Gould, J. L. 1986, 'Pattern learning by honeybees', *Animal Behaviour*, vol. 34, pp. 991–7.

Gould, J. L. 1987, 'Landmark learning by honey bees', *Animal Behaviour*, vol. 35, pp. 26–34.

Gould, J. L. and Gould, C. G. 1988, *The Honey Bee*, Scientific American Library, Freeman, New York.

Gray, J. and Lissmann, H. W. 1946, 'The coordination of limb movements in the amphibia', *Journal of Experimental Biology*, vol. 23, pp. 133–42.

Gregory, R. L. 1981, *Mind in Science*, Penguin Books, London.

Grenacher, H. 1879, *Untersuchungen über das Sehorgan der Arthropoden, insbesondere der Spinnen, Insecten und Crustaceen*, Vandenhoeck and Ruprecht, Göttingen.

Gross, H. J., Pahl, M., Si, A., Zhu, H., Tautz, J. et al. 2009, 'Number-based visual generalisation in the honeybee', *PLoS ONE*, vol. 4, no. 1.

Guerten, R. H., Nordström, K., Sprayberry, J. D. H., Bolzon, D. M. and O'Carroll, D. C. 2007, 'Neural mechanisms underlying target detection in a dragonfly centrifugal neuron', *Journal of Experimental Biology*, vol. 210, pp. 3277–84.

Hardie, R. C. 1985, 'Functional organisation of the fly retina', in D. Ottoson (ed.), *Progress in Sensory Physiology 5*, Springer, Berlin, pp. 1–79.

Hardie, R. C. 1986, 'The photoreceptor array of the dipteran retina', *Trends in Neurosciences*, vol. 9, pp. 419–23.

Hardie, R. C. 1987, 'Is histamine a neurotransmitter in insect photoreceptors?', *Journal of Comparative Physiology*, vol. 161, pp. 201–13.

Hardie, R. C. 1988a, 'The use of local ionophoresis to identify neurotransmitter candidates in the housefly *Musca domestica*', *Journal of Physiology*, vol. 396, p. 7.

Hardie, R. C. 1988b, 'Neurotransmitters in compound eyes', in D. G. Stavenga and R. C. Hardie (eds), *Facets of Vision*, Springer, Berlin.

Hassenstein, B. 1951, 'Ommatidienraster und afferente Bewegungsintegration', *Zeitschrift für vergleichende Physiologie*, vol. 33, pp. 301–26.

Hassenstein, B. and Reichardt, W. 1956, 'Systemtheoretische analyse der Zeit-, Reihenfolgen- und Vorzeichenauswertung bei der Bewegungsperzeption des Rüsselkäfers *Chlorophanus*', *Zeitschrift für Naturforschung*, vol. 31c, pp. 629–33.

Hateren, J. H. van, 1989, 'Photoreceptor optics, theory and practice', in D. G. Stavenga and R. C. Hardie (eds), *Facets of Vision*, Springer, Berlin, pp. 74–89.

Hateren, J. H. van, 1992, 'Theoretical predictions of spatiotemporal receptive fields of fly LMCs, and experimental validation', *Journal of Comparative Physiology*, vol. 171, pp. 157–70.

Hateren, J. H. van, Srinivasan, M. V. and Wait, P. B. 1990, 'Pattern recognition in bees: orientation discrimination', *Journal of Comparative Physiology*, vol. 167, pp. 649–54.

Hausen, K. 1982, 'Motion-sensitive interneurons in the optomotor system of the fly. I–II', *Biological Cybernetics*, vol. 45, pp. 143–56; and vol. 46, pp. 67–79.

Hausen, K. 1984, 'The lobula-complex of the fly: structure, function and significance in visual behavior', in M. A. Ali (ed.), *Photoreception and Vision in Invertebrates*, Plenum, New York, pp. 523–59.

Hausen, K. and Egelhaaf, M. 1989, 'Neural mechanisms of visual course control in insects', in D. G. Stavenga and R. C. Hardie (eds), *Facets of Vision*, Springer, Berlin, pp. 391–424.

Hebb, D. O. 1949, *The Organization of Behavior*, Wiley, New York.

Hecht, S. and Wald, G. 1934, 'The visual acuity and intensity discrimination of *Drosophila*', *Journal of General Physiology*, vol. 17, pp. 517–47.

Hecht, S. and Wolf, E. 1929, 'The visual acuity of the honeybee', *Journal of General Physiology*, vol. 12, pp. 727–60.

Heinze, S. and Homberg, U. 2007, 'Map-like representation of celestial E-vector orientations in the brain of an insect', *Science*, vol. 315, pp. 995–7.

Heisenberg, M. 1995, 'Pattern recognition in insects', *Current Opinion in Neurobiology*, vol. 5, pp. 475–81.

Heisenberg, M. and Wolf, R. 1984, *Vision in Drosophila: Genetics of microbehavior*, Springer, Berlin.

Heisenberg, M. and Wolf, R. 1988, 'Reafferent control of optomotor yaw torque in *Drosophila melanogaster*', *Journal of Comparative Physiology*, vol. 163, pp. 373–88.

Heisenberg, M. and Wolf, R. 1990, 'Visual control of straight flight in *Drosophila melanogaster*', *Journal of Comparative Physiology*, vol. 167, pp. 269–83.

Heisenberg, M. and Wolf, R. 1992, 'The sensory motor link in motion-dependent flight control of flies', in J. Wallman and F. A. Miles (ed.), *Visual Motion and Its Role in the Stabilization of Gaze*, Elsevier, Amsterdam.

Helversen, O. von, 1972, 'Zur spektralen Unterscheidsempfindlichkeit der Honigbiene', *Journal of Comparative Physiology*, vol. 80, pp. 439–72.

Hempel de Ibarra, N. and Giurfa, M. 2003, 'Discrimination of closed coloured shapes by honeybees requires only contrast to the long wavelength receptor type', *Animal Behaviour*, vol. 66, pp. 903–10.

Hengstenberg, R. 1982, 'Common visual response properties of giant vertical cells in the lobula plate of the blowfly *Calliphora*', *Journal of Comparative Physiology*, vol. 149, pp. 179–93.

Hensler, K. and Rowell, C. H. F. 1990, 'Control of optomotor responses by descending deviation detector neurons in intact flying locusts', *Journal of Experimental Biology*, vol. 149, pp. 191–205.

Hertel, H. 1980, 'Chromatic properties of identified interneurons in the optic lobes of the bee', *Journal of Comparative Physiology*, vol. 137, pp. 215–31.

Hertel, H. and Maronde, U. 1987a, 'Processing of visual information in the honeybee brain', in R. Menzel and A. Mercer (eds), *Neurobiology and Behaviour in Honeybees*, Springer, Berlin, pp. 141–57.

Hertel, H. and Maronde, U. 1987b, 'Processing of visual information in the centrally projecting visual interneurons in the honeybee brain', *Journal of Experimental Biology*, vol. 133, pp. 301–15.

Hertel, H., Schäfer, S. and Maronde, U. 1987, 'The physiology and morphology of visual commissures in the honeybee brain', *Journal of Experimental Biology*, vol. 133, pp. 283–300.

Hertz, M. 1929–31, 'Die Organisation des optischen Feldes bei der Biene', *Zeitschrift für vergleichende Physiologie*, vol. 8, pp. 693–748; vol. 11, pp. 107–45; vol. 14, pp. 629–74.

Hertz, M. 1933, 'Über figurale Intensität und Qualitäten in der optische Wahrnehmung der Biene', *Biologische Zentralblatte*, vol. 53, pp. 10–40.

Hertz, M. 1934, 'Die Untersuchungen über den Formensinn der Honigbiene', *Naturwissenschaften*, vol. 23, pp. 618–24.

Hinde, R. A. 1990, 'Nikolaas Tinbergen', *Biographical Memoirs of the Fellows of the Royal Society*, vol. 36, pp. 549–65.

Hinton, G. E., McClelland, J. L. and Rumelhart, D. E. 1986, 'Distributed representations', in D. E. Rummelhart and J. L. McClelland (eds), *Parallel Distributed Processing*, MIT Press, Cambridge, Mass., pp. 77–109.

Holmes, W., Pumphrey, R. J. and Young, J. Z. 1941, 'The structure and conduction velocity of the medullated nerve fibres of prawns', *Journal of Experimental Biology*, vol. 18, pp. 50–4.

Holst, E. von, and Mittelstaedt, H. 1950, 'Das Reafferenzprinzip. Wechselwirkungen zwischen Zentralnervensytem und Peripherie', *Naturwiss*, vol. 37, pp. 464–76.

Hooke, R. 1665, *Micrographia or Some Physiological Descriptions of Minute Bodies Made by Magnifying Glasses*, J. Martyn and J. Allestry, London.

Horridge, G. A. 1962, 'Learning of leg position by the ventral nerve cord in headless insects', *Proceedings of the Royal Society of London*, vol. 157, pp. 33–52.

Horridge, G. A. 1966a, 'Perception of edges versus areas by the crab *Carcinus*', *Journal of Experimental Biology*, vol. 44, pp. 247–54.

Horridge, G. A. 1966b, 'The retina of the locust', in C. G. Bernhard (ed.), *The Functional Organization of the Compound Eye*, Pergamon Press, Oxford, pp. 513–42.

Horridge, G. A. 1968, *Interneurons: Their origin, action, specificity, growth and plasticity*, Freeman and Co., London and San Francisco.

Horridge, G. A. (ed.) 1975, *The Compound Eye and Vision of Insects*, Oxford University Press, Oxford.

Horridge, G. A. 1977a, 'The compound eye of insects', *Scientific American*, vol. 237, pp. 108–20.

Horridge, G. A. 1977b, 'Insects which turn and look', *Endeavour*, [new series], vol. 1, pp. 7–17.

Horridge, G. A. 1978, 'The separation of visual axes in apposition compound eyes', *Philosophical Transactions of the Royal Society of London*, vol. 285, pp. 1–59.

Horridge, G. A. 1980, 'Apposition eyes of large diurnal insects as organs adapted to seeing', *Proceedings of the Royal Society of London*, vol. 207, pp. 287–309.

Horridge, G. A. 1987, 'The evolution of visual processing and the construction of seeing systems', *Proceedings of the Royal Society of London*, vol. 220, pp. 279–92.

Horridge, G. A. 1994, 'Bee vision of pattern and 3D', *Bioessays*, vol. 16, pp. 877–84.

Horridge, G. A. 1996a, 'Vision of the honeybee *Apis mellifera* for patterns with two pairs of equal orthogonal bars', *Journal of Insect Physiology*, vol. 42, pp. 131–8.

Horridge, G. A. 1996b, 'The relation between pattern and landmark vision of the honeybee (*Apis mellifera*)', *Journal of Insect Physiology*, vol. 42, pp. 373–81.

Horridge, G. A. 1996c, 'Pattern vision of the honeybee (*Apis mellifera*): the significance of the angle subtended by the target', *Journal of Insect Physiology*, vol. 42, pp. 693–703.

Horridge, G. A. 1996d, 'The honeybee (*Apis mellifera*) detects bilateral symmetry and discriminates its axis', *Journal of Insect Physiology*, vol. 42, pp. 755–64.

Horridge, G. A. 1997a, 'Pattern discrimination by the honeybee, disruption as a cue', *Journal of Comparative Physiology*, vol. 181, pp. 267–77.

Horridge, G. A. 1997b, 'Vision of the honeybee *Apis mellifera* for patterns with one pair of equal orthogonal bars', *Journal of Insect Physiology*, vol. 43, pp. 741–8.

Horridge, G. A. 1997c, 'Spatial and non-spatial coding of patterns by the honeybee', in M. V. Srinivasan and S. Venkatesh (eds), *From Living Eyes to Seeing Machines*, Oxford University Press, Oxford, pp. 52–79.

Horridge, G. A. 1998a, 'Spatial coincidence of cues in visual learning by the honeybee', *Journal of Insect Physiology*, vol. 44, pp. 343–50.

Horridge, G. A. 1998b, 'Pattern vision by the honeybee (*Apis mellifera*): training on two pairs of patterns alternately', *Journal of Insect Physiology*, vol. 45, pp. 349–55.

Horridge, G. A. 1999a, 'Two-dimensional pattern discrimination by the honeybee', *Physiological Entomology*, vol. 24, pp. 1–17.

Horridge, G. A. 1999b, 'Pattern vision by the honeybee (*Apis mellifera*) is colour blind for radial/tangential cues', *Journal of Insect Physiology*, vol. 184, pp. 413–22.

Horridge, G. A. 1999c, 'Pattern vision of the honeybee (*Apis mellifera*). The effect of pattern on the discrimination of location', *Journal of Comparative Physiology*, vol. 185, pp. 105–13.

Horridge, G. A. 2000a, 'Pattern vision of the honeybee (*Apis mellifera*). What is an oriented edge?', *Journal of Comparative Physiology*, vol. 186, pp. 521–34.

Horridge, G. A. 2000b, 'Seven experiments on pattern vision of the honeybee, with a model', *Vision Research*, vol. 40, pp. 2589–603.

Horridge, G. A. 2000c, 'Visual discrimination of radial cues by the honeybee (*Apis mellifera*)', *Journal of Insect Physiology*, vol. 46, pp. 629–45.

Horridge, G. A. 2000d, 'Pattern vision of the honeybee (*Apis mellifera*): the discrimination of location by the blue and green receptors', *Neurobiology of Learning and Memory*, vol. 74, pp. 1–16.

Horridge, G. A. 2003a, 'Discrimination of single bars by the honeybee (*Apis mellifera*)', *Vision Research*, vol. 43, pp. 1257–71.

Horridge, G. A. 2003b, 'The visual system of the honeybee (*Apis mellifera*): the maximum length of the orientation detector', *Journal of Insect Physiology*, vol. 49, pp. 621–8.

Horridge, G. A. 2003c, 'Visual resolution of gratings by the compound eye of the bee (*Apis mellifera*)', *Journal of Experimental Biology*, vol. 206, pp. 2105–10.

Horridge, G. A. 2003d, 'Visual discrimination by the honeybee (*Apis mellifera*): the position of the common centre as the cue', *Physiological Entomology*, vol. 28, pp. 132–43.

Horridge, G. A. 2003e, 'The effect of complexity on the discrimination of oriented bars by the honeybee (*Apis mellifera*)', *Journal of Comparative Physiology*, vol. 189, pp. 703–14.

Horridge, G. A. 2003f, 'Visual resolution of the orientation cue by the honeybee (*Apis mellifera*)', *Journal of Insect Physiology*, vol. 49, pp. 1145–52.

Horridge, G. A. 2005a, 'The spatial resolutions of the apposition compound eye and its neurosensory feature detectors: observation versus theory', *Journal of Insect Physiology*, vol. 51, pp. 243–66.

Horridge, G. A. 2005b, 'What the honeybee sees: a review of the recognition system of *Apis mellifera*', *Physiological Entomology*, vol. 30, pp. 2–13.

Horridge, G. A. 2006a, 'Visual discrimination of spokes, sectors, and circles by the honeybee (*Apis mellifera*)', *Journal of Insect Physiology*, vol. 52, pp. 984–1003.

Horridge, G. A. 2006b, 'Some labels that are recognized on landmarks by the honeybee (*Apis mellifera*)', *Journal of Insect Physiology*, vol. 52, pp. 1254–71.

Horridge, G. A. 2006c, 'Visual processing of pattern', in E. Warrant and D.-E. Nilsson (eds), *Invertebrate Vision*, Cambridge University Press, Cambridge, pp. 494–525.

Horridge, G. A. 2007, 'The preferences of the honeybee (*Apis mellifera*) for different visual cues during the learning process', *Journal of Insect Physiology*, vol. 53, pp. 877–89.

Horridge, G. A. 2009a, 'Generalization in visual recognition by the honeybee (*Apis mellifera*). A review and explanation', *Journal of Insect Physiology*, vol. 55, pp. 499–511.

Horridge, G. A. 2009b, 'What does the honeybee see?', in O. Lazareva, T. Shimizu and E. Wasserman (eds), *How Animals See the World*, Oxford University Press, Oxford.

Horridge, G. A. 2009c, 'What does an insect see?', *Journal of Experimental Biology*, vol. 212, pp. 2721–2729.

Horridge, G. A. and Marčelja, L. 1992, 'On the existence of "fast" and "slow" directionally sensitive motion detector neurons in insects', *Proceedings of the Royal Society of London*, vol. 248, pp. 47–54.

Horridge, G. A. and Meinertzhagen, I. A. 1970, 'The exact neural projection of the visual fields upon the first and second ganglia of the insect eye', *Zeitschrift für vergleichende Physiologie*, vol. 66, pp. 369–78.

Horridge, G. A. and Zhang, S. W. 1995, 'Pattern vision of bees, flower-like patterns with no predominant orientation', *Journal of Insect Physiology*, vol. 41, pp. 681–8.

Horridge, G. A., Duniec, J. and Marčelja, L. 1981, 'A 24-hour cycle in single locust and mantid photoreceptors', *Journal of Experimental Biology*, vol. 91, pp. 307–22.

Horridge, G. A., Marčelja, L., Jahnke, R. and McIntyre, P. 1983, 'Daily changes in the compound eye of a beetle (*Macrogyrus*)', *Proceedings of the Royal Society of London*, vol. 217, pp. 265–85.

Horridge, G. A., Mimura, K. and Hardie, R. C. 1976, 'Fly photoreceptors III. Angular sensitivity as a function of wavelength and the limits of resolution', *Proceedings of the Royal Society of London*, vol. 194, pp. 151–77.

Horridge, G. A., Scholes, J. H., Shaw, S. and Tunstall, J. 1965, 'Extracellular recordings from single neurons in the optic lobe and brain of the locust', in J. E. Treherne and J. S. C. Beament (eds), *The Physiology of the Insect Central Nervous System*, Academic Press, London, pp. 165–202.

Horridge, G. A., Zhang, S. W. and Lehrer, M. 1992, 'Bees can combine range and visual angle to estimate absolute size', *Philosophical Transactions of the Royal Society of London*, vol. 337, pp. 49–57.

Horridge, G. A., Zhang, S. W. and O'Carroll, D. 1992, 'Insect perception of illusory contours', *Philosophical Transactions of the Royal Society of London*, vol. 337, pp. 59–64.

Horridge, G. A., Zhang, S. W. and Srinivasan, M. V. 1992, 'Pattern recognition in honeybees: local and global analysis', *Proceedings of the Royal Society of London*, vol. 248, pp. 55–61.

Howard, J. and Snyder, A. W. 1983, 'Transduction as a limitation on compound eye function and design', *Proceedings of the Royal Society of London*, vol. 217, pp. 287–307.

Hubel, D. H. and Wiesel, T. N. 1959, 'Receptive fields of single neurons in the cat's striate cortex', *Journal of Physiology*, vol. 148, pp. 574–91.

Ibbotson, M. R. 1991a, 'Wide-field motion-sensitive neurons tuned to horizontal movement in the honeybee, *Apis mellifera*', *Journal of Comparative Physiology*, vol. 168, pp. 91–102.

Ibbotson, M. R. 1991b, 'A motion-sensitive visual descending neuron in *Apis mellifera* monitoring translatory flow-fields in the horizontal plane', *Journal of Experimental Biology*, vol. 157, pp. 1–5.

Ibbotson, M. R., Maddess, T. and Dubois, R. 1991, 'A system of insect neurons sensitive to horizontal and vertical image motion connects the medulla and midbrain', *Journal of Comparative Physiology*, vol. 169, pp. 355–67.

Ichikawa, T. 1990, 'Spectral sensitivities of elementary colour-coded neurons in butterfly larva', *Journal of Neurophysiology*, vol. 64, pp. 1861–72.

Ichikawa, T. 1991, 'Integration of colour signals in the medulla of the swallowtail butterfly larva', *Journal of Experimental Biology*, vol. 155, pp. 127–45.

Ichikawa, T. and Tateda, H. 1980, 'Cellular patterns and spectral sensitivity of larval ocelli in the swallowtail butterfly *Papilio*', *Journal of Comparative Physiology*, vol. 139, pp. 41–7.

Ichikawa, T. and Tateda, H. 1982a, 'Receptive field of the stemmata in the swallowtail butterfly *Papilio*', *Journal of Comparative Physiology*, vol. 146, pp. 191–9.

Ichikawa, T. and Tateda, H. 1982b, 'Distribution of color receptors in the larval eyes of four species of *Lepidoptera*', *Journal of Comparative Physiology*, vol. 149, pp. 317–24.

Jacobs-Jessens, U. F. 1959, 'Zur Orientierung der Hummeln und einiger anderer Hymenopteren', *Zeitschrift für vergleichende Physiologie*, vol. 41, pp. 597–641.

James, A. C. 1992, 'Non-linear operator network models of processing in the fly lamina', in R. B. Pinter and B. Nabet (eds), *Nonlinear Vision*, CRC Press, Boca Raton, pp. 39–73.

James, A. C. and Osorio, D. 1996, 'Characterization of columnar neurons and visual signal processing in the medulla of the locust optic lobe by system identification techniques', *Journal of Comparative Physiology*, vol. 178, pp. 183–99.

Jander, R. 1964, 'Die Detektortheorie optischer Auslösemechanismen von Insekten', *Zeitschrift für Tierpsychologie*, vol. 21, pp. 302–7.

Jander, R. and Volk-Heinrichs, I. 1980, 'Das strauchspezifische visuel Perceptorsystem der Stabheuschrecke (*Carausius morosus*)', *Zeitschrift für vergleichende Physiologie*, vol. 70, pp. 425–77.

Jander, R. and Voss, C. 1963, 'Die Bedeutung von Streifenmustern für das Formensehen der Roten Waldameise (*Formica rufa L.*)', *Zeitschrift für Tierpsychologie*, vol. 20, pp. 1–9.

Jander, R., Fabritius, M. and Fabritius, M. 1970, 'Die Bedeutung von gliederung und Kantenrichtung für die visuelle Formunterscheidung der Wespe *Dolichovespula saxonica* am Flugloch', *Zeitschrift für Tierpsychologie*, vol. 27, pp. 881–93.

Jander, U. and Jander, R. 2002, 'Allometry and resolution of bee eyes (Apoidea)', *Arthropod Structure and Development*, vol. 30, pp. 179–93.

Järvilehto, M. 1985, 'The eye, vision and perception', in G. A. Kerkut and L. I. Gilbert (eds), *Comprehensive Insect Physiology, Biochemistry and Pharmacology. Volume 6. Nervous System, Sensory*, Pergamon Press, Oxford, pp. 355–429.

Jawlowski, H. 1958, 'Nerve tracts in bee (*Apis mellifera*) running from the sight and antennal organs to the brain', *Annales of the Université, M. Curie-Sklodowska C*, vol. 12, pp. 307–23.

Jennings, H. S. 1905, *The Behaviour of the Lower Organisms*, Columbia University Press, New York.

Jones, C. E. and Buchmann, S. L. 1974, 'Ultraviolet floral patterns as functional orientation cues in hymenopterous pollination systems', *Animal Behaviour*, vol. 22, pp. 481–5.

Kennedy, J. S. 1940, 'The visual responses of flying mosquitoes', *Proceedings of the Zoological Society of London*, vol. 109, pp. 221–42.

Kenyon, F. C. 1986, 'The brain of the bee. A preliminary contribution to the morphology of the nervous system of the Arthropoda', *Journal of Comparative Neurology*, vol. 6, pp. 133–210.

Kien, J. 1975, 'Motion detectors in locusts and grasshoppers', in G. A. Horridge (ed.), *The Compound Eye and Vision of Insects*, Oxford University Press, Oxford, pp. 410–22.

Kirschfeld, K. 1966, 'Discrete and graded receptor potentials in the compound eye of the fly (*Musca*)', in C. G. Bernhard (ed.), *The Functional Organization of the Compound Eye*, Pergamon Press, Oxford, pp. 291–308.

Kirschfeld, K. 1967, 'Die projektion der optischen Umwelt auf das Raster der Rhabdomeren im Komplexauge von *Musca*', *Experimental Brain Research*, vol. 3, pp. 248–70.

Kirschfeld, K. 1972, 'The visual system of *Musca*. Studies on optics, structure and function', in R. Wehner (ed.), *Information Processing in the Visual System of Arthropods*, Springer, Berlin, pp. 63–74.

Kirschfeld, K. 1976, 'The resolution of lens and compound eyes', in F. Zettler and R. Weiler (eds), *Neural Principles in Vision*, Springer, Berlin, pp. 354–70.

Kirschfeld, K. and Franceschini, N. 1969, 'Ein Mehanismus zur Steuerung des Lichtflusses in den Rhabdomeren des Komplexauges von *Musca*', *Kybernetic*, vol. 6, pp. 13–22.

Kirschfeld, K. and Lutz, B. 1974, 'Lateral inhibition in the compound eye of the fly, *Musca'*, *Zeitschrift für Naturforschung*, vol. 29c, pp. 95–6.

Koehler, W. 1925, *The Mentality of Apes*, Kegan Paul, London and New York.

Koffka, K. 1924, *The Growth of the Mind*, Translated by R. M. Ogden, Kegan Paul, London.

Koffka, K. 1935, *Principles of Gestalt Psychology*, Kegan Paul, London.

Kolb, G. and Autrum, H. 1972, 'Die Feinstruktur im Auge der Biene bei Hell- und Dunkeladaption', *Journal of Comparative Physiology*, vol. 77, pp. 113–25.

Kuffler, S. 1953, 'Discharge patterns and functional organization of mammalian retina', *Journal of Neurophysiology*, vol. 16, pp. 37–68.

Kühn, A. 1919, *Die Orientierung der Tiere im Raum*, Fischer, Jena.

Kuhn, T. S. 1970, *The Structure of Scientific Revolutions*, University of Chicago Press, Chicago.

Kuiper, J. W. 1962, 'The optics of the compound eye', *Symposium of the Society for Experimental Biology*, vol. 16, pp. 58–71.

Kuiper, J. W. 1966, 'On the image formation in a single ommatidium of the compound eye in Diptera', in C. G. Bernhard (ed.), *The Functional Organization of the Compound Eye*, Pergamon Press, Oxford, pp. 35–50.

Kunze, P. 1961, 'Untersuchungen des Bewegungssehens fixiert fliegender Bienen', *Zeitschrift für vergleichende Physiologie*, vol. 44, pp. 656–84.

Labhart, T. 1980, 'Specialized photoreceptors at the dorsal rim of the honey bee's compound eye: polarization and angular sensitivity', *Journal of Comparative Physiology*, vol. 141, pp. 19–30.

Labhart, T. 1988, 'Polarization-opponent interneurons in the insect visual system', *Nature*, vol. 331, pp. 435–7.

Land, M. F. 1975, 'Head movements and fly vision', in G. A. Horridge (ed.), *The Compound Eye and Vision of Insects*, Oxford University Press, Oxford, pp. 469–89.

Land, M. F. 1989, 'Variations in the structure and design of compound eyes', in D. G. Stavenga and R. C. Hardie (eds), *Facets of Vision*, Springer, Berlin, pp. 90–111.

Land, M. F. 1997a, 'Visual acuity in insects', *Annual Review of Entomology*, vol. 42, pp. 147–77.

Land, M. F. 1997b, 'The resolution of insect compound eyes', *Israel Journal of Plant Sciences*, vol. 45, pp. 79–91.

Land, M. F. and Collett, T. S. 1974, 'Chasing behaviour of houseflies (*Fannia cannicularis*)', *Journal of Comparative Physiology*, vol. 89, pp. 331–57.

Land, M. F. and Eckert, H. 1985, 'Maps of the acute zones of fly eyes', *Journal of Comparative Physiology*, vol. 156, pp. 525–38.

Lashley, K. S. 1938, 'Conditional reactions in the rat', *Journal of Psychology*, vol. 6, pp. 311–24.

Laughlin, S. B. 1975, 'The function of the lamina ganglionaris', in G. A. Horridge (ed.), *The Compound Eye and Vision of Insects*, Clarendon Press, Oxford.

Laughlin, S. B. 1981, 'A simple coding procedure enhances a neuron's information capacity', *Zeitschrift für Naturforschung*, vol. 36c, pp. 910–12.

Laughlin, S. B. 1989, 'Coding efficiency and design in visual processing', in D. G. Stavenga and R. C. Hardie (eds), *Facets of Vision*, Springer, Berlin, pp. 213–34.

Laughlin, S. B. 1994, 'Matching coding, circuits, cells and molecules to signals. General principles of retinal design in the fly's eye', *Progress in Retinal and Eye Research*, vol. 13, pp. 165–96.

Laughlin, S. B. and Hardie, R. C. 1978, 'Common strategies for light adaptation in the peripheral visual systems of fly and dragonfly', *Journal of Comparative Physiology*, vol. 128, pp. 319–40.

Laughlin, S. B. and Horridge, G. A. 1972, 'Angular sensitivity of the retinula cells of dark-adapted worker bee', *Zeitschrift für vergleichende Physiologie*, vol. 74, pp. 329–35.

Laughlin, S. B. and Weckström, M. 1993, 'Fast and slow photoreceptors—a comparative study of the functional diversity of coding and conductances in the Diptera', *Journal of Comparative Physiology*, vol. 172, pp. 593–609.

Laughlin, S. B., Howard, J. and Blakeslee, B. 1987, 'Synaptic limitations to contrast coding in the retina of the blowfly *Calliphora*', *Proceedings of the Royal Society of London*, vol. 231, pp. 437–67.

Lehrer, M. 1990, 'How bees use peripheral eye regions to localize a frontally positioned target', *Journal of Comparative Physiology*, vol. 167, pp. 173–85.

Lehrer, M. 1993, 'Why do bees turn back and look?', *Journal of Comparative Physiology*, vol. 172, pp. 544–63.

Lehrer, M. and Bischof, S. 1995, 'Detection of model flowers by honeybees: the role of chromatic and achromatic contrast', *Naturwissenschaften*, vol. 82, pp. 145–7.

Lehrer, M. and Campan, R. 2004, 'Shape discrimination by wasps (*Paravespula germanica*) at the food source: generalization among various types of contrast', *Journal of Comparative Physiology*, vol. 190, pp. 651–63.

Lehrer, M. and Campan, R. 2006, 'Generalizatoin of convex shapes by bees: what are shapes made of?', *Journal of Experimental Biology*, vol. 208, pp. 3233–3247.

Lehrer, M. and Srinivasan, M. V. 1992, 'Freely flying bees discriminate between stationary and moving objects: performance and possible mechanisms', *Journal of Comparative Physiology*, vol. 171, pp. 457–67.

Lehrer, M. and Srinivasan, M. V. 1993, 'Object detection by honeybees: why do they land on edges?', *Journal of Comparative Physiology*, vol. 173, pp. 23–32.

Lehrer, M., Horridge, G. A., Zhang, S. W. and Gadagkar, R. 1995, 'Shape vision in bees' innate preference for flower-like patterns', *Philosophical Transactions of the Royal Society of London*, vol. 347, pp. 123–37.

Lehrer, M., Srinivasan, M. V. and Zhang, S. W. 1990, 'Visual edge detection in the honeybee, and its chromatic properties', *Proceedings of the Royal Society of London*, vol. 238, pp. 321–30.

Lehrer, M., Srinivasan, M. V., Zhang, S. W. and Horridge, G. A. 1988, 'Motion cues provide the bee's visual world with a third dimension', *Nature*, vol. 332, pp. 356–7.

Lehrer, M., Wehner, R. and Srinivasan, M. V. 1985, 'Visual scanning behaviour in honeybees', *Journal of Comparative Physiology*, vol. 157, pp. 405–15.

Lehrer, M., Wunderli, M. and Srinivasan, M. V. 1993, 'Perception of heterochromatic flicker by honeybees: a behavioural study', *Journal of Comparative Physiology*, vol. 172, pp. 1–6.

Lehrman, D. S. 1953, 'A critique of Konrad Lorenz's theory of instinctive behaviour', *Quarterly Review of Biology*, vol. 28, pp. 337–63.

Lehrman, D. S. 1970, 'Semantic and conceptual issues in the nature–nurture problem', in L. R. Aronson, E. Tobach, D. S. Lehrman and J. S. Rosenblatt (eds), *Development and Evolution of Behavior: Essays in memory of T. C. Schneirla*, Freeman, San Francisco, pp. 17–52.

Lettvin, J. Y., Maturana, H. R., McCulloch, W. S. and Pitts, W. H. 1959, 'What the frog's eye tells the frog's brain', *Proceedings of the Institute of Radio Engineers*, vol. 47, pp. 1940–51.

Lillywhite, P. G. 1977, 'Single photon signals and transduction in an insect eye', *Journal of Comparative Physiology*, vol. 122, pp. 189–200.

Lillywhite, P. G. and Dvorak, D. R. 1981, 'Responses to single photons in a fly optomotor neuron', *Vision Research*, vol. 21, pp. 279–90.

Lindauer, M. 1978, *Communication Among Social Bees*, Harvard University Press, Cambridge, Mass.

Lindauer, M. and Martin, P. 1968, 'Die Schwereorientierung der Bienen unter dem Einfluss des Erdmagnetfeldes', *Zeitschrift für vergleichende Physiologie*, vol. 60, pp. 219–43.

Lubbock, J. 1865, *Prehistoric Times*, Williams and Norgate, London.

Lubbock, J. 1871, *The Origin of Civilisation and the Primitive Condition of Man*, Longmans/Green, London.

Lubbock, J. 1881 [1898], *Ants, Bees and Wasps*, 13th edn, Kegan Paul, London.

McCann, G. D. and Dill, J. C. 1969, 'Fundamental properties of intensity, form and motion perception in the visual nervous system of *Calliphora phaenicia* and *Musca domestica*', *Journal of General Physiology*, vol. 53, pp. 385–413.

Maddess, T. 1986, 'After-image-like effects in the motion-sensitive neuron H1', *Proceedings of the Royal Society of London*, vol. 228, pp. 433–59.

Maddess, T. and Laughlin, S. B. 1985, 'Adaptation of the motion-sensitive neuron H1 is generated locally and governed by contrast frequency', *Proceedings of the Royal Society of London*, vol. 225, pp. 251–75.

Maldonado, H. 1970, 'The deimatic reaction in the praying mantis *Stagmatoptera biocellata*', *Zeitschrift für vergleichender Physiologie*, vol. 68, pp. 60–71.

Mallock, A. 1894, 'Insect sight and the defining power of composite eyes', *Proceedings of the Royal Society of London*, vol. 55, pp. 85–90.

Maronde, U. 1991, 'Common projection areas of antennae and visual pathways in the honeybee brain, *Apis mellifera*', *Journal of Comparative Physiology*, vol. 309, pp. 328–40.

Marr, D. 1982, *Vision*, Freeman, San Francisco.

Maturana, R., Lettvin, J., Pitts, W. and McCulloch, W. 1960, 'Anatomy and physiology of vision in the frog (*Rana pipiens*)', *Journal of General Physiology*, vol. 43, pp. 129–75.

Mazokhin-Porshnyakov, G. A. 1969, *Insect Vision*, Plenum Press, New York.

Meinertzhagen, I. A. 1976, 'The organisation of perpendicular fibre pathways in the insect optic lobe', *Philosophical Transactions of the Royal Society of London*, vol. 274, pp. 555–94.

Meinertzhagen, I. A. and Sorra, K. E. 1976, 'Synaptic organization in the fly's optic lamina: few cells, many synapses and divergent microcircuits', *Progress in Brain Research*, vol. 131, pp. 53–69.

Menzel, R. 1979, 'Spectral sensitivity and colour vision in invertebrates', in H. Autrum (ed.), *Handbook of Sensory Physiology. Volume VII. Part 6A. Invertebrate Visual Centres and Behaviour*, Springer, Berlin, pp. 503–80.

Menzel, R. 2008, 'Insect minds for human minds', in A. S. Benjamin, J. S. de Belle and T. A. Polk (eds), *Human Learning*, Elsevier, London, pp. 271–85.

Menzel, R. 2009, 'Working memory in bees, also in flies?', *Journal of Neurogenetics*, vol. 8, pp. 1–8.

Menzel, R and Giurfa, M. 2006, 'Dimensions of cognition in an insect: the honeybee', *Behavioral and Cognitive Neuroscience Reviews*, vol. 5, pp. 24–40.

Menzel, R. and Greggers, U. 1992, 'Temporal dynamics and foraging behaviour in honeybees', in T. Billen (ed.), *Biology and Evolution of Social Insects*, Leuven University Press, Leuven, Belgium, pp. 303–18.

Menzel, R. and Mercer, A. (eds) 1987, *Neurobiology and Behavior of Honeybees*, Springer, Berlin.

Menzel, R., Chyittka, L., Eichmuller, S., Geiger, K., Peitsch, D. and Knoll, P. 1990, 'Dominance of celestial cues over landmarks disproves map-like orientation in honey bees', *Zeitschrift für Naturforschung*, vol. 45c, pp. 723–6.

Menzel, R., Greggers, U., Smith, A., Berger, S., Brandt, R., Brunke, S., Bundrock, G., Huelse, S., Pluempe, T., Schaupp, F., Schuettler, E., Stach, S., Stind, J., Stollhoff, N. and Watzl, S. 2005, 'Honeybees navigate according to a map-like spatial memory', *Proceedings of the National Academy of Sciences of the USA*, vol. 102, pp. 3040–5.

Meyer, H. W. 1971, 'Visuelle Schlüsselreize für die Aulösung der Beutefanghandlung beim Bachwasserläufer *Velia capria* (Hemiptera, Heteroptera)', *Zeitschrift für vergleichende Physiologie*, vol. 72, pp. 260–342.

Meyer, H. W. 1974, 'Geometrie und funktionelle Specialisierung des optischen Abtastrasters beim Bachwasserläufer (*Velia capria*)', *Journal of Comparative Physiology*, vol. 92, pp. 85–103.

Mill, J. S. 1843, *A System of Logic, Ratiocinative and Inductive, Being a Connected View of the Principal Evidence, and the Methods of Scientific Investigation*, 2 vols, John W. Parker, London.

Mill, J. S. 1873, *Autobiography*, Penguin Classics, United States.

Mobbs, P. G. 1982, 'The brain of the honeybee *Apis mellifera I*. The connections and spatial organization of the mushroom bodies', *Philosophical Transactions of the Royal Society of London*, vol. 298, pp. 309–54.

Møller, A. P. 1995, 'Bumblebee preference for symmetrical flowers', *Proceedings of the National Academy of Science of the USA*, vol. 92, pp. 2288–92.

Mollon, J. D. 1997, 'On the basis of velocity clues alone': some perceptual themes, 1946–1996', *Quarterly Journal of Experimental Psychology*, vol. 50, pp. 859–78.

Morgan, C. L. 1890, *Animal Life and Intelligence*, Edward Arnold, London.

Müller, J. 1826, *Zur vergleichende Physiologie des Gesichtssinnes*, Cnobloch, Leipsig.

Naka, K. 1961, 'Recording of retinal action potentials from single cells in the insect compound eye', *Journal of General Physiology*, vol. 44, pp. 571–84.

Naka, K. and Eguchi, E. 1962, 'Spike potentials recorded from the insect photoreceptor', *Journal of General Physiology*, vol. 45, pp. 663–80.

Nelson, R. C. and Aloimonos, J. 1988, 'Finding motion parameters from spherical motion fields (or the advantages of having eyes in the back of your head)', *Biological Cybernetics*, vol. 58, pp. 261–73.

Neumann, J. von, 1958, *The Computer and the Brain*, Yale University Press, Newhaven, Conn.

Nilsson, D. E. 1989, 'Optics and evolution of the compound eye', in D. G. Stavenga and R. C. Hardie (eds), *Facets of Vision*, Springer, Berlin, pp. 30–73.

Nordström, K. and O'Carroll, D. C. 2006, 'Small object detection neurons in female hoverflies', *Proceedings of the Royal Society of London*, vol. 273, pp. 1211–16.

Nordström, K., Barnett, P. D. and O'Carroll, D. C. 2006, 'Insect detection of small targets moving in visual clutter', *PloS Biology*, vol. 4, no. 3.

Northrop, R. B. 1975, 'Information processing in the insect compound eye', in G. A. Horridge (ed.), *The Compound Eye and Vision of Insects*, Oxford University Press, Oxford, pp. 378–409.

Olberg, R. M. 1981, 'Object and self-movement detectors in the ventral nerve cord of the dragonfly', *Journal of Comparative Physiology*, vol. 141, pp. 327–34.

Olberg, R. M. 1986, 'Identified target-selective visual interneurons descending from the dragonfly brain', *Journal of Comparative Physiology*, vol. 159, pp. 827–40.

Osborne, J. L., Williams, I. H., Carreck, N. L., Poppy, G. M., Riley, J. R., Smith, A. D., Reynolds, D. R. and Edwards A. S. 1996, 'Harmonic radar: a new technique for investigating bumblebee and honey bee foraging flight', *VII International Symposium on Pollination. ISHS Acta Horticulturae*, vol. 43.

Osorio, D. 1986, 'Directionally selective cells in the locust medulla', *Journal of Comparative Physiology*, vol. 159, pp. 841–7.

Osorio, D. 1987a, 'Temporal and spectral properties of sustaining cells in the medulla of the locust', *Journal of Comparative Physiology*, vol. 161, pp. 441–8.

Osorio, D. 1987b, 'The temporal properties of non-linear transient cells in the locust medulla', *Journal of Comparative Physiology*, vol. 161, pp. 431–40.

Osorio, D., Snyder, A. W. and Srinivasan, M. V. 1987, 'Bi-partitioning and boundary detection in natural scenes', *Spatial Vision*, vol. 2, pp. 191–8.

Palka, J. 1965, 'Diffraction and visual acuity of insects', *Science*, vol. 149, pp. 551–3.

Palka, J. and Pinter, R. B. 1975, 'Theoretical and experimental analysis of visual acuity in insects', in G. A. Horridge (ed.), *The Compound Eye and Vision of Insects*, Oxford University Press, Oxford, pp. 321–37.

Paulk, A. C., Dacks, A. M. and Gronenberg, W. 2009, 'Color processing in the medulla of the bumblebee (*Apidae: Bombus impatiens*)', *Journal of Comparative Neurology*, vol. 513, pp. 441–56.

Pick, B. and Buchner, E. 1979, 'Visual movement detection under light- and dark-adaptation in the fly, *Musca domestica*', *Journal of Comparative Physiology*, vol. 134, pp. 45–54.

Pièron, H. 1904, 'Du rôle de sens musculaire dans l'orientation des fourmis', *Bulletin de l'Institute génerale de Psychologie*, vol. 45, pp. 221–9.

Pinter, R. B. 1979, 'Inhibition and excitation in the locust DCMD receptive field: spatial frequency, temporal and spatial characteristics', *Journal of Experimental Biology*, vol. 80, pp. 191–216.

Plateau, F. 1885–99, 'Comment les fleurs attirent les insects. Recherches expérimentales', *Bulletin Academie, Société royale belge*, vol. 30, pp. 466–88, [see papers listed by Forel (1908:142)].

Poggio, T. and Reichardt, W. 1976, 'Visual control of orientation in the fly. Part II. Towards the underlying neural interactions', *Quarterly Review of Biophysics*, vol. 9, pp. 377–438.

Poggio, T. and Reichardt, W. 1981, 'Visual fixation and tracking in flies. Mathematical properties of simple control systems', *Biological Cybernetics*, vol. 40, pp. 101–12.

Popper, K. R. 1935 [1959], *Logik der Forschung*, Springer, Vienna [Translated as *The Logic of Scientific Discovery*, Hutchinson, London].

Popper, K. 1972, *Objective Knowledge: An evolutionary approach*, Oxford University Press, Oxford.

Praagh, J. P. van, Ribi, W., Wehrhahn, C. and Wittmann, D. 1980, 'Drone bees fixate the queen with the dorsal front part of their compound eyes', *Journal of Comparative Physiology*, vol. 162, pp. 159–72.

Preiss, R. 1987, 'Motion parallax and figural properties of depth control and flight speed in an insect', *Biological Cybernetics*, vol. 57, pp. 1–9.

Preiss, R. 1992, 'Set point of retinal velocity of ground images in the control of swarming flight of desert locusts', *Journal of Comparative Physiology*, vol. 171, pp. 251–6.

Preiss, R. and Kramer, E. 1984, 'Control of flight speed by minimization of the apparent ground pattern movement', in D. Varjú and H. U. Schnitzler (eds), *Localization and Orientation in Biology and Engineering*, Springer, Berlin, pp. 140–2.

Pringle, J. W. S. 1938, 'Proprioception in insects. I–III', *Journal of Experimental Biology*, vol. 15, pp. 101–31, 467–73.

Pumphrey, R. J. and Young, J. Z. 1938, 'The rates of conduction of nerve fibres of various diameters in cephalopods', *Journal of Experimental Biology*, vol. 15, pp. 453–66.

Pumphrey, R. J., Schmit, O. H. and Young, J. Z. 1940, 'Correlation of local excitability with local physiological response in the giant axon of the squid (*Loligo*)', *Journal of Physiology*, vol. 98, pp. 47–72.

Pyza, E. and Meinertzhagen, I. A. 2003, 'The regulation of circadian rhythms in the fly's visual system', *Neuropeptides*, vol. 37, pp. 227–89.

Rabaud, E. 1928, *How Animals Find Their Way About*, Translated by H. Myers, Kegan Paul, London.

Reichardt, W. 1961, 'Autocorrelation: a principle for evaluation of sensory information by the central nervous system', in W. A. Rosenblith (ed.), *Principles of Sensory Communication*, Wiley, New York, pp. 303–17.

Reichardt, W. 1962, 'Nervous integration in the facet eye', *Journal of Biophysics*, vol. 2, pp. 121–43.

Reichardt, W. (ed.) 1969, *Processing of Optical Data by Organisms and by Machines*, Academic Press, New York.

Reichardt, W. 1970, 'The insect eye as a model for analysis of uptake, transduction and processing of optical data in the nervous system', in F. O. Schmitt (ed.), *The Neurosciences: Second study program*, Rockefeller University Press, New York, pp. 494–511.

Reichardt, W. 1986, 'Processing of optical information by the visual system of the fly', *Vision Research*, vol. 26, pp. 113–26.

Reichardt, W. 1987a, 'Computation of optical motion by movement detectors', *Biophysics and Chemistry*, vol. 26, pp. 263–78.

Reichardt, W. 1987b, 'Evaluation of optical motion information by movement detectors', *Journal of Comparative Physiology*, vol. 161, pp. 533–47.

Reichardt, W. and Poggio, T. 1976, 'Visual control of orientation behavior in the fly. Part I. A quantitative analysis', *Quarterly Review of Biophysics*, vol. 9, pp. 311–75.

Reichardt, W. and Poggio, T. 1979, 'Figure-ground discrimination by relative movement in the visual system of the fly. Part I. Experimental results', *Biological Cybernetics*, vol. 35, pp. 81–100.

Reichardt, W., Poggio, T. and Hausen, K. 1983, 'Figure-ground discrimination by relative movement in the visual system of the fly. Part II. Towards the neural circuitry', *Biological Cybernetics*, vol. 46 (Supplement), pp. 1–30.

Reichert, H. and Rowell, C. H. F. 1986, 'Neuronal circuits controlling flight in the locust: how sensory information is processed for motor control', *Trends in Neurosciences*, vol. 9, pp. 281–3.

Ribi, W. 1975–79, 'The first optic ganglion of the bee. I–III', *Cell and Tissue Research*, vol. 165, pp. 103–11; vol. 171, pp. 359–73; vol. 200, pp. 345–57.

Rind, F. C. 1990, 'A directionally selective motion-detecting neuron in the brain of the locust: physiological and morphological characterization', *Journal of Experimental Biology*, vol. 149, pp. 1–19.

Robert, D. and Rowell, C. H. F. 1992, 'Locust flight steering', *Journal of Comparative Physiology*, vol. 171, pp. 41–51.

Romanes, G. J. 1885, 'Homing faculty of Hymenoptera', *Nature*, vol. 32, p. 630.

Ronacher, B. 1979, 'Äquivalenz zwischen Größen- und Helligkeitsunterschieden im Rahmen der visuellen Wahrnehmung der Honigbiene', *Biological Cybernetics*, vol. 32, pp. 63–75.

Ronacher, B. and Duft, U. 1996, 'An image matching mechanism describes a generalization task in honeybees', *Journal of Comparative Physiology*, vol. 178, pp. 803–12.

Rose, A. 1973, *Vision, Human and Electronic*, Plenum Press, New York and London.

Rossel, S. 1979, 'Regional differences in photoreceptor performance in the eye of the praying mantis', *Journal of Comparative Physiology*, vol. 131, pp. 95–112.

Rossel, S. and Wehner, R. 1987, 'The bee's E-vector compass', in R. Menzel and A. Mercer (eds), *Neurobiology and Behavior of the Honeybee*, Springer, Berlin, pp. 76–93.

Rowell, C. H. F. 1971, 'The orthopteran descending movement-detector (DMD) neurons: a characterization and review', *Zeitschrift für vergleichende Physiologie*, vol. 73, pp. 167–94.

Rowell, C. H. F. and Reichert, H. 1991, 'Mesothoracic interneurons involved in flight steering in the locust', *Tissue and Cell*, vol. 23, pp. 75–139.

Rowell, C. H. F., O'Shea, M. and Williams, J. L. D. 1977, 'The neuronal basis of a sensory analyser; the acridid movement detector system. I. The preference for small-field stimuli', *Journal of Experimental Biology*, vol. 68, pp. 157–85.

Rummelhart, D. E. and McClelland, J. L. (eds) 1986, *Parallel Distributed Processing*, MIT Press, Cambridge, Mass.

Ryback, J. and Menzel, R. 1993, 'Anatomy of the mushroom bodies in the honeybee brain: the neuronal connections of the alpha lobe', *Journal of Comparative Neurology*, vol. 334, pp. 444–65.

Sandeman, D. C., Kien, J. and Erber, J. 1975, 'Optokinetic eye movements in the crab, *Carcinus maenas*. II. Responses of optokinetic interneurons', *Journal of Comparative Physiology*, vol. 101, pp. 259–74.

Sanders J. S. (ed.) 1996, *Selected Papers on Natural and Artificial Compound Eye Sensors*, SPIE Optical Engineering Press, Bellingham, Washington, DC.

Santschi, F. 1911, 'Observations et remarques critiques sur le mécanisme de l'orientation chez les fourmis', *Revue Suisse de Zoologie*, vol. 19, pp. 303–38.

Santschi, F. 1923, *Memoires de la Societe Vaudoise des Sciences Naturelles*, vol. 137.

Schnetter, B. 1968, 'Visuelle Formunterscheidung der Honigbiene im Bereich von Vier- und Sechs-strahlsternen', *Zeitschrift für vergleichende Physiologie*, vol. 59, pp. 90–109.

Schnetter, B. 1972, 'Experiments on pattern discrimination in honey bees', in R. Wehner (ed.), *Information Processing in the Visual Systems of Arthropods*, Springer, Berlin, pp. 195–200.

Scholes, J. H. 1964, 'Discrete subthreshold potentials from the dimly-lit insect eye', *Nature*, vol. 202, pp. 572–3.

Scholes, J. 1965, 'Discontinuity of the excitation process in locust visual cells', *Cold Spring Harbor Symposium on Quantitative Biology*, vol. 30, pp. 517–27.

Schwind, R. 1984, 'Evidence for true polarization vision based on a two-channel analyzer system in the eye of the water bug *Notonecta glauca*', *Journal of Comparative Physiology*, vol. 154, pp. 53–7.

Seidl, R. 1982, Die Sehfelder und Ommatidien Divergenzwinkel von Arbeiterin, Königin und Drohne der Honigbiene (*Apis mellifera*), PhD thesis, Technische Hochschule, Darmstadt.

Seidl, R. and Kaiser, W. 1981, 'Visual field size, binocular domain and ommatidial array of the compound eyes in worker honey bees', *Journal of Comparative Physiology*, vol. 143, pp. 17–26.

Shaw, S. R. 1968, 'Organisation of the locust retina', *Symposia of the Zoological Society of London*, vol. 23, pp. 135–63.

Shaw, S. R. 1984, 'Early visual processing in insects', *Journal of Experimental Biology*, vol. 112, pp. 225–51.

Shaw, S. R. 1989, 'The retina-lamina pathway in insects, particularly Diptera, viewed from an evolutionary perspective', in D. G. Stavenga and R. C. Hardie (eds), *Facets of Vision*, Springer, Berlin, pp. 186–212.

Shepheard, P. R. B. 1966, 'Optokinetic memory and the perception of movement by the crab, *Carcinus*', in C. G. Bernhard (ed.), *The Functional Organization of the Compound Eye*, Pergamon Press, Oxford, pp. 543–57.

Sherrington, C. S. 1906, *The Integrative Action of the Nervous System*, Yale University Press, New Haven, Conn.

Smakman, J. G. J., van Hateren, J. H. and Stavenga, D. G. 1984, 'Angular sensitivity of blowfly photoreceptors, intracellular measurements and wave-optical predictions', *Journal of Comparative Physiology*, vol. 155, pp. 239–47.

Snyder, A. W. 1973, 'Structure and function of the fused rhabdom', *Journal of Comparative Physiology*, vol. 87, pp. 99–135.

Snyder, A. W. 1975, 'Optical properties of invertebrate photoreceptors', in G. A. Horridge (ed.), *The Compound Eye and Vision of Insects*, Oxford University Press, Oxford, pp. 179–235.

Snyder, A. W. 1979, 'The physics of vision in compound eyes', in H. Autrum (ed.), *Handbook of Sensory Physiology. Volume VII. Part 6A. Vision in Invertebrates*, Springer, Berlin, pp. 255–314.

Snyder, A. W. and Menzel, R. (eds) 1975, *Photoreceptor Optics*, Springer, Berlin.

Snyder, A. W., Stavenga, D. G. and Laughlin, S. B. 1977, 'Spatial information capacity of compound eyes', *Journal of Comparative Physiology*, vol. 116, pp. 183–207.

Sobel, E. C. 1990, 'The locust's use of motion parallax to measure distance', *Journal of Comparative Physiology*, vol. 167, pp. 579–88.

Sobey, P., Sasaki, S., Nagle, M., Toriu, T. and Srinivasan, M. V. 1992, 'A hardware system for computing image velocity in real time', *Proceedings SPIE, Boston*, vol. 1823, pp. 334–41.

Spaethe, J. and Chittka, L. 2003, 'Inter-individual variation of eye optics and single object resolution in bumblebees', *Journal of Experimental Biology*, vol. 206, pp. 3447–53.

Srinivasan, M. V. 1983, 'The impulse response of a movement-detecting neuron and its interpretation', *Vision Research*, vol. 23, pp. 659–63.

Srinivasan, M. V. 1985, 'Shouldn't directional movement detection necessarily be "colour-blind"?', *Vision Research*, vol. 25, pp. 997–1000.

Srinivasan, M. V. 1992, 'How insects exploit optic flow: behavioural experiments and neural models', *Philosophical Transactions of the Royal Society of London*, vol. 337, pp. 253–9.

Srinivasan, M. V. 1994, 'Pattern recognition in the honeybee: recent progress', *Journal of Insect Physiology*, vol. 40, pp. 183–94.

Srinivasan, M. V. 2006, 'Small brains, smart computations: vision and navigation in honeybees, and applications to robotics', *International Congress Series*, Elsevier, vol. 1291, pp. 30–7.

Srinivasan, M. V. and Bernard, G. D. 1977, 'The pursuit response of the housefly and its interaction with the optomotor response', *Journal of Comparative Physiology*, vol. 115, pp. 101–17.

Srinivasan, M. V. and Dvorak, D. R. 1980, 'Spatial processing of visual information in the movement detecting pathway of the fly', *Journal of Comparative Physiology*, vol. 140, pp. 1–23.

Srinivasan, M. V. and Lehrer, M. 1984a, 'Temporal acuity of honeybee vision: behavioural studies using moving stimuli', *Journal of Comparative Physiology*, vol. 155, pp. 297–312.

Srinivasan, M. V. and Lehrer, M. 1984b, 'Temporal acuity of honeybee vision: behavioural studies using flickering stimuli', *Physiological Entomology*, vol. 9, pp. 447–57.

Srinivasan, M. V. and Lehrer, M. 1988, 'Spatial acuity of honeybee vision, and its spectral properties', *Journal of Comparative Physiology*, vol. 162, pp. 159–72.

Srinivasan, M. V. and Venkatesh, S. (eds) 1997, *From Living Eyes to Seeing Machines*, Oxford University Press, New York.

Srinivasan, M. V., Laughlin, S. B. and Dubs, A. 1982, 'Predictive coding: a fresh view of inhibition in the retina', *Proceedings of the Royal Society of London*, vol. 216, pp. 427–59.

Srinivasan, M. V., Lehrer, M. and Horridge, G. A. 1990, 'Visual figure-ground discrimination in the honeybee: the role of motion parallax at boundaries', *Proceedings of the Royal Society of London*, vol. 238, pp. 331–50.

Srinivasan, M. V., Lehrer, M., Kirchner, W. H. and Zhang, S. W. 1991, 'Range perception through apparent image speed in freely flying honeybees', *Visual Neurosciences*, vol. 6, pp. 519–35.

Srinivasan, M. V., Lehrer, M., Zhang, S. W. and Horridge, G. A. 1989, 'How honeybees measure their distance from objects of unknown size', *Journal of Comparative Physiology*, vol. 165, pp. 605–13.

Srinivasan, M. V., Zhang, S. W. and Bidwell, N. J. 1997, 'Visually mediated odometry in honeybees', *Journal of Experimental Biology*, vol. 200, pp. 2513–22.

Srinivasan, M. V., Zhang, S. W. and Chahl, J. S. 2001, 'Landing strategies in honeybees, and possible applications to autonomous airborne vehicles', *Biological Bulletin*, vol. 200, pp. 216–21.

Srinivasan, M. V., Zhang, S. W. and Chandrashekara, K. 1993, 'Evidence for two distinct movement-detecting mechanisms in insect vision', *Naturwissenschaften*, vol. 80, pp. 38–41.

Srinivasan, M. V., Zhang, S. W. and Rolfe, B. 1993, 'Is pattern vision in insects mediated by "cortical" processing?', *Nature*, vol. 362, pp. 539–40.

Srinivasan, M. V., Zhang, S. W. and Witney, K. 1994, 'Visual discrimination of pattern orientation by honeybees', *Philosophical Transactions of the Royal Society of London*, vol. 343, pp. 199–210.

Srinivasan, M. V., Zhang, S. W. and Zhu, H. 1998, 'Honeybees link sights to smells', *Nature*, vol. 396, pp. 637–8.

Srinivasan, M. V., Zhang, S. W., Altwein, A. and Tautz, J. 2000, 'Honeybee navigation: nature and calibration of the "odometer"', *Science*, vol. 287, pp. 851–3.

Srinivasan, M. V., Zhang, S. W., Lehrer, M. and Collett, T. S. 1996, 'Honeybee navigation en route to the goal, visual flight control and odometry', *Journal of Experimental Biology*, vol. 199, pp. 237–44.

Stach, S. and Giurfa, M. 2005, 'The influence of training length on generalization of visual feature assemblies in honeybees', *Behavioural Brain Research*, vol. 161, pp. 8–17.

Stach, S., Benard, J. and Giurfa, M. 2004, 'Local feature assembling in visual pattern recognition and generalization in honeybees', *Nature*, vol. 429, pp. 758–61.

Stange, G. 1981, 'The ocellar component of flight equilibrium control in dragonflies', *Journal of Comparative Physiology*, vol. 141, pp. 335–47.

Stange, G., Stowe, S., Chahl, J. S. and Massaro, A. 2002, 'Anisotropic imaging in the dragonfly median ocellus: a matched filter for horizon detection', *Journal of Comparative Physiology*, vol. 188, pp. 455–67.

Stavenga, D. G. 1979, 'Pseudopupils of compound eyes', in H. Autrum (ed.), *Invertebrate Photoreceptors. Handbook of Sensory Physiology. VII/6A*, Springer, Berlin, pp. 357–439.

Stavenga, D. G. 2003, 'Angular and spectral sensitivity of fly photoreceptors. Parts I, II, III', *Journal of Comparative Physiology*, vol. 189, pp. 1–17; vol. 189, pp. 189–202; vol. 190, pp. 115–29.

Stavenga, D. G. and Hardie, R. C. (eds) 1989, *Facets of Vision*, Springer, Berlin.

Strausfeld, N. J. 1976, *Atlas of an Insect Brain*, Springer, Berlin.

Strausfeld, N. J. 1989, 'Beneath the compound eye. Neuroanatomical analysis and physiological correlates in the study of insect vision', in D. G. Stavenga and R. C. Hardie (eds), *Facets of Vision*, Springer, Berlin, pp. 317–59.

Strausfeld, N. J. 2002, 'Organization of the honey bee mushroom body: representation of the calyx within the vertical and gamma lobes', *Journal of Comparative Neurology*, vol. 450, pp. 4–33.

Strausfeld, N. J. and Lee, J. K. 1991, 'Neuronal basis for parallel visual processing in the fly', *Visual Neuroscience*, vol. 7, pp. 13–33.

Strausfeld, N. J. and Seyan, H. S. 1985, 'Convergence of visual, haltere and prosternal inputs at neck motor neurons of *Calliphora erythrocephala*', *Cell and Tissue Research*, vol. 240, pp. 601–15.

Strausfeld, N., Douglass, J. K., Campbell, H. and Higgins, C. M. 2006, 'Parallel processing in the optic lobes of flies and the occurrence of motion computing circuits', in E. Warrant and D.-E. Nilsson (eds), *Invertebrate Vision*, Cambridge University Press, Cambridge, pp. 349–98.

Tatler, B., O'Carroll, D. C. and Laughlin, S. B. 2000, 'Temperature and the temporal resolving power of fly photoreceptors', *Journal of Comparative Physiology*, vol. 186, pp. 399–407.

Tautz, J., Rohrseitz, K. and Sandeman, D. C. 1996, 'One-strided waggle dance in bees', *Nature*, vol. 382, p. 32.

Tautz, J., Zhang, S. W., Spaethe, J., Brockman, A., Si, A. and Srinivasan, M. V. 2004, 'Honeybee odometry: performance in varying natural terrain', *PloS Biology*, vol. 2, pp. 915–23.

Thorpe, W. H. 1956 [1963], *Learning and Instinct in Animals*, 2nd edn, Methuen, London.

Thorson, J. 1966a, 'Small-signals analysis of a visual reflex in locust', *Kybernetik*, vol. 3, pp. 54–66.

Thorson, J. 1966b, 'Small-signal analysis of a visual reflex in locust. I. Input parameters', *Kybernetik*, vol. 3, pp. 41–53.

Tinbergen, N. and Kruyt, W. 1938, 'Über die Orientierung des Bienenwolfes (*Philanthus triangulum Fabr.*). III. Die Bevorzugung bestimmter Wegmarken', *Zeitschrift für vergleichende Physiologie*, vol. 25, pp. 292–334.

Tunstall, J. and Horridge, G. A. 1967, 'Electrophysiological investigation of the optics of the locust retina', *Zeitschrift für vergleichende Physiologie*, vol. 55, pp. 167–82.

Turner, C. H. 1911, 'Experiments on pattern vision of the honeybee', *Biological Bulletin, Woods Hole*, vol. 21, pp. 249–64.

Tye, M. 1997, 'The problem of simple minds: is there anything it is like to be a honey bee?', *Philosophical Studies*, vol. 88, pp. 289–317.

Uexkull, J. von, 1908, *Umwelt und Innenwelt*, J. Springer, Berlin.

Vallet, A. M. and Coles, J. A. 1993, 'The perception of small objects by the drone honeybee', *Journal of Comparative Physiology*, vol. 172, pp. 183–8.

Varjú, D. 1959, 'Anwendung der Systemtheorie auf Experimente am Rüsselkäfer *Chlorophanus viridis*', *Zeitschrift für Naturforschung*, vol. 14b, pp. 724–6.

Varjú, D. and Schnitzler, H. U. (eds) 1984, *Localization and Orientation in Biology and Engineering*, Springer, Berlin.

Verlaine, L. 1927, 'L'instinct et l'intelligence chez les Hyménoptères. VII L'abstraction', *Annales de la Societe Royale Zoologique de Belgique*, vol. 55, pp. 58–88.

Victor, J. D. and Shapley, R. M. 1980, 'A method of non-linear analysis in the frequency domain', *Biophysical Journal*, vol. 29, pp. 459–84.

Vigier, P. 1907, 'Sur les terminations photoréceptrices dans les yeux composés des Muscides', *Comptes Rendues, Academie des Sciences, Paris*, vol. 63, pp. 532–36.

Vigier, P. 1909, 'Mécanisme de la synthèse des impressions lumineuses recueilles par les yeux composés des Diptères', *Comptes Rendues, Academie des Sciences, Paris*, vol. 65, pp. 1221–3.

Vladusich, T., Hemmi, J. M. and Zeil, J. 2006, 'Honeybee odometry and scent guidance', *Journal of Experimental Biology*, vol. 209, pp. 1367–75.

Vladusich, T., Hemmi, J. M., Srinivasan, M. V. and Zeil, J. 2005, 'Interactions of visual odometry and landmark guidance during food search in honeybees', *Journal of Experimental Biology*, vol. 208, pp. 4123–35.

Vorobyev, M. and Osorio, D. 1998, 'Receptor noise as a determinant of colour thresholds', *Proceedings of the Royal Society of London*, vol. 265, pp. 351–8.

Vorobyev, M., Brandt, R., Peitsch, D., Laughlin, S. B. and Menzel, R. 2001, 'Colour thresholds and receptor noise, behaviour and physiology compared', *Vision Research*, vol. 41, pp. 639–53.

Voss, C. 1967, 'Das Formensehen der Roten Waldameise *Formica rufa*', *Zeitschrift für vergleichende Physiologie*, vol. 55, pp. 225–54.

Walcott, B. 1975, 'Anatomical changes during light adaptation in insect compound eyes', in G. A. Horridge (ed.), *The Compound Eye and Vision of Insects*, Oxford University Press, Oxford, pp. 20–33.

Wallace, D. F. 2008, 'It all gets quite tricky', Harpers Magazine, 317, 31.

Wallace, G. K. 1959, 'Visual scanning in the desert locust *Schistocerca gregaria Forskal*', *Journal of Experimental Biology*, vol. 36, pp. 512–25.

Warrant, E. and Nillson, D.-E. (eds) 2006, *Invertebrate Vision*, Cambridge University Press, Cambridge.

Warrant, E., Kelber, A., Gislén, A., Greiner, B., Ribi, W. and Wcislo, T. 2004, 'Nocturnal vision and landmark orientation in a tropical halictid bee', *Current Biology*, vol. 14, pp. 1309–18.

Warrant, E., Porombka, T. and Kirchner, W. 1996, 'Neural image enhancement allows honeybees to see at night', *Proceedings of the Royal Society of London*, vol. 263, pp. 1521–6.

Wehner, R. 1967, 'Pattern recognition in bees', *Nature*, vol. 215, pp. 1244–8.

Wehner, R. 1968, 'Die Bedeutung der Streifenbreite für die optische Winkelmessung der Biene (*Apis mellifica*)', *Zeitschrift für vergleichende Physiologie*, vol. 58, pp. 322–43.

Wehner, R. 1969, 'Die Mechanismus der optischen Winkelmessung bei der Biene (*Apis mellifera*)', *Zoologische Anzeiger*, vol. 33 (Supplement), pp. 586–92.

Wehner, R. 1971, 'The generalization of directional visual stimuli in the honeybee, *Apis mellifera*', *Journal of Insect Physiology*, vol. 17, pp. 1579–91.

Wehner, R. 1972a, 'Dorsoventral asymmetry in the visual field of the bee, *Apis mellifica*', *Journal of Comparative Physiology*, vol. 77, pp. 256–77.

Wehner, R. 1972b, 'Pattern modulation and pattern detection in the visual systems of Hymenoptera', in R. Wehner (ed.), *Information Processing in the Visual Systems of Arthropods*, Springer, Berlin, pp. 183–94.

Wehner, R. (ed.) 1972c, *Information Processing in the Visual Systems of Arthropods*, Springer, Berlin.

Wehner, R. 1975, 'Pattern recognition', in G. A. Horridge (ed.), *The Compound Eye and Vision of Insects*, Oxford University Press, Oxford, pp. 75–114.

Wehner, R. 1981, 'Spatial vision in arthropods', in H. Autrum (ed.), *Handbook of Sensory Physiology. Volume VII/6C. Vision in Invertebrates*, Springer-Verlag, Berlin, pp. 287–616.

Wehner, R. 1987, '"Matched filters": neural models of the external world', *Journal of Comparative Physiology*, vol. 161, pp. 511–31.

Wehner, R. 1989, 'The hymenopteran skylight compass: matched filtering and parallel coding', *Journal of Experimental Biology*, vol. 146, pp. 63–85.

Wehner, R. and Lindauer, M. 1966a, 'Zur Physiologie des Formensehens bei der Honigbiene. I Winkelunterscheidung an vertikal orientierten Streifenmustern', *Zeitschrift für vergleichende Physiologie*, vol. 52, pp. 290–324.

Wehner, R. and Lindauer, M. 1966b, 'Die optische Orientierung der Honigbiene (*Apis mellifica*) nach der Winkelrichtung frontal gebotener Streifenmuster', *Zoologische Anzeiger*, vol. 30 (Supplement), pp. 239–46.

Wehner, R. and Menzel, R. 1990, 'Do insects have cognitive maps?', *Annual Review of Neurosciences*, vol. 13, pp. 403–14.

Wehner, R. and Müller, M. 1985, 'Does interocular transfer occur in visual navigation by ants?', *Nature*, vol. 315, pp. 228–9.

Wehner, R. and Rossel, S. 1985, 'The bee's celestial compass—a case study in behavioural neurobiology', *Fortschritt für Zoologie*, vol. 31, pp. 11–53.

Wehner, R. and Srinivasan, M. V. 1984, 'The world as the insect sees it', in T. Lewis (ed.), *Insect Communication*, Academic Press, New York, pp. 29–47.

Wehner, R., Bleuler, S., Nievergelt, C. and Shah, D. 1990, 'Bees navigate by using vectors and routes rather than maps', *Naturwissenchaften*, vol. 77, pp. 479–82.

Wehrhahn, C. 1985, 'Visual guidance of flies during flight', in G. A. Kerkut and L. I. Gilbert (eds), *Comprehensive Insect Physiology, Biochemistry and Pharmacology. Volume 6. Nervous System, Sensory*, Pergamon Press, Oxford, pp. 673–84.

Weiss, K. 1953, 'Versuche mit Bienen und Vespen in farbigenlabryrinthen', *Zeitschrift für Tierpsychologie*, vol. 10, pp. 29–44.

Wells, P. H. and Wenner, A. M. 1973, 'Do honey bees have a language?', *Nature*, vol. 241, pp. 171–5.

Wenner, A. M. 1967, 'Honey bees: do they use the distance information contained in their dance maneuver?', *Science*, vol. 155, pp. 847–9.

Wenner, A. M. [see Munz, T. 2005, 'The bee battles: Karl von Frisch, Adrian Wenner and the honey bee dance language controversy', *Journal of the History of Biology*, vol. 38, pp. 535–70].

Wenner, A. M. and Wells, P. H. 1990, *Anatomy of a Controversy: The question of a 'language' among bees*, Columbia University Press, New York.

Wertheimer, M. 1912, 'Experimentelle Studien über das Sehen von Bewegung', *Zeitschrift für Psychologie und Physiologie der Sinnesorgane*, vol. 61, pp. 161–265.

Wertheimer, M. 1924 [1938], Über Gestalttheorie [an address before the Kant Society, Berlin, 7 December 1924, Translated by Willis D. Ellis, published in his *Source Book of Gestalt Psychology*, Harcourt, Brace and Co. New York. Reprinted in 1997 by Gestalt Journal Press, New York].

Westaway, F. W. 1937, *Scientific Method*, Revised 5th edn, Blackie, London.

Whewell, W. 1837, *History of the Inductive Sciences*, Parker, London.

Whewell, W. 1840, *The Philosophy of the Inductive Sciences, Founded Upon Their History*, Parker, London.

Whewell, W. 1858, *Novum Organon Renovatum*, Parker, London.

Whitaker, D., Bradley, A., Barrett, B. T. and McGraw, P. V. 2002, 'Isolation of stimulus characteristics contributing to Weber's law for position', *Vision Research*, vol. 42, pp. 1137–48.

Wiechert, E. 1938, 'Zur Frage der Koordinaten des subjectiven Sehraumes der Biene', *Zeitschrift für vergleichende Physiologie*, vol. 25, pp. 455–93.

Wiersma, C. A. G. 1958, 'On the functional connections of single units in the central nervous system of the crayfish *Procambarus clarkii Girard*', *Journal of Comparative Neurology*, vol. 110, pp. 421–71.

Wiersma, C. A. G. 1966, 'Integration in the visual pathway of crustacea', *Symposia of the Society of Experimental Biology*, vol. 20, pp. 151–78.

Wigglesworth, V. B. [1965], *The Principles of Insect Physiology*, Revised 6th edn, Methuen, London.

Wilson, M. 1975, 'Angular sensitivity of light and dark adapted locust retinula cells', *Journal of Comparative Physiology*, vol. 97, pp. 323–8.

Winsor, F. 1958 [2001], *The Space Child's Mother Goose*, Simon Schuster, New York [Reprinted by Purple House Press].

Wolf, E. 1931, 'Sehschärfeprüfung an Bienen im Freilandversuch', *Zeitschrift für vergleichende Physiologie*, vol. 14, pp. 746–62.

Wolf, E. 1933, 'The visual intensity discrimination of the honeybee', *Journal of General Physiology*, vol. 16, pp. 407–22.

Wolf, E. 1935, 'An analysis of the visual capacity of the bee's eye', *Cold Spring Harbor Symposium on Quantitative Biology*, vol. 3, pp. 255–60.

Wolf, E. and Zerrahn-Wolf, G. 1935, 'The dark adaptation of the eye of the honeybee', *Journal of General Physiology*, vol. 19, pp. 229–37.

Wolf, E. and Zerrahn-Wolf, G. 1936, 'Flicker and the reactions of bees to flowers', *Journal of General Physiology*, vol. 20, pp. 511–18.

Wolf, R. and Heisenberg, M. 1986, 'Visual orientation in motion-blind flies is an operant behaviour', *Nature*, vol. 323, pp. 154–6.

Wolf, R. and Heisenberg, M. 1990, 'Visual control of straight flight in *Drosophila melanogaster*', *Journal of Comparative Physiology*, vol. 167, pp. 269–83.

Wolf, R. and Heisenberg, M. 1991, 'Basic organization of operant behaviour as revealed in *Drosophila* flight orientation', *Journal of Comparative Physiology*, vol. 169, pp. 699–705.

Wolf, R., Wittig, T., Li, L., Wustmann, G., Eyding, D. and Heisenberg, M. 1998, '*Drosophila* mushroom bodies are dispensable for visual, tactile, and motor learning', *Learning and Memory*, vol. 5, pp. 166–78.

Yang, E.-C. and Maddess, T. 1997, 'Orientation-sensitive neurons in the brain of the honey bee (*Apis mellifera*)', *Journal of Insect Physiology*, vol. 43, pp. 329–36.

Yang, E.-C. and Osorio, D. 1996, 'Spectral responses and chromatic processing in the dragonfly lamina', *Journal of Comparative Physiology*, vol. 178, pp. 543–50.

Yang, E.-C., Lin, H.-C. and Yung, Y.-S. 2004, 'Patterns of chromatic information processing in the lobula of the honeybee', *Journal of Insect Physiology*, vol. 50, pp. 913–25.

Young, J. Z. 1939, 'Fused neurons and synaptic contacts in the giant nerve fibres of cephalopods', *Philosophical Transactions of the Royal Society of London*, vol. 229, pp. 465–503.

Zawarzin, A. 1913, 'Histologische Studien über Insekten IV. Die optischen Ganglien der *Aeschna* Larven', *Zeitschrift für wissenschaftlich Zoologie*, vol. 108, pp. 175–257.

Zeil, J. 1993, 'Orientation flights of solitary wasps (*Cerceris*; Sphecidae; Hymenoptera). Parts I and II', *Journal of Comparative Physiology*, vol. 172, pp. 189–205, 207–22.

Zeil, J., Nalbach, G. and Nalbach, H. O. 1989, 'Spatial vision in a flat world: optical and neural adaptations in arthropods', in R. N. Singh and N. Strausfeld (eds), *Neurobiology of Sensory Systems*, Plenum, New York, pp. 123–36.

Zerrahn, G. 1933, 'Formdressur und Formunterscheidung bei der Honigbiene', *Zeitschrift für vergleichende Physiologie*, vol. 20, pp. 117–50.

Zettler, F. and Weiler, R. 1976, *Neural Principles in Vision*, Springer, Berlin.

Zhang, S. W. and Horridge, G. 1992, 'Pattern recognition in bees, size of regions in spatial layout', *Transactions of the Royal Society of London*, vol. 337, pp. 65–71.

Zhang, S. W. and Srinivasan, M. V. 1994a, 'Prior experience enhances pattern discrimination in insect vision', *Nature*, vol. 368, pp. 330–2.

Zhang, S. W. and Srinivasan, M. V. 1994b, 'Pattern recognition in honeybees: analysis of orientation', *Philosophical Transactions of the Royal Society of London*, vol. 346, pp. 399–406.

Zhang, S. W. and Srinivasan, M. V. 2004, 'Exploration of cognitive capacity in honeybees: higher functions emerge from a small brain', in F. R. Prete (ed.), *Complex Worlds From Simpler Nervous Systems*, MIT Press, Cambridge, Mass., pp. 41–74.

Zhang, S. W., Lehrer, M. and Srinivasan, M. V. 1998, 'Eye-specific route learning and interocular transfer in walking honeybees', *Journal of Comparative Physiology*, vol. 182, pp. 745–54.

Zhang, S. W., Srinivasan, M. V. and Collett, T. 1995, 'Convergent processing in honeybee vision: multiple channels for the recognition of shape', *Proceedings of the National Academy of Sciences of the USA*, vol. 92, pp. 3029–31.

Zhang, S. W., Srinivasan, M. V. and Horridge, G. A. 1992, 'Pattern recognition in honeybees: local and global analysis', *Proceedings of the Royal Society of London*, vol. 248, pp. 55–61.

Zhang, S. W., Wang, X., Liu, Z. and Srinivasan, M. V. 1990, 'Visual tracking of moving targets by freely flying honeybees', *Visual Neuroscience*, vol. 4, pp. 379–86.

www.ingramcontent.com/pod-product-compliance
Lightning Source LLC
Chambersburg PA
CBHW061238270326
41928CB00037B/3353